T0258761

MOLECULAR
ELECTRONICS

MOLECULAR ELECTRONICS

*Properties, Dynamics,
and Applications*

Edited by

GÜNTER MAHLER
University of Stuttgart
Stuttgart, Germany

VOLKHARD MAY
Humboldt University of Berlin
Berlin, Germany

MICHAEL SCHREIBER
Technical University
Chemnitz, Germany

Marcel Dekker, Inc.   **New York • Basel • Hong Kong**

Library of Congress Cataloging-in-Publication Data

Mahler, Günter.
   Molecular electronics : properties, dynamics, and applications /
Günter Mahler, Volkhard May, Michael Schreiber.
      p. cm.
   Includes index.
   ISBN 0-8247-9526-1 (alk. paper)
   1. Molecular electronics. I. May, Volkhard. II. Schreiber,
Michael. III. Title.
TK7874.8.M34   1996
621.381—dc20
                                                                    95-52749
                                                                    CIP

The publisher offers discounts on this book when ordered in bulk quantities. For more information, write to Special Sales/Professional Marketing at the address below.

This book is printed on acid-free paper.

Marcel Dekker, Inc.
270 Madison Avenue, New York, New York 10016

Current printing (last digit):
10 9 8 7 6 5 4 3 2 1

PRINTED IN THE UNITED STATES OF AMERICA

Preface

During the 1950s and 1960s, increasing understanding was reached with respect to the structure, dynamics, and reactivity of chemical compounds. This more and more improved view of chemistry at the atomic level even included complex organic molecules up to very sophisticated biomolecules. Along with this accumulation of knowledge in the field of chemistry, concepts of using molecules and molecular complexes in a very new way entered scientific discussion. Neither their special structure nor their properties to form new compounds were at the center of interest: Instead, their ability to conduct and transfer "particles" or energy was envisaged as the basis for a novel technology.

Since microelectronics achieved a certain high point in the late 1970s, we have hoped to realize information processing and information storage on an even shorter scale by using single molecules. This dream has been called molecular electronics. Let us recall in this connection the "molecular rectifier" proposed by Aviram and Ratner in 1974 (1).

Keeping this idea in mind and drawing on experiences in organic chemistry and molecular and solid-state physics, a large number of proposals for molecular wires, molecular switches, and so on emerged. Molecular electronics has been strongly shaped by the personality of Forrest L. Carter, who succeeded in popularizing it not only in the scientific community but also throughout society (2,3).

This early state of molecular electronics might be characterized as a phase of formulating wishes and almost unconstrained dreams. Molecular structures that might allow for different functions have been suggested (2-8). The main guideline in this early phase was the simulation of functions on a molecular level that are already well established in microelectronics. For example, molecular structures such as polymeric chains have been discussed for the conduction of single electrons, and a number of molecular candidates have been offered for rectification of electron transport.

A preliminary understanding of charge transport in conjugated polymers (9) let polyacetylene become most popular. Different molecular architectures of polymeric strands to switch charge motion or to split off charge motion were proposed. These structures would be completed by additional bistable molecular structures that would act as the active part and allow us to switch charge motion or the motion of other types of quasi-particles. A popular proposal has been a photochromic molecule that, after absorption of a light quantum, would reorganize the single and double bonds in polyacetylene, as is typical for the solitonic mechanism of charge transport (see e.g. (6)).

These few examples clearly point to the two main difficulties of molecular electronics, which appeared not only in its early phase of development: (1) It is not hard to suggest molecular structures that might fulfill simple bistable or multistable functions. There is no lack of structures or mechanisms, at least in principle, to realize such functions on the microscopic level. The central problem is the synthesis of stable molecular arrangements with an error rate comparable to that of traditional semiconductor microelectronics. In contrast to microelectronics, structurization for molecular electronics has to be done in the range of nanometers; moreover, structurization has to be done from below (i.e., single molecules or molecular complexes have to be "put together" step by step to form a macroscopic device). (2) Techniques are not yet available to realize stable structurization on the nanometer scale up to a macroscopic level, to help us learn what kind of structures to build up and what kind of functions to get.

As the 1980s ended, the shortcomings of the somewhat naive conception of molecular electronics were more fully realized. Both of the abovementioned concerns entered scientific discussion. The dreaming phase of molecular electronics was over, and some pessimism concerning the field was expressed (10). Even today, molecular electronics is not a strictly defined field of science. And, as is typical for new fields that start with a vague idea rather than an experiment (that challenges the known context of science), a weakness of concepts can be stated. (It might be appropriate in this context to remember the idea of thermonuclear synthesis in the early 1940s and the long way to recently announced promising experimental results.) Nevertheless, it is necessary for a field of science that tries to transfer a rough idea into reality to have the basic conception in mind and to redefine the conception again and again in the course of its realization. Therefore, let us explain what we think molecular electronics should be. This definition is not given to reflect the points of view exemplified in the various chapters of this book. It is offered as a starting point for further reading and to trigger controversial but, we hope, productive discussions.

We believe that molecular electronics is a conception for information storage and processing that (1) introduces nanometer-scaled systems such as single atoms, molecules, clusters, complexes, or aggregates as basic structural components, (2) uses microscopic particles such as electrons, protons, small chemical groups or molecular degrees of freedom e.g. excitons, intramolecular excitations, conformational transitions as information carriers, and (3) applies (with some exceptions) the laws of quantum mechanics and quantum dynamics such as the uncertainty principle, the probabilistic aspect of motion, and wave function interference for information processing.

In connection with this proposed definition we can locate three main problems that have already been tackled in the past and have to be solved in the future to go a step toward molecular electronic devices.

The first basic problem concerns the laws of particle dynamics in the basic units of molecular electronic devices and thus the detailed understanding of dissipative quantum dynamics in molecular systems. Let us assume that the well-accepted conception of using single microscopic particles, e.g., electrons, protons, small chemical groups, or quasi-particles such as excitons or conformational transitions, as information carriers is the right one. Then we have to understand all details of the quantum motion inside molecules or nanostructures—namely the interaction with the molecular structure and the determination of the type of motion by the structure, the influence of a dissipative environment and the determination of the type of motion by the environment, and the action of external fields.

As the second basic problem we underline the complications of structurization on a nanometer level. The first promising steps have been done. We mention here the techniques for the self-organization of molecular structures (chemical synthesis, Langmuir–Blodgett techniques), the techniques based on the scanning tunneling microscope to allow structurization from below the molecular level, and the more traditional structurization from the macroscopic level (cf. the arsenal of structurization techniques of microelectronics). Nevertheless, a new type of material science has to be built up.

The last set of problems concerns the adequate development of system architecture. This part is still the most speculative. Here, one has to think about appropriate molecular electronic device architecture: In what manner should different microscopic parts communicate? What about local versus delocalized information storage? Does the concept of neural networks help? Is there any need for an alternative approach, sometimes called quantum system theory?

As a by-product one can use ensemble properties of systems of potential interest for molecular electronics.

Despite serious objections against the concept of molecular electronics, molecular biology clearly demonstrates the possibility of cooperativity on a molecular level, which we usually call life. The new insights into the organization of life on a molecular level are so promising that we feel assured that the conception of molecular electronics is also basically sound. This as well as greater understanding of biological processes on the molecular level has brought researchers interested in molecular electronics to think about the direct application of biomolecules. With increasing understanding of tunneling processes in biological systems (see (11-13)) and speculations on solitonic mechanisms of energy transduction (14), the catchword biomolecular electronics has become popular.

The clarification of the structure and function of the bacterial photosynthetic reaction center (15) represents the outstanding example in this connection. The reaction center, a protein complex, is the prototype of a biological system able to realize directed particle transfer initiated by light absorption. An artificial copy of the main ingredients of the reaction center has been proposed (8) and used to speculate on an electron shift register based on a polymeric system. A more practical output of all these speculations is the application of bacteriorhodopsin, another special protein complex, in the field of photonics.

At the beginning of the 1990s it was realized that it may be more appropriate to speak about nanoelectronics than about molecular electronics [see e.g. (16,17)]. Alternative conceptions have been introduced based on semiconducting and metallic systems of some nanometers. Nevertheless, the basic problems are the same whether one uses an organic molecule of some hundred or thousand different atoms or an "artificial" complex of atoms forming a semiconducting or metallic cluster. (Quantum dots are a prominent example.)

The present collection of chapters tries to reflect this view of molecular electronics, which thus differs somewhat from the traditional one. The book has been separated into three main sections each starting with an introduction by the editors. In the first, we discuss "Physical Systems and Information Theory", the basis of molecular electronics. This concerns the hardware as well as the software of molecular electronics, i.e., the material basis and the necessary theoretical concepts. H. Mühlenbein offers results of automata theory that are of interest for the main concern of the book. G. Mahler discusses what theoretical physics and quantum theory can tell us about these basic questions. How to handle molecular systems and how to form an artificial molecular network up to the macroscopic length scale are described in the contribution of I. R. Peterson. The study of *single* molecules by light fields is explained by M. Orrit and J. Bernard. The description of *single* electron tunneling in mesoscopic structures can be found in the contribution of M. H. Devoret, D. Esteve, and C. Urbina.

The second section is concerned with the "Dynamical Repertoire of Interacting Networks." What is already available in the field of semiconductor microstructures is presented by K. Ensslin, W. Hansen, and J. P. Kotthaus. In the contribution of M. Mehring and H. P. Knoll one finds experimental examples and some general concepts of the controllability of electron motion in molecular systems like donor–acceptor complexes. A more general view of the cooperativity of particle dynamics in networks of a mesoscopic length scale is offered by H. Körner and G. Mahler.

After discussion of the basics of molecular electronics and some cooperative properties of artificial mesoscopic systems, the third section addresses "Ensemble Properties and Applications." Of course, these examples do not offer real molecular electronic devices in the strict sense. Nevertheless, they use properties of distinct molecules, but joined together to a macroscopic ensemble of (more or less) identical units. To give an adequate view of this promising development, a number of research groups contributed to this chapter. The main interest is in the optical addressability of different states in bacteriorhodopsin, a natural protein complex. The basic features of the related experimental technique, i.e., spectral hole burning, are presented in separate contributions by U. Bogner and K. K. Rebane and A. Rebane. C. De Caro, S. Bernet, A. Renn, and U. P. Wild explain the application of holographic techniques to realize three-dimensional data storage. The specifics of bacteriorhodopsin in its natural form and the properties of gene-manipulated molecules as well as the consequences for information processing systems are discussed by R. R. Birge, R. B. Gross, and A. F. Lawrence, and by C. Bräuchle, N. Hampp, and D. Oesterhelt.

In the last chapter, entitled "Prospects of Molecular Electronics," we present our own view as editors, on what may be important for further development of molecular electronics, i.e., what may be related to some basic directions as well as lead us toward real device applications.

We hope that this collection of articles, and our particular view of molecular electronics, will encourage scientists in this field and investigators from neighboring fields to contribute further to the utilization of single molecules and their concerted action in an ordered structure and to improve our possibilities of dealing with information processing.

It is a pleasure for us to thank all the contributors for their effort in presenting a comprehensive and up-to-date view of molecular electronics. Finally it remains to thank Marcel Dekker, Inc., for its cooperation in producing this book.

Günter Mahler
Volkhard May
Michael Schreiber

REFERENCES

1. Aviram, A., and Ratner, M. A., *Chem. Phys. Lett.*, *29*, 281 (1974).
2. Carter, F. L., ed., *Molecular Electronic Devices*, Marcel Dekker, New York, 1982.
3. Carter, F. L., ed., *Molecular Electronic Devices II*, Marcel Dekker, New York, 1987.
4. Adey, W. R., and Lawrence, A. F., eds., *Nonlinear Electrodynamics in Biological Systems*, Plenum Press, New York, 1984.
5. Aviram, A., *J. Am. Chem. Soc.*, *110*, 5687 (1988).
6. Launay, J. P., and Joachim, C., *J. Chimie Phys.*, *85*, 1135 (1988).
7. Joachim, C., *J. Mol. Electr.*, *4*, 125 (1988).
8. Hopfield, J. J., Onuchic, J. N., and Beratan, D. N., *Sciences*, *241*, 817 (1988).
9. Eteman, S., Heeger, A. J., and MacDiarmid, A. G., *Ann. Rev. Phys. Chem.*, *33*, 443 (1982).
10. Haddon, R. C., and Lamola, A. A., *Proc. Natl. Acad. Sci. USA*, *82*, 1874 (1985).
11. Chance, B., ed., *Tunneling in Biological Systems*, Academic Press, New York, 1979.
12. DeVault, D., ed., *Quantum Mechanical Tunneling in Biological Systems*, Cambridge Univ., New York, 1974.
13. Jortner, J., and Pullman, B., eds., *Perspectives in Photosynthesis*, Kluwer Academic, New York, 1990.
14. Davydov, A. S., *Solitonij v Bioenergetikje*, Naukova Dumka, Kiev, 1986.
15. Michel–Beyerle, M. E., ed., *Antenna and Reaction Centers of Photosynthetic Bacteria*, Springer Series in Chemical Physics 42, Berlin, 1985.
16. Proceedings of the First European Conference on Molecular Electronics, *Mol. Cryst. Liq. Cryst.*, *234* (1993).
17. Abstracts of the Second European Conference on Molecular Electronics, Kloster Banz, Germany, 1994.

Contents

Contributors

J. Bernard   CNRS and University of Bordeaux, Talence, France

Stefan Bernet   Swiss Federal Institute of Technology, Zürich, Switzerland

Robert R. Birge   Syracuse University, Syracuse, New York

C. Bräuchle   University of Munich, Munich, Germany

U. Bogner   University of Regensburg, Regensburg, Germany

Cosimo De Caro   Swiss Federal Institute of Technology, Zürich, Switzerland

Michel H. Devoret   CEA-Saclay, Gif-sur-Yvette, France

K. Ensslin   Ludwig-Maximilians-University, Munich, Germany

Daniel Esteve   CEA-Saclay, Gif-sur-Yvette, France

Richard B. Gross   Syracuse University, Syracuse, New York

N. Hampp   University of Marburg, Marburg, Germany

W. Hansen   Ludwig-Maximilians-University, Munich, Germany

Hanspeter Knoll   ComTech GmbH, Waiblingen, Germany

H. Körner   University of Stuttgart, Stuttgart, Germany

J. P. Kotthaus   Ludwig-Maximilians-University, Munich, Germany

Albert F. Lawrence   Syracuse University, Syracuse, New York

Günter Mahler   University of Stuttgart, Stuttgart, Germany and Santa Fe Institute, Santa Fe, New Mexico

Volkhard May   Humboldt University of Berlin, Berlin, Germany

Michael Mehring   University of Stuttgart, Stuttgart, Germany

Heinz Mühlenbein   GMD Schloss Birlinghoven, Sankt Augustin, Germany

D. Oesterhelt   Max-Planck-Institute of Biochemistry, Martinsried, Germany

M. Orrit   CNRS and University of Bordeaux, Talence, France

I. R. Peterson   Coventry University, Coventry, England

Alexander Rebane   Swiss Federal Institute of Technology, Zürich, Switzerland

Karl K. Rebane   Institute of Physics, Tartu, Estonia

Alois Renn   Swiss Federal Institute of Technology, Zürich, Switzerland

Michael Schreiber   Technical University, Chemnitz, Germany

Cristian Urbina   CEA-Saclay, Gif-sur-Yvette, France

Urs P. Wild   Swiss Federal Institute of Technology, Zürich, Switzerland

MOLECULAR ELECTRONICS

1

Physical Systems and Information Theory

Günter Mahler

University of Stuttgart
Stuttgart, Germany

Volkhard May

Humboldt University of Berlin
Berlin, Germany

Michael Schreiber

Technical University
Chemnitz, Germany

This section addresses the interplay between the *description* of nature (in terms of mathematical models *defining* algorithms) and the *use* of physical systems for *representing* algorithms according to some preselected tasks. The objective of an explicit model is always a particular and usually very small part of the universe, thus defining a "cut" between the system proper and its environment. This holds for physical modeling as well as for abstract computational schemata. The machine metaphor stresses this very openness of the system and the need for an external reference, a certain context.

Computers are special machines that implicitly define computability: computable operations are operations that can be represented physically. This implies a strange cyclic relationship between physical models and their mathematical codification: to be practically useful, physics should be formulated in terms of computable functions, which in turn are defined via physical models.

For any physical system under consideration there are various levels of description and thus, so it seems, largely different computational modes. This conclusion, to be sure, has been severely constrained by the well-known Church–Turing thesis telling us that so-called universal computational systems are equivalent in the sense that they all can simulate each other. This conjecture appears to render the question of physical implementation almost irrelevant. On the other hand, progress in computer science has been based on improved algorithms and more and more sophisticated implementations: not only what a system does is important but also how it is done. One key for appreciating this issue appears to be the conception of complexity in its various meanings.

In the following, Mühlenbein discusses complexity in the context of computer science. Here, the problems arise in various complexity classes including simple polynomial time algorithms and nondeterministic polynomial problems. This computational complexity is intimately related to the efficiency of the algorithm used and of its implementation. Both these efficiencies can be improved, as argued by Mühlenbein, in particular by analyzing and—to some extent—copying nature. Machines are specified by their tasks; typical computational tasks are optimization problems, which can be approached by learning procedures. Collective learning models can be designed to simulate Darwinian evolution and are then often known as genetic algorithms. It is fairly obvious that these algorithms correspond to a high level description of nature including aspects of genotypes and phenotypes. Nevertheless, the corresponding abstract rules can be transcribed into a conventional electronic computer operating on completely different physical objects and on largely different time scales. One may wonder whether such a flexibility even holds down in the quantum world.

Superficially, quantum dynamics is just another level of description, which should eventually show up in any physical system. There are design principles to make this level more relevant for actual observation: pertinent dynamical models are, according to Mahler, hierarchical networks of (quasi) molecular subsystems. They still allow for different modes of operation (different levels of description) starting from ensemble properties (based on essentially independent subunits as used for spectral hole burning) and proceeding to incoherent and, possibly, to (partly) coherent network dynamics. Incoherent networks are, typically, molecular realizations of classical machine concepts like cellular automata; they exploit nanometer scale organization and massive parallelism, but they reflect quantum behavior only to the extent that the local rules are fundamentally stochastic. These systems would thus continue the down-scaling trend of today's computer fabrication into a physical regime, where they can no longer be based on the transistor and electric wiring. Such systems, if realized, could be fascinating prototypes also for information processing in biological systems. On the other hand, they do not constitute any new "computational paradigm."

Quantum computation is still a rather controversial subject. That a quantum system might do something that a classical computer could not simulate is a misconception: the algorithms by which we describe quantum objects are routinely carried out by classical machines without any known exception. However, this does not tell us anything about the respective efficiency: specific computational tasks might most appropriately be represented by nonclassical dynamical modes. As briefly mentioned by Mahler, coherent networks would live in a much larger state space (compared to incoherent ones) characterized by nonlocal correlations or so-called entanglements. These could be represented also classically but at the cost of significantly increased requirements for memory and interconnects. We shall briefly return to this issue in the last section of this book.

An important part of the desription of nature is concerned with structure. Structure defines and selects dynamical paths, which necessarily underlie any machine function. Various material classes of possible interest to molecular electronics are reviewed by Mahler. However, as argued by Petersen, the unbiased assessment of different materials requires reference to clearly defined operational modes specified on the local network-node level as well as with respect to internal and external interactions. The choice of a molecular structure is therefore by itself an optimization problem. But not only that: as systems must be composed in the real world, fabrication even represents a kind of directed molecular dynamics (which, in a different context, might establish a chemical computer). Petersen carefully discusses pertinent problems as exemplified by the Langmuir–Blodgett technique for organic films; severely limited control also exists for (molecular beam) epitaxy, lithography, or structuring based on the scanning–tunneling microscope. New methods are evolving that should eventually improve the structural control on molecular length scales.

The paper by Orrit and Bernard is concerned with single-molecule spectroscopy, a technique by which ensemble averaging on the molecular level is avoided. This kind of fascinating experiment demonstrates that individual quantum objects (defects in the present case) can indeed be isolated (to some extent) even in a macroscopic environment. This isolation allows novel modes of observation: the structure in the 2-time correlation function, e.g., would be washed out for an ensemble; the dynamical influence of the local microscopic environment, which clearly shows up here, would otherwise disappear within an apparently stationary broadening. These findings underline that under different structural conditions qualitatively *new types* of information can become available; it is not just the "old stuff" reduced to a much smaller scale.

Devoret, Esteve, and Urbina, finally, introduce the reader into the world of single-electron transfer in custom-made nanostructures. These effects emerge as tiny isolated islands (which could even be molecules) and become accessible to controlled changes of their charge state. These discrete charge states would thus play a similar role to the discrete energy states of Orrit and Bernard. In either case, quantum effects are responsible for the discretization of state space; such a discretization is, of course, highly welcome for the interpretation of those states as information carriers; it is a kind of fundamental digitalization. In the former case, the addressing is via optical fields (nonlocal on molecular scales), while for the charge states the addressing is through local junctions that can be biased by corresponding voltage sources. As the authors clearly point out, the building of larger networks faces severe and even conceptional problems: the transfer of individual electrons would allow only for very constrained step-by-step changes of local charge states within the net, an inherently stochastic process. To what extent here also coherence and/or entanglement (interesting for quantum computation) might be reached remains yet unclear. Furthermore, a clearly defined charge reference state (like the ground state in energy states) is obviously hard to achieve. Nevertheless, it should be quite rewarding to compare such quasi-molecular schemes based on local charge states with those based on local energy states.

2

Algorithms, Data, and Hypotheses: Learning in Open Worlds

Heinz Mühlenbein

GMD Schloss Birlinghoven
Sankt Augustin, Germany

I. INTRODUCTION

With the introduction of computers, automata have been playing a continuously increasing role in the natural sciences. In this paper I focus on special automata called *learning systems*. A learning system has a learning procedure by which it can develop methods that cannot be deduced trivially from its learning procedure. The learning system tries out hypotheses (methods) and selects the better ones. It has *a priori* a well defined *universe of hypotheses* from which it must choose those to be tried. If this universe is small, then the "inventiveness" of the machine is severely limited, and the value of the methods that it develops depends more on the astuteness of the programmer in choosing a universe containing good hypotheses than an ability of the learning system to pick the best hypothesis from among those in the universe. In order to give the learning system a "free hand", it should have a universe which, although well-defined, is so large and varied that the user of the system is not even acquainted with the forms of all the methods it contains.

Artificial learning systems need huge processing capabilities. New physical concepts of information processing have to be developed to meet these requirements. A promising research direction is molecular electronics. But I would like to mention a second reason why molecular electronics might be interesting for the design of artificial learning systems. Learning systems, natural or artificial, face the problem of finding good hypotheses that explain the past data and that can be used for predictions. But a fundamental theorem states that a hypothesis cannot be assigned a probability in the classical sense of being true (Popper, 1972). There is a similar problem in quantum mechanics, the theoretical foundation of molecular electronics. Here also researches are looking for an extension of classical probability theory. It is a general feature of quantum mechanics that one needs a rule to determine which of the alternative "histories" can be assigned probabilities (Gell-Mann and Hartle, 1992).

A rigorous treatment of the above problems seems to be out of reach at this moment. Therefore I shall concentrate on some important aspects of the general problem. The outline of the paper is as follows. In section II, data fitting will be described from the mathematical point of view. In section III, I extend the basic model. The general task of data modelling and its connection to Occam's razor is described in a Bayesian framework. Section IV introduces the concept of complexity as defined in computer science and in probability theory. The problem of discrete vs. continuous representations is discussed in section V. Some principal limitations of artificial automata are discussed in section VI. In the final section, two learning systems are discussed that have been implemented by my research group. One system models collective learning of populations in a similar way to *Darwinian evolution*; the other system models learning of an artificial organism equipped by something like a *brain*. The paper ends with a short discussion of the question: Can quantum mechanics make contributions towards finding a new computing paradigm needed for systems operating in the real world?

II. APPROXIMATION OF FUNCTIONS

Many learning procedures can be formulated as approximation problems. Let X be the input space and Y the output space of an unknown process $s := X \rightarrow Y$. The problem is to find an approximation $f: X \rightarrow Y$ in a search space (the universe) F such that

$$f(x) \approx s(x) \qquad x \in X \tag{1}$$

The approximation problem can be precisely defined if a norm is given in the search space F. In this case one looks for an ϵ-*approximation* such that

$$\| f - s \| < \epsilon \tag{2}$$

In order to solve the problem, some information about s has to be used. I shall investigate the case where the unknown solution can be computed using a finite *data set* D, where

$$D = \{(x_j, y_j = s(x_j)), \quad j = 1, \ldots, n\}$$

Many learning procedures determine an approximation f by fitting the data according to some criterion. The most popular criterion is called *least mean square error (LMSE)*, which minimizes the sum of the squared errors between the data and the model predictions:

$$Err_n = \frac{1}{n} \sum_{j=1}^{n} \| y_j - f(x_j) \|^2 \rightarrow MIN \tag{3}$$

This minimization problem is investigated in different scientific disciplines. If the search space is a space of functions, then the problem belongs to *mathematical approximation theory*. If the search space consists of nonnumeric elements such as rules or program components, then the problem belongs to *artificial intelligence*. It is called *program synthesis by examples*.

In mathematical approximation theory, many results have been obtained. A distinction is made between the *interpolation problem* and the *approximation problem*. In interpolation, the function f has to fit the data points exactly. In approximation one looks for a function f that approximates the unknown function s best according to some norm in the function space. By definition, the error of the best approximation function is not more than that of the best interpolation function. But the best interpolation function can be constructed for many subspaces of functions, whereas no method has been discovered for constructing the best approximation function.

Most of the results from mathematical approximation have been obtained for the one-dimensional case only. I shall summarize some well-known results for the maximum norm and search spaces F consisting of polynomials. Here the best interpolation polymomial was computed by Chebychev. The optimal interpolation points $\{x_i\}$ are given by the zeros of the Chebychev polynomials. It could be shown that the error of the best interpolation polynomial is only a factor of $O(log(n))$ larger than the error of the best approximation polynomial, where n is the order of the polynomial. For the Euclidian norm, the optimal interpolation polynomial is defined at the zeros of the Legendre polynomials.

An extension of the classical mathematical approximation problem was developed by Traub and Wozniakowski (1980). The generality and power of the extension is a result of the fact that *information and problem complexity* play a central role in this approach. In classical approximation theory, optimal algorithms are computed by making various technical assumptions about the class of algorithms and the class of problem elements. These assumptions are often not verifiable. Furthermore, depending on the assumptions, many different optimal approximations might exist. The concept of problem complexity deals with the metaproblem: how to find the best of the optimal approximations.

Problem complexity is a measure of the intrinsic difficulty of obtaining the solution to a problem regardless of how this solution is obtained. It can be defined with respect to a model of computation and a class of "permissible" information operators. Unfortunately, the determination of problem complexity is very difficult; it has been completely solved for only very few problems. I shall not discuss this approach further. The interested reader is referred to Traub and Wozniakowski (1980).

The theory of optimal algorithms closes the gap between the mathematical and the statistical approach to the data-driven learning problem. In the statistical approach, the data is not reliable but corrupted by noise. Therefore the approximation problem consists of two subproblems: first to estimate the amount of noise and then to approximate the corrected data. I shall discuss this problem with a simple example. Consider the problem of a nonparametric estimation of a regression function s from observations of the form $y_i = s(x_i) + \xi_i$, $i = 1, \ldots, n$, where $x_i = i/n$, and ξ_i are random variables such that

$$E(\xi_i) = 0, \qquad E(\xi_i\xi_j) = \sigma^2\delta_{ij} \qquad \sigma^2 > 0 \tag{4}$$

It is assumed that s is defined on $[0, 1]$ and can be represented as a Fourier series: $s(x) = \sum_{j=1}^{\infty} c_j\phi_j(x)$. Let the approximation be defined for $N \leq n$ as

$$f_N(x) = \sum_{j=1}^{N} \hat{c}_j\phi_j(x) \qquad \hat{c}_j = \frac{1}{n}\sum_{m=1}^{n} y_m\phi_j(x_m) \tag{5}$$

What is the optimal order N of the approximation? The answer depends on the optimality criterion to be used. The usual C_p-criterion leads to

$$N_{\text{opt}} = \arg_{N \leq n} \min(Err_N + 2\sigma^2 Nn^{-1}) \tag{6}$$

where

$$Err_N = \frac{1}{n}\sum_{i=1}^{n} (f_N(x_i) - y_i)^2$$

For $\sigma \ll 1$ we have the mathematical approximation problem with the solution $N = n$. With a very large amount of noise, i.e. $\sigma \gg 1$, the optimal N can be much smaller. If we interpret N as an indicator for the complexity of the approximation model, we see that the C_p-criterion tries to balance model complexity with interpolation error.

In the next section I will treat the above problem in a general Bayesian framework. This model is able to handle noisy data as well as active data selection.

III. DATA SELECTION, MODEL SELECTION, AND OCCAM'S RAZOR

In science, a central task is to develop and compare models to account for data. Two levels of inference are involved in the task of data-driven modelling. At the first level of inference, one assumes that one of the models that was invented is true: that model is then fitted to the data. Typically a model includes some free parameters; *fitting the model to the data* involves inferring what values those parameters should probably take, given the data. This is the approach of mathematical approximation theory. The results of this inference are often summarized by the most probable parameter values and hopefully some error bars on those parameters. The second level of inference is the task of *model comparison*. Here, one wishes to compare the models in the light of the data, and assign some sort of preference or ranking to the alternatives.

Model comparison is a difficult task because it is not possible simply to choose the model that fits the data best: more complex models can always fit the data better, so the maximum likelihood model choice would lead us inevitably to implausible over-parameterized models that generalize poorly. *Occam's razor* states that unnecessarily complex models should not be prefered to simpler ones.

In this section I survey the Bayesian approach to Occam's razor. This survey is based on MacKay (1992).

Let us write down the Bayes rule for the two levels of inference described above. Each model H_i is assumed to have a vector of parameters w. A model is defined by its functional form and two probability distributions: a prior distribution $P(w \mid H_i)$, which states what values the model's parameters might plausibly take; and the predictions $P(D \mid w, H_i)$ that the model makes about the data D when its parameters have particular values w. Note that models with the same parameterization but different priors over the parameters are defined to be different models.

In *model fitting* it is assumed that one model H_i is true, and the model's parameters w are then inferred from the given data. Using the Bayes rule, the *posteriori probability* of the parameter w is

$$P(w \mid D, H_i) = \frac{P(D \mid w, H_i)\, P(w \mid H_i)}{P(D \mid H_i).} \tag{7}$$

In words,

$$Posterior = \frac{Likelihood * Prior}{Evidence}$$

For model fitting, the normalizing constant $P(D \mid H_i)$ is commonly ignored. It will be important in the second level of inference, and it is named *evidence* for H_i. For *model comparison* one wishes to infer which model is most plausible given the data. The posterior probability of each model is

$$P(H_i \mid D) \propto P(D \mid H_i)\, P(H_i) \tag{8}$$

The second term, $P(H_i)$, is, a *subjective* prior over our hypothesis space that expresses how plausible we thought the alternative models were *before* the data arrived. This subjective part of the inference will typically be overwhelmed by the objective term, the evidence. Assuming that there is no reason to assign strongly differing priors $P(H_i)$ to the alternative models, models H_i are ranked by evaluating the evidence.

A. Model Fitting

Let us now explicitly study the evidence in order to gain insight into how the Bayesian Occam's razor works. The evidence is defined as

$$P(D \mid H_i) = \int P(D \mid w, H_i) \, P(w \mid H_i) \, dw \tag{9}$$

For many problems, including interpolation, it is common that the integrand has a strong peak at the most probable parameters w^*. Then the evidence can be approximated by the height of the peak of the integrand times its width, $\Delta w = w - w^*$.

$$P(D \mid H_i) \simeq P(D \mid w^*, H_i) P(w^* \mid H_i) \, \Delta w \tag{10}$$

If w is k-dimensional, and if the posterior is well approximated by a gaussian, the above equation can be computed. The factor Δw is given by the determinant of the gaussian covariance matrix (MacKay, 1992):

$$P(D \mid H_i) \simeq P(D \mid w^*, H_i) P(w^* \mid H_i)(2\pi)^{k/2}(\det C)^{-1/2} \tag{11}$$

where

$$C = -\nabla\nabla \log P(w \mid D, H_i)$$

Let us apply this framework to the noisy interpolation problem. For simplicity, let us assume that x and y are scalars. To define a linear interpolation model, a set of k fixed basis functions $A = \{\phi_j(x)\}$ is chosen. The interpolated function is assumed to have the form

$$y(x) = \sum_{j=1}^{k} w_j \phi_j(x)$$

The data set is modeled as deviating from this mapping under some additive noise process:

$$y_i = y(x_i) + \xi_i$$

If the ξ have a zero-mean gaussian distribution whose standard deviation is o_y, then the probability of the data given the parameters is

$$P(D \mid w, \beta, A, N) = \frac{\exp(-\beta/2 Err_D(D \mid w, A))}{Z_D \beta} \tag{12}$$

where $\beta = 1/\sigma_\nu^2$, $Err_D = \Sigma(y(x_i) - y_i)^2$, and $Z_D(\beta) = (2\pi/\beta)^{N/2}$. Under these assumptions, finding the maximum likelihood parameters w^* is identical to minimizing the quadratic error Err_D. This is just the least mean square error (LMSE) criterion mentioned in section II. It is well known that this may be an "ill-posed" problem. That is, the w that minimizes Err_D is underdetermined and/or depends sensitively on the details of the noise in the data. Thus it is clear that to complete our interpolation model we need a prior R that expresses the sort of smoothness we expect the interpolation $y(x)$ to have. I shall not discuss this extension here. Strictly one should write

$$P(D \mid w, \beta, a, N) = P(\{y_i\}|\{xi\}, w, \beta, A, N),$$

since interpolation models do not predict the distribution of inputs $\{x_i\}$. But with the Bayesian framework this problem, often called *active learning* or *sequential design*, can also be addressed. There are two scenarios in which one would like actively to select training data. In the first, data measurements are expensive or slow, and the researcher wants to know where to look next so as to learn as much as possible. In the second scenario, there is an immense amount of data, and one has to select a subset of points that are the most useful. For active data selection, objective functions have to be defined that measure the *expected informativeness* of candidate measurements. At least three different criteria are possible: maximizing the total information gain, maximizing the information gain in a region of interest, and maximizing the discrimination between two models. All these criteria depend on the assumption that the hypothesis space is correct. This is their main weakness. Paaß and Kindermann (1995) used the variance of the predictions of a population of models. Data is selected in areas where the variance is highest.

B. Model Comparison

I now proceed with the second level of inference, model comparison. To rank alternative basis sets A, noise models N, and regularizers R in the light of the data D, the posterior probabilities for alternative models $H = \{A, N, R\}$ are examined:

$$P(H \mid D) \propto P(D \mid H)P(H) \tag{13}$$

Assuming that there is no reason to assign strongly differing priors $P(H)$, alternative methods H are ranked just by examining the evidence $P(D \mid H)$.

A slightly different approach to the model selection problem uses the *minimal description length* of Rissanen (1992). It is restricted to binary problems. Let C be an injective coding function from a discrete set X into the set of all binary strings B^*. Let $L(x)$ be the length of $C(x)$, i.e., the number of binary digits in $C(x)$. A code C is said to be a prefix code, if

$$\sum_{x \in X} 2^{-L(x)} \leq 1 \tag{14}$$

Thus a prefix code defines a distribution on X. Shannon's fundamental coding theorem states that for a given distribution P(x), all prefix codes must have a mean length bounded below by the entropy

$$\sum_{x} P(x)L(x) \geq - \sum_{x} P(x) \; log_2 \; P(x) \tag{15}$$

The lower bound can be reached only if the lengths satisfy the equality $Lx) = -log_2 P(x)$ for every x. In this sense one could call -$log_2 P(x)$ the Shannon complexity of x relative to the ''model'' P.

The above analysis can be extended to a whole class $M = \{P(y \mid x, \theta)\}$, where θ ranges over some subset of the k-dimensional Euclidean space. In this case the minimum description length criterion can be computed, which combines model complexity, the number of parameters, and the precision of the data n.

$$MDL \; (y \mid x, k) = -log_2[P(y \mid x, \theta)] + \frac{k}{2} \; log_2 \; n \tag{16}$$

MDL has to be minimized over k to get the optimal model complexity.

C. Some Remarks

Bayesian model selection is a simple extension of maximum-likelihood model selection: the evidence is obtained by multiplying the best fit likelihood by a model complexity factor. The evidence is a measure of a model's *plausibility*. The amount of CPU time required to run a model is not addressed. Choosing between models on the basis of how many operations they need can be seen as an exercise in *decision theory*. This needs further study.

The Bayesian framework does not lead to new learning procedures, but it is very useful in clarifying the many implicit assumptions hidden in the specific learning procedures.

The framework presented in this section is currently one of the most advanced methods for data-driven learning. Its application depends on many assumptions. The crucial question is whether these assumptions are fulfilled for an unknown data set.

I shall now discuss other measures of the complexity of a problem.

IV. INFORMATION, COMPLEXITY, AND UNCERTAINTY

In computer science the complexity of a problem is measured by the length of the *shortest program* written in some standard language (e.g., a program for a Turing machine) by which the problem can be solved. This information is called the *algorithmic complexity* of the problem. Often this measure cannot be computed. It is therefore of limited practical use. Furthermore, it does not take into account how many operations have to be executed by the program to solve the problem. Such a measure is the *computational complexity*.

Computational complexity is a characterization of the time or space requirements for solving a problem by a particular algorithm. Both of these requirements are usually expressed in terms of a single parameter that represents the size of the problem.

Definition: *The time complexity function f (n) of an algorithm is the largest amount of time required to solve a problem of size n.*

It has been very useful to distinguish between two classes of algorithms by the rate of growth of their time complexity function. One class is called *P*. It consists of *polynomial time algorithms*. Here the time complexity can be expressed in terms of a polynomial. The second class of algorithms consists of *exponential time algorithms*. This class is called "nonpolynomial", *NP*. More precisely the *NP* classes of problems are defined as follows. If an individual problem has a solution, then the algorithm will find that solution in *exponential time*. But it must be possible to check in polynomial time that the proposed solution is indeed a solution.

NP problems arise in many contexts. A very popular problem is the "travelling salesman problem." Here one seeks a tour that visits each city exactly once for which the distance is a minimum. The number of possible tours grows exponentially with the number of cities. In fact, this problem is not only *NP*, but what is named as *NP-complete*. This means that any other *NP* problem can be converted into it in polynomial time. It is commonly believed by computer scientists that it is impossible, with a Turing-machine-like device, to solve an *NP-complete* problem in polynomial time.

Conjecture: $P \neq NP$.

This conjecture remains the most important unsolved problem in complexity theory. *NP* problems are the hard ones. For large problem sizes, they are *transcomputational*. This term was coined by Bremermann (1962). An algorithm that needs more than 10^{93} operations is transcomputational. It cannot run until completion on any real computational system. The exact number is not so important, but it shows that there are definite limits to the computational power of any system in our universe. This bound has implications for *NP* problems. The travelling salesman problem for instance has approximately 10^{90} tours for 66 cities. In real life one is interested in good solutions for problems with more than 1000 cities.

Bremermann (1962) computed his bound by simple considerations based on quantum theory. It is surely an upper bound.

Bremermann's bound: *No data processing system, whether artificial or living, can process more than $2 * 10^{47}$ bits per second per gram of its mass.*

Bremermann derives the limit from the following considerations based on quantum physics. The phrase "processing x bits" means the transmission of that many bits over one or several communication channels within the computing system. Now assume that information is encoded in terms of energy levels within the interval $[O, E]$. Assume further that energy levels can be measured with an accuracy of only ΔE. The most refined encoding is defined in terms of markers by which the whole interval is divided into $N = E/\Delta E$ equal subintervals, each associated with the amount of energy ΔE. In order to represent more information with the same amount of energy, it is desirable to reduce ΔE. The extreme case is represented by the Heisenberg principle of uncertainty: energy can be measured to the accuracy of ΔE if the inequality

$$\Delta E \; \Delta t \geqslant h \tag{17}$$

is satisfied. This means that

$$N \leqslant \frac{E \; \Delta t}{h}$$

Now by Einstein's formula

$$E = mc^2$$

If we take the upper (most optimistic) bound of N we get

$$N = \frac{mc^2 \; \Delta t}{h}$$

Substituting numerical values for c and h, one obtains $N = 1.36 \cdot 10^{47} \, m \, \Delta t$. Using this bound, Bremermann calculated the total number of bits processed by a hypothetical computer the size of the earth within a time period equal to the estimated age of the earth. He computed 10^{93} bits. This number is referred to as *Bremermann's limit*. Problems that require processing more than 10^{93} bits of information are called *transcomputational problems*. It is obvious that exponential time algorithms are already transcomputational for fairly small problem sizes.

Recent research in fuzzy sets and probability theory has taken a different approach in trying to define complexity. It is not absolutely defined, but relative to the knowledge of a given observer. A good survey about the different definitions is given by Klir and Folger (1988).

Two general methods of defining system complexity can be distinguished: one is based on *information*, the other on *uncertainty*. In the first one the complexity is proportional to the amount of information required to *describe the system*. In the second one, system complexity is proportional to the amount of information needed to *resolve any uncertainty* associated with the system.

To the neurophysiologist, for instance, the brain consists of a network of fibers and a soup of enzymes. Therefore the transmission of a detailed description of it requires much time and space. To a butcher, in contrast, the brain is simple, for he has to distinguish it from only about thirty other types of "meat."

Both definitions of complexity are relative to an observer and its knowledge. They are related to each other under the *closed world assumption*. With this assumption the universe of discourse can be divided into two sets: the set of events known as possible and the set of events known as impossible. The set of unknown events is assumed to be empty.

The above definitions have been primarily used for the purpose of developing computational methods by which systems that seem incomprehensible can be simplified to an acceptable level of complexity. There is a major problem with this approach. Even if one has found an algorithm that reduces the complexity of the given system, the computational complexity associated with the simplification algorithm has to be taken into account. If the resulting algorithm is transcomputational, it is of no practical use.

Another severe problem is the closed-world assumption. The real world is open for any system operating in it. For any system the set of unknown events is infinite. Unfortunately, the scientific understanding of *open worlds* is in its infancy. I believe that the development of a calculus for dealing with open systems is one of the most important problems in epistemology, probability theory, and also quantum physics. I shall just mention the work of Jaynes (1992). He raises the question of whether probability theory is a "physical" theory of phenomena governed by "chance" or "randomness" or whether it should be considered as an extension of logic, showing how to reason in situations of incomplete information. Jaynes remarks: "We then see the possibility of a future quantum theory in which the role of incomplete information is recognized: for any variable F, the dispersion $(\Delta F)^2 = <F^2> - <F>^2$ represents only the accuracy with which the theory is able to predict the value of F . . . When ΔF is infinite, it means only that the theory is completely unable to predict F. The only thing that is infinite is the uncertainty of prediction."

In summary: the concepts of information, complexity, and uncertainty are used differently in different disciplines. In the future a common framework of these concepts is needed. Quantum mechanics can play a major role in this development. I now turn to another important topic for any system, that of the representation.

V. DISCRETE VS. CONTINUOUS REPRESENTATIONS

The theory of computing has been centered on the binary, all-or-none type. It has been, from the mathematical point of view, combinatorial rather than analytical. Rigid, all-or-none concepts have little connection to the continuous concept of real or complex numbers, on which mathematical analysis is based. John von

Neumann (1948), one of the founders of today's computers, warned: "Formal logic is, by the nature of its approach, cut off from the best cultivated portions of mathematics, and forced onto the most difficult part of the terrain, into combinatorics." Therefore von Neumann predicted that a powerful theory of automata will differ from the present system of formal logic in two relevant aspects.

1. The actual length of "chains of reasoning," that is, of the chains of operations, will have to be considered.
2. The operations of logic will all have to be treated by procedures that allow exceptions. All of this will lead to theories that are less rigid than past and present formal logic.

Von Neumann continued: "There are numerous indications to make us believe that this new system of formal logic will move closer to another discipline which has been little linked in the past with logic. This is thermodynamics, primarily in the form it was received from Boltzmann."

In my opinion, von Neumann's predictions turned out to be right. The importance of the length of the chain of operations was first recognized in computer science. It lead to the theory of computational complexity discussed in the previous section. Boltzmann's thermodynamics approach, especially the concept of entropy, is becoming increasingly popular in the design of new learning systems. I like to call this new emerging field "quantitative artificial intelligence."

New learning systems now under development for robotics do not use just one learning procedure: they frequently employ different learning procedures at different levels of the system architecture. The learning systems try to process both discrete and continuous information. Some promising new architectures consist of three levels. An overview and a semantic description of the three levels is shown in Table 1.

Table 1

Semantics	Characteristics
Plans	Discrete processes
Relations	Discrete values
Objects	
Features	Discrete processes
	Continuous values
Signals	Continuous processes
	Continuous values

At the most abstract level, there are *discrete values and discrete processes*. The idealization of this representation is that its members can be characterized abstractly as a set of discrete elements. At the lowest level, the information is in the form of *continuous values and continuous processes*. The constraints at this level are captured by Shannon's information and his sampling theorem. The idealization of this representation is that of a continuous function of a set of variables, e.g., $y = f(x, t)$. In between these extremes there is an intermediate level that can be characterized as requiring *continuous values of discrete processes*. For example, the rotation of the visual field can be characterized by rotational values of a single rigid body motion process. There is only one process, but the actual parameter values that describe that process are continuous.

After describing some advanced methods to synthesize reasonable hypotheses from data, I shall discuss the question, What are the principal limits of such an approach? I shall show that there are limitations, which follow from the general *induction problem*, discussed intensively by Popper (1972).

VI: PRINCIPAL LIMITATIONS OF ARTIFICIAL AUTOMATA

In 1943 McCulloch and Pitts proved this remarkable theorem: *Anything that can be defined at all logically, strictly, and unambiguously in a finite number of words can also be realized by an artificial neural network.* At first this theorem seems to indicate that an artificial system is able to solve any clearly defined problem. But the content of the theorem has to be interpreted differently. This was shown by von Neumann (1948) who raised the two questions:

1. Can the network be realized within practical limits, e.g., is the required number of connections less than the number of atoms in the universe?
2. Can every existing mode of behavior be put completely and unambiguously into words?

Let us discuss both questions with a specific example, the classification of geometrical entities as performed by humans. There have been three approaches to this central problem of vision. I call them the *theoretical*, the *learning-from-example* and the *copy-the-brain* approach.

In the theoretical approach researchers try to find a computational calculus that solves the classification problem. Up to now, a calculus has only been developed for very restricted idealized geometric objects. The limitations of the learning-by examples approach was already discussed by von Neumann (1948). He argued as follows: There seems to be no difficulty in describing how automata might be able to identify any two rectilinear triangles. The classification of more general kinds of triangles—triangles whose sides are curved, triangles that are indicated by shading, etc. can also easily be done. Next we want the system to recognize handwritten objects and letters. Von Neumann remarks: "At this point we should have the vague and uncomfortable feeling that a complete catalogue along such lines would not only be exceedingly long, but also unavoidably indefinite.'' These problems, however, constitute only a small fragment of the more general concept of identification of analogous geometrical entities. This, in turn, is only a microscopic piece of the general concept of analogy. ''Nobody would attempt to describe and define within any practical amount of space and time the general concept of analogy which dominates human vision.'' Learning from examples just by enumeration is not effective for large problem domains. The number of examples goes to infinity.

Therefore, a bottom-up approach was tried—replicating the brain, which obviously solves the classification problem—instead of solving the problem. The only way to define what constitutes a visual analogy may be a description of the connections of the visual cortex of the human brain. Any attempt to describe it by literal and formal-logical methods may lead to something less manageable. But this means that *the connections of the brain might be the simplest description of the functions it can perform*. Von Neumann remarks: ''In fact, results in modern logic indicate that phenomena like this have to be expected when we deal with really complicated entities.''

But the brain consists of about 10^{12} neurons and 10^{16} connections. How long will it take to produce a description it? Furthermore, the structure of the brain only partly defines visual analogy. The data flow, i.e., the processing of the data, must also be described. Such a description might be finite, but it is obviously *transcomputational*. It cannot be expressed by using all the atoms in the universe. ''Obviously, there is on this problem no more profit in the McCulloch–Pitts result.'' Von Neumann concludes his discussion of the theorem with the remarks: ''It may be, however, that in the process of understanding the central nervous system, logic will have to undergo a pseudomorphosis to neurology to a much greater extent than the reverse. One of the relevant things we can do at this moment with respect to the theory of the central nervous system is to point out the directions in which the real problem does not lie.''

VII. LEARNING FROM NATURE

In the previous section I showed some general limitations of artificial automata. Nevertheless, each system, whether natural or artificial, can and should *improve its capabilities*. Learning is a dominant feature of living beings. Therefore, my research group at the GMD concentrates on learning methods used in nature. Currently we model two different natural learning methods. One method models collective learning of populations similar to *Darwinian evolution*, the other method models individual learning done by organisms equipped with a *brain*. Learning and adaptation is one of the most important features of nature. Therefore it seems that *learning from nature* is a good strategy. This was already advocated by John von Neumann (1948). He wrote: "Some of the regularities which we observe in the organization of natural organisms may be instructive in our thinking and planning of artificial automata. Conversely, a good deal of our experiences with our artificial automata can be to some extent projected on our interpretations of natural organisms."

In learning by simulating evolution, we distinguish between two models. One model is based on *natural evolution* without any central control (Mühlenbein (1991), and the other model is based on *artificial selection* as carried out by human breeders. This model, the *breeder genetic algorithm*, has been used successfully for large-scale optimization problems. The theory of this algorithm is based on the equation for the response to selection, which is also used by breeders. An overview can be found in Mühlenbein and Schlierkamp-Voosen (1993, 1994).

The second learning method models learning in individuals. The emphasis is on real-world applications. The learning method is surprisingly similar to learning by evolution. It can be called the *Darwinian model of individual learning*. The model is based on the philosophy of Popper (1972). From the data seen so far, the learning system generates hypotheses explaining the data. The hypotheses are used to predict the outcome for new data. All hypotheses are preliminary; they are evaluated according to how well they explain the data. A fundamental problem is that *hypotheses cannot be assigned a probability of being true*. This was most clearly stated by Popper (1972). Hypotheses cannot be ranked according to a probability; two hypotheses can only be compared according to their likelihood of explaining the data. This is a general formulation of the classical *induction problem*.

The induction problem can easily be shown. Let the unknown function generating the data be a Boolean function of input size n. If $2^n - 1$ inputs are given, two hypotheses are left which explain all the data. Each hypothesis will correctly predict the output with a probability of only 0.5 for the very last input. If a smaller input set is given, the probability of correct prediction goes to zero rapidly.

Quantum mechanics is faced with a surprisingly similar problem. Not every "history" in quantum mechanics can be assigned a probability of being true. In order to derive an understandable calculus, Gell-Mann (1992) proposes a decoherence functional. It is a complex functional on any pair of histories in the set of alternative histories. Decoherence is also critical to molecular electronics. Two quantum systems that have interacted in the past (which is necessary to process information) and evolve coherently in time cannot be separated again. In order to assure the independent preparation and measurement of the subsystems, it is necessary to include dissipation, which destroys the coherence between the two subsystems. To make my point clear: I am not saying that researchers in quantum mechanics are working on the induction problem for learning systems in general. But within their smaller domain of research, they seem to have similar methodological problems as a designer of a learning system.

Our current approach to learning in an open-world problem is based on the idea of *reflection*. The learning system continuously observes and assesses its own behavior. It tries at every step to be aware of *what it knows and what it does not know*. A system meeting this claim is able to learn incrementally, and, furthermore, actively to explore its environment. Our first implementation is a hand-eye robot which consists of two arms and sensors. Design principles and some applications can be found in Beyer and Smieja (1995) and Smieja (1995).

VIII. A QUANTUM NEURAL COMPUTER

So far, all computers have been designed based on rigid all-or-none concepts. I have shown the limitations of this approach. Currently, more flexible concepts are emulated by software on the otherwise rigid hardware. Intelligence has been taken by many scientists to emerge from the complexity of the interconnections between the neurons of the brain. I have argued in this paper that it seems to be impossible to model this interconnection scheme on a computer. It is transcomputational. It seems therefore fruitless to build an intelligent system by a ''copy-the-brain'' approch.

The most promising way is the ''learning-from-examples'' approach. Unfortunately this approach suffers from the induction problem, which has been discussed most vividly by Popper (1972). I hope that this paper has shown that molecular electronics people and computer scientists should work together for two reasons. The conventional one is just to increase the speed of computation and leave the computational model as it is. The theoretical one is to investigate new models of computation based on quantum mechanics. A promising approach is the ''many histories'' view of Gell-Mann.

REFERENCES

Beyer, U., and Smieja, F. J., Learning from examples, agent teams and the concept of reflection, *International Journal of Pattern Recognition and Artificial Intelligence*, to be published, 1995.

Bremermann, H. J., *Optimizing* Through Evolution and Recombination, in *Self-Organizing Systems* (M. C. Yovits, ed.), Spartan Books, Washington, 1962, pp. 93–106.

Gell-Mann, M., and Hartle, J. B., Quantum mechanics in the light of quantum cosmology, in Complexity, Entropy and the Physics of Information (W. H. Zurek, ed.), Addison-Wesley, New York, 1992, pp. 425–458.

MacKay, D. J. C., Bayesian methods of adaptive models. Ph.D thesis, California Institute of Technology, Pasadena, 1992.

Jaynes, E. T., *Probability in Quantum theory*, in *Complexity, Entropy and the Physics of Information* (W. H. Zurek, ed), Addison-Wesley, New York, 1992, pp. 381–404.

Klir, G. J., and Folger, T. A., *Fuzzy Sets, Uncertainty, and Information*, Prentice Hall, London, 1988.

McCulloch, W. S., and Pitts, W.: A logical calculus of the ideas immanent in nervous activity, *Bull. of Mathematical Biophysics 9*, 127–147 (1943).

Mühlenbein, H., Evolution in time and space: The parallel genetic algorithm, in *Foundations of Genetic Algorithms* (G. Rawlins, ed.), Morgan Kaufmann, San Mateo, 1991, pp. 316–337.

Mühlenbein, H., and Schlierkamp-Voosen, D., *Predictive models for the breeder genetic algorithm: Continuous parameter optimization*, Evolutionary Computation 1, 1–26 (1993).

Mühlenbein, H., and Schlierkamp-Voosen, D., The science of breeding and its application to the breeder genetic algorithm,'' *Evolutionary Computation 1*, 335–360 (1994).

Paaß, G., and Kindermann, J., Bayesian query construction for neural network models, in *Advances in Neural Information Processing Systems 7* (G. Tesauro, D. S. Touretzky, and T. K. Leen, eds.), MIT Press, 1995.

Popper, K. R., *Objective Knowledge*, Clarendon Press, Oxford, 1972.

Rissanen, J.: *Complexity of models.*, in *Complexity, Entropy and the Physics of Information* (W. H. Zurek, ed.), Addison-Wesley, New York, 1992, pp. 117–126.

Smieja, F. J., The Pandemonium system of reflective agents, *IEEE Transactions on Neural Networks*, to be published, 1995.

Traub, J. F., and Wozniakowski, H., *A General Theory of Optimal Algorithms*, Academic Press, New York, 1980.

Von Neumann, J.: On the logical and mathematical theory of automata, in *Collected Works of John von Neumann V*, Pergamon Press, London, 1965, pp. 288–328.

3

Synthetic Nanostructures as Quantum Control Systems

Günter Mahler

Santa Fe Institute
Santa Fe, New Mexico
and University of Stuttgart
Stuttgart, Germany

I. INTRODUCTION

Is molecular electronics simply the "logical" culmination point in the ever continuing miniaturization? As physics is not scale invariant, this cannot be taken for granted, and old questions call for a fresh look: computers are physical systems (machines) (1), but what makes a physical system a computer? Are machines processing information in a well-defined special sense?

It is pretty obvious what computation means in everyday life. So it would seem natural to say that any system capable of performing such a calculation (i.e. "mapping" input into the appropriate output according to a prescribed rule) should be called a computer. Unfortunately, this leads to rather unsatisfying consequences.

For given systems, mathematical functions characterize the relationship between physical observables. Knowing such a function and based on appropriate coding would thus allow us to "use" the physical system as a representation of this very function. For given input field x_j, $(j = 1,2 \ldots)$, *we could read off (i.e. measure) the output field y_i, $(i = 1,2, \ldots)$:*

$$y_i, = f_i(x_j) \tag{1}$$

There are many and, under some modest requirements, nontrivial examples: it is well known, e.g., that under suitable conditions the output of a biconvex lens is equivalent to a spatial Fourier transform of the input. Holographic image reconstruction is another optical realization. Similarly, the relationship between a classical initial state (in terms of generalized coordinates) and the state at some other time t can be seen as a mapping generated by the corresponding Hamilton function. All these examples indicate a kind of information processing, but most people would be unwilling to call such a device already a computer.

What could be missing? The "program" (mapping) is given with the system (lens, Hamiltonian). Selecting other observables (e.g., in terms of the previous ones) would formally change the function f, but unless these new observables are actually measured (prepared), this transformation is not at all *represented* by the physical system. Rather, there should be external control parameters P, by which we can select *different* functions,

$$y_i = f^P_i (x_j) \tag{2}$$

Examples also of this type abound; they include idealized ones like the many-particle Hamilton system with elastic short-range pair interactions and single-particle potentials like hard-wall segments, which has been interpreted as a "reversible billiard ball computer" by ambiguously treating part of it as external parameters P (2). More realistic computing systems (prototypes exist already) are based on spectral hole burning with the electric field as a parameter (3).

Should the definition of a computing device be still more restrictive? Being concerned with various mappings it would be highly inefficient to ignore possible relations between them: typically, they can be thought of as being composed of a set of elementary maps, a property that at the same time simplifies the verification of each mapping. Even complicated ones could thus be realized via sequential applications of elementary steps, which, on the appropriate level of description, might then be identified with the individual computational step n:

$$x_i^{(n)} = f_i^{P(n)} (x_j^{n-1})) \tag{3}$$

Of this type are "abstract machines" like the Turing machine, which employs local operations controlled by the position of the so-called Turing head. For the cellular automaton the step n would comprise updating of the whole array of "cells," usually with fixed P (fixed rules). Unsynchronized versions with changing P appear to be closer to physical realizations in the microworld.

Finally, based on such recursive operations are "computation universal" systems. Specific idealized versions of Turing machines and cellular automata have this formal property (4): Such machines can simulate any other machine (given infinite state space). They are the centerpieces of computation theory.

In the following we shall require a computing system to allow for *function composition*, i.e., different sequences of "elementary processes" (in time and/ or space); we shall not insist on universal computation. The external cost of computation is not discussed here; lower bounds have been proposed by a number of researchers (5). Contrary to most other machines, though, such computers should be able to work even on the molecular level (i.e., exploiting even quantum dynamics proper). At this level the "elementary process" receives its clearest and most fundamental meaning; it can, by any physical means, no longer be decomposed into more fundamental steps; it is, typically, no longer deterministic, and it is no longer a property of the system proper alone, but reflects the embedding into the environment.

II. IMPLEMENTATION OF COMPUTATION

A. On Making Machines

D. Deutsch (6) remarks: "The reason why we find it possible to construct, say, electronic calculators, and indeed why we can perform mental arithmetic, cannot be found in mathematics or logic. The reason is that the laws of physics 'happen to' permit the existence of physical models for the operations of arithmetic such as addition, substraction and multiplication. If they did not, these familiar operations would be noncomputable functions."

Condensed matter physics has traditionally been concerned with large and—in some sense—homogeneous systems. Though models do start on the microscopic level of atoms and electrons, one is eventually interested in macroscopic phenomena emerging on the many-particle level. Correspondingly, applications have primarily been based on the control of macrostates. Even one of the most advanced technologies—ultra large scale integrated circuits—still defines its function on the hydrodynamical level.

It is convenient to visualize nanostructures (whether synthetic or natural) as networks made up of certain "nodes" and "interconnects" ("edges"). Such networks, of course, also exist in the macroscopic world (e.g., traffic nets, telephone nets, etc). Here we restrict ourselves to nodes specified on a molecular level, i.e., on a nanometer scale. The "compartmentalization" on various internal length scales allows for states of various localization types, which carries over to the type of dynamics: various localizations imply different time scales with respect to local coupling operators (like the dipole operator). As a result, dynamical phenomena, which would otherwise all happen on nearly the same time scale, now disentangle into diverse hierarchies of "slow," "fast," and "very fast" ones, etc. This gives rise to new phenomena depending on the time scale of observation and the level of description. A measure of isolation is the respective coherence time. The network states of such a nanostructured system may represent fairly isolated microstates that merely differ with respect to one or a few effective particles. In this sense they become similar to simple atomic states. Molecular electronics will most probably not rely on individual molecules but rather on such networks.

We will thus focus on (quasi-)molecular machinery. It is fairly obvious that such machines exist in the biological world (7). However, as the user is free to formulate a specific task, he cannot expect that the systems of his natural environment will have any tendency to evolve in the desired direction. Contrary to the forces governing biological evolution, human needs or mere wishful thinking have no selective power on the construction level. Machines or tools are therefore either *selected* by its user from a given or evolving ensemble of systems, or are intentionally *designed* (manufactured). Building a "useful" machine will always require some direct interaction and manipulation.

Synthesis and characterization are based on certain rules. It is unlikely that *any* such rule can be imposed on *any* level of description: besides efficiency limits there are severe physical constraints, in particular on the atomic level. Here, the local growing rules can, in principle, be taken care of by self-organization (like atomic layer epitaxy), while nonlocal rules usually require direct manipulation (like lithography). The physical basis of those self-organizational processes appears to be energy minimization constrained by geometry, but also by intentionally blocking unwanted reaction paths etc., which again means manipulation. Anyway, without letting self-organization do the main job, one would be restricted to very small systems, if any. Self-repair—often attributed to so-called "intelligent" materials (8)—is another variant of self-organized synthesis. It can only happen, if the necessary environmental conditions are met even during active "use" of the system. This is obviously the case for some biological systems; it will be hard to mimic in technical systems, as the repair rules tend to be even more complicated than the mere growth rules.

B. Embedding

Machines need to be controlled; control systems are necessarily open systems, i.e., systems embedded in a specific way into an environment. Though this environment cannot be the object of detailed investigation, we will need information about it in terms of boundary conditions. These boundary conditions might be "passive" (like "infinite wall" conditions) or "active" in terms of driving forces (like electrical, thermal, optical, or chemical contacts). The total of conditions defines the physical "interface," through which the system exchanges particles, energy, etc. This exchange constitutes "indirect measurements" of some sort and thus provides communication with the outside world. Of course, this interface is specified in an abstract space, which might include aspects of the material surface of the system; it, again, will depend on the level of description.

C. Control and Complexity

Complexity physics (9), though still not well-defined, is becoming a popular branch of science. This complexity conception differs from that used in information theory (10), where one distinguishes feasible and computable algorithms from those that are not, i.e., are "complex." Most people will expect that "simple" systems and "complex" systems should constitute different dynamical classes; however, we do not want complex systems to mean that they are not accessible by theory. If complexity implied unpredictability or absence of control, complex systems would certainly be "useless" as machines.

For the present purpose it seems most adequate to define complex systems as systems having a rich repertoire of "functions" (set of rules), which can be distinguished (measured) and addressed (controlled). Of course, virtually any individual behavioral mode of this system could be simulated by a much simpler system (with only a few functions or just a single function): it is these selected modes that allow for intuitive any simple theoretical models. So one can, e.g., understand elastic waves without being concerned about atomic structure and electron distributions.) What counts here, however, are not those individual modes but rather the total of inherent possibilities.

It is not guaranteed, though, that *any* function can be realized on *any* level of description: the types of functions will necessarily be restricted if the "molecular" or quantum level should really become important. Applications still possible should include (quantum) computation (based on symbolic assignments, encoding), sensorics (molecular recognition), heterogeneous catalysis (induced molecular reactions), and molecular energy conversion (microactuators).

III. MATERIAL OPTIONS: NETWORKS IN REAL SPACE

A. Geometries

Structural networks are primarily based on some skeleton keeping the nodes in place. This can be achieved in various geometries, which, in turn, influence the network dynamics. The combinations of node types and such network geometries—provided they can be realized—make up a large class of possible materials. Pertinent categories are network dimension, space curvature and boundary conditions (torus, spheres, etc.), node positioning (random, ordered), and set of allowed nodes (nodes of single type A, pairs A,B, continuous distribution of local properties).

Compartmentalization, though being a very common design principle of nature, does not necessarily show up dynamically: any crystal is composed of unit cells, but its properties usually draw on nonlocalized elementary excitations (and nonlocal elementary events). One will thus have to look for more specific material candidates including homogeneous, mixed, and hybrid systems. It appears that dynamical compartmentalization requires *hierarchical structures* with a minimal number of internal length scales of 4 or larger (see Ref. (11); compare (12) for biological systems). A typical example is shown in Fig. 1.

B. Molecular Crystals

Molecular crystals are composed of (organic) molecules, which retain, to some extent, their individuality. It is a natural network, often strongly anisotropic. However, the network character is likely to show up only for larger molecules, if at all; the hierarchical complexity is apparently too small otherwise. Most protein macromolecular crystals have lattice constants between 50 and 150 Å.

C. Langmuir–Blodgett Films

The Langmuir–Blodgett technique is a well-established way to transfer molecules in solution (e.g., water) onto a substrate (13). Molecules to be used may vary from small synthetic polymers up to large natural proteins. These molecules will form a "film." One can make monolayers, multilayers, or heterolayers. It is a simple and rather cheap method, which nevertheless offers much room for manipulations (custom-made structures); it refers to a small scale: the node is identical with a single molecule. However, properties of the node may change with the type of assemblage, and films are far from perfect, being often chemically and structurally unstable. Stabilization is possible, though.

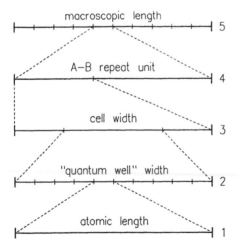

Figure 1   Hierarchical structure. Quantum dot array with alternating dots *A* and *B*.

D. Semiconductor Quantum Dot Arrays

This is a recent development and still in progress. Semiconductor physics becomes a kind of "molecular physics" (14). Quantum dots are mesoscopic "quasi molecules," built of some 10^6 atoms embedded in other semiconductor material (see Fig. 2). This technique is very flexible, being based on well-known material properties and on chemically stable, well defined rigid structures. Disadvantages are related to the complicated processing (combination of epitaxy and lithography); furthermore, the use of mixed crystals implies statistical fluctuations, in particular in small subunits; the node size is typically larger than 100 Å.

E. Inclusions

Inclusion of droplets of one material within another (then denoted as the matrix) are relatively easy to manufacture. The matrix is usually taken as "passive." Such systems have been realized also with semiconductors (see Fig. 2 and Ref. 15). The simple technique is advantageous; the droplet size can be made rather small. On the other hand, position and size of the droplets (= nodes) vary statistically; properties of droplets cannot be manipulated at will: there is no internal structure.

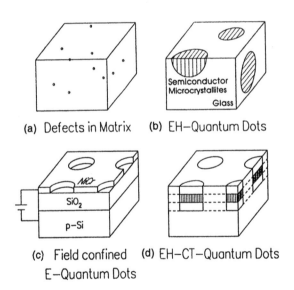

(a) Defects in Matrix   (b) EH–Quantum Dots

(c) Field confined   (d) EH–CT–Quantum Dots
   E–Quantum Dots

Figure 2   Nanostructured semiconductors. (a) Randomly positioned defects. (b) Spherical inclusions. (c) Field-confined quantum dots. (d) Charge-transfer quantum dots.

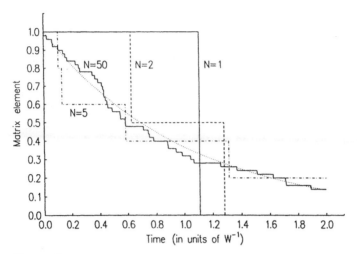

Figure 3   Stochastic simulations. Exponential decay of the matrix element ρ_{22} (occupation of excited state) for $N = 1,5,10,50$ equally prepared two-level-systems.

F.   Zeolites

Zeolites have been studied already for a long time (16). They represent sponge-type skeletons, the cages of which may be doped with "active" atomic groups (nodes). An obvious advantage is that the structure is built by self-organization rather than via step-by-step processing. The structure is thus fairly precise and reproducible. However, the cage size limits the complexity of the node that can be brought in; the doping (with those nodes) is likely to be statistical.

G.   Fullerenes

The recently discovered crystals of large carbon-based spheres ("footballs") may also be of potential use. As the cage size can be varied, it could host a number of different-size active centers (17). Experiments have shown that doping is indeed possible. This is another variant of materials in which the passive geometric positioning and the active degrees of freedom are assigned to different material classes. There are many other examples of molecular container compounds (18). At the same time these systems represent molecular crystals with a rather large and complex unit cell.

H.   Organic Macromolecules on Substrate

An interesting approach to organic macromolecules is by genetic engineering. Bacteriorhodopsin is an example of manipulated macromolecules, which, when randomly fixed on a substrate, have been used for spectral hole burning and even a kind of optical information processing (19): two-dimensional arrays of protein molecules are built on (solid or fluid) substrate. Other variants might envolve DNA or RNA molecules. For these systems additional dynamical levels become available: selection and mutation. The advantage is that systems are stable and optimized by nature. Within some limits they can be manipulated, e.g., with respect to optical properties. The disadvantage is that due to the complicated structure it is difficult to bring these subunits in controlled geometric relation (for the purpose of direct interaction), except by exploiting self-organization (lattice structures; see, e.g., (20)).

IV. NETWORKS IN ELECTRONIC STATE SPACE

We now approach the connection between structure and dynamics: the structural network gives rise to a network in state space. We restrict ourselves to electronic states, and first analyze individual nodes and their physical means of mutual coupling.

A. "Nodes": Models of Eigenstates

The network concept (N nodes, $n = 1, 2, \ldots, N$, with I states each) is based on a "local picture," not far away from the chemist's view of matter. Here, to be sure, we are not concerned with any chemical details of the nodes, which may represent simple atoms, molecules, or mesoscopic clusters like semiconductor quantum dots. The electronic eigenstates emerge from the local states by including the respective node-node interaction (24).

$$H^{(1)} = \sum_{n=1}^{N} H_n + \frac{1}{2} \sum_{\substack{n,m=1 \\ n \neq m}}^{N} H_{nm} \tag{4}$$

$$H_n = \sum_{i_n=1}^{I} E_{i_n}^0 |i_n> < i_n| \tag{5}$$

Here, $|i_n>$ are the respective eigenfunctions of H_n with eigenvalue $E_{i_n}^0$. The interaction might be of single-particle type (overlap)

$$H_{nm} = \sum_{i_n \cdot i_m} U_{i_n i_m}^{nm} |i_n> < i_m| \tag{6}$$

which would lead to conventional single-particle bands, or of two-particle type

$$H_{nm} = \sum_{\substack{i_n,j_n,i_m,j_m \\ i_n \neq j_n, i_m \neq j_m}} W_{i_n j_n i_m j_m}^{nm} |i_n i_m> < j_n j_m| \tag{7}$$

leading to pair bands. In either case the eigenstates would become delocalized superpositions of the original ones. This delocalization trend is quenched by disorder: "diagonal disorder" is characterized by a distribution of the E_{in}^o for fixed i with width ΔE_i.

Finally, there is the "diagonal" pair-interaction

$$H_{nm}^d = \sum_{i_n, i_m} V_{i_n i_m}^{nm} |i_n i_m > < i_n i_m| \tag{8}$$

which leaves the original eigenstates unchanged, i.e., the eigenstates remain simple product states

$$|J> = |\{i_n\}> = |i_1, i_2, \ldots > = |i_1 > \otimes |i_2 > \ldots \tag{9}$$

This interaction is not suppressed by disorder. Note that the nodes are "distinguishable" by their classical index n. The resulting eigenvalues are

$$E_{\{i_n\}} = \sum_{n=1}^{N} E_{i_n}^o + \frac{1}{2} \sum_{\substack{n,m=1 \\ n \neq m}}^{N} V_{i_n i_m}^{nm} \tag{10}$$

Any of these coupling functions, U^{nm}, W^{nm}, and V^{nm}, will go to zero as the distance between the centers n, m approaches infinity. A rough measure for the strength of interaction is thus the maximum nearest neighbor value U^{max}, W^{max}, V^{max}, respectively. Interactions of the type of Eq. (6) are basic to single-particle tunnelling, and, supplemented by electrical contacts for external control, are being investigated as one version of future quantum devices. Interactions of the type of Eq. (8) are basic to "optical wiring," as will be discussed below. Meanwhile, many researchers agree that the latter case of energy transfer—rather than internode charge transfer—should be advantageous for control.

B. Information Feedback

In the strong damping limit coherence effects can be neglected (on time scales large compared to the coherence time), and the master equation for the material system in the eigenbasis of its total Hamiltonian, $|J>$, reduces to

$$\frac{d\rho_{ii}}{dt} = \sum_{K \neq J} \{R_{JK}\rho_{KK} - R_{KJ}\rho_{jj}\} \tag{11}$$

where the ρ_{jj} denote the occupation probabilities of state $|J>$. In this case the open dynamics is easily interpreted. If the $\rho_{jj} > 0$ are all different, there exists a unique decomposition into orthogonal pure states (21). The individual quantum network will be found in any of those eigenstates with a probability as given by its corresponding eigenvalue. When simulated for given external parameters, the history trace of this process,

$$|K > \to |J > \qquad \text{with transition rate } R_{KJ} \tag{12}$$

provides the missing information flow to render the network state a pure state at any instant of time (limited resolution). Any elementary transition is correlated with the emission of a photon (if this is the only damping channel).

The sequence of events represents information. The information gain (either simulated or actually measured) has to be fed back into the subsequent dynamical description. This is exemplified in Fig. 3, where we simulate measurement records by using the stochastic process underlying the rate equation (11) for a network consisting of a single node: a 2-level-system is prepared in state 2 and radiatively decays into state 1 at constant rate. We thus know that the expectation value for the system being still in 2 will decay exponentially. However, as we continuously monitor the individual system (if no luminescence photon has been seen, the system is definitely still in 2; detection of a single photon indicates the system is in 1), we cannot reasonably stick to our original forecast: the new information has to be worked in as it becomes available. Also shown are the results for finite ensembles ($N = 5, 10, 50$). One should note that in the very large ensemble limit no information feedback is necessary any more: the deviation from the time-dependent expectation value (noise) tends to zero. In this way the macroscopic deterministic behavior emerges from the underlying microscopic stochastic rules.

C. Noninteracting Nodes

Let us restrict ourselves to a node consisting of an electronic three-level system ($i = -1, 0, 1; -1$ = ground state, $+1$ = metastable state, 0 = transient excited state). The decay rate between $+1$ and -1 is assumed to be small compared with that between state 0 and $+1$, -1. Such a system might be realized by any of the material classes as discussed in section III. It is likely that semiconductor materials will be the first choice (cf. Fig. 4), though other classes may overtake them in the long run.

The spectral density of the applied electromagnetic field fieled determines the effective transition rates from -1 to $+1$ and vice versa. It is this pattern in time that can be controlled from the outside even under stationary conditions. If the decay of state $+1$ can be neglected, the node can be made to operate as a reversible switch. This is the operation mode for pulsed excitation.

The various states of the node might be associated with a (reversible) structural reorganization (22). This reorganization usually stabilizes the excited states, thus improving metastability. In principle, these structural changes might be sensed via short range distortions by the neighboring nodes.

The most primitive network dynamics is due to a noninteracting ensemble of such nodes with varying paramters (excitation energies). Suppose these nodes have (at least) two discernible metastable states, which can selectively be addressed in frequency space. A pertinent model could be the three-level system

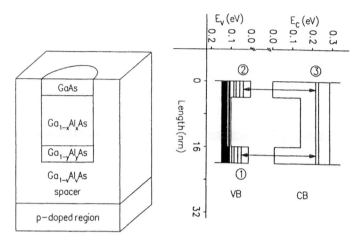

Figure 4 Charge transfer quantum dot. (a) Structure. (b) Energy spectrum and localization of states; Fermi-level between state 1 and 2, lowest conduction band state 3 can selectively be reached from both these states.

discussed above. Depending on the spectral density and polarization of the electromagnetic field (constant in time) certain transitions will be activated.

Mainly the fast (i.e., resonant) transitions are addressed by light pulses: if the final states are long-lived, the transitions are bleached and will then appear as "holes" in a subsequent low-intensity absorption profile (23). These holes, to be sure, are due to ensembles of bleached nodes with almost identical paramters. Spatial resolution is limited by the wavelength of the light source. This persistent spectral hole burning (which is a "one-shot" dynamics) allows for information storage in extended parameter space (i.e., as a static pattern in the frequency domain in addition to the spatial coordinates). Technological applications (on the ensemble level) have long since been demonstrated. Instead of simply storing information, one may use the system as an optical processor. One can, e.g., imprint various holograms (the only step involving direct molecular dynamics), which can then be brought to interact depending on the electric field as a parameter P. For these interactions the network acts as a passive filter according to Eq. (2).

D. Local Charge Transfer and Dipole Interaction

Let us now consider a local few-level system, for which excitations are connected not only with a change of energy but also with a change of charge distribution, which might be approximated by a static electric dipole moment d_{in}. This means that the eigenstates of H_n are eigenstates also of the local dipole operator. If $R_{nm} = R_n - R_m$ is the separation between the two node centers considered, the dipole-dipole interaction implements the diagonal interaction (24) introduced in Eq. (8):

$$V_{inim} = \frac{1}{4\pi\epsilon_0 |R_{nm}|^3} \left(d_{in} d_{im} - \frac{3}{|R_{nm}|^2} (R_{nm} d_{in})(R_{nm} d_{im}) \right) \tag{13}$$

A similar form of interaction could also be induced by local strain fields (elastic dipole-dipole interaction). As a simple example we show in Fig. 4 the idealized structure of a charge transfer quantum dot, consisting of three relevant states with different localization.

E. "Edges": Optical Wiring

Quasi-static electrical fields as driving forces abound in conventional electronics. On the molecular level, however, "wiring" is strongly restricted if not impossible ("molecular wires" have often been discussed but usually in an ill-defined manner). Electromagnetic fields, being independent of material support, are ideal to connect different material sections in space. Generation and guiding of those fields might require material devices on appropriate length scales.

The "nodes" of the state space network are the respective local eigenstates E_{in}^0. "Edges" can then be implemented at will by external driving fields ("optical wiring"), which might be continuous or pulsed.

Let us restrict ourselves to diagonal interactions (cf. Eq. (8)), with or without disorder, and one-particle transitions (i.e., transitions corresponding to the transfer of exactly one electron: this means that the connected states, Eq. (9), differ in only one position). This coupling is selective in frequency space and, possibly, in additional quantum numbers pertaining to a complete set of operators. The latter are important to distinguish degenerate transitions and to formulate selection rules with respect to a specific coupling Hamiltonian (like optical dipole transitions).

Without loss of generality we shall focus on nodes with $I = 2$ local states only. We briefly consider finite 1-dimensional networks with N nodes and periodic boundary conditions (rings). Depending on the parameters in the Hamiltonian, characterized by ΔE_i, U^{max}, V^{max}, W^{max}, and the external (spatially homogeneous) driving field, characterized by its respective amplitudes in frequency space, various state space networks result.

For $I = 2$ there are $n = 2^N$ eigenstates, each directly connectable to $c = N$ others, resulting in a total of $e = cn/2$ allowed connections ("edges") out of $n(n - 1)$ possible ones; in this sense the networks are sparsely connected. In Fig. 5 we see how, under these constraints, the structural network translates into a state space network: for $N = 2$ we get a ring, for $N = 3$ a connected double ring (as a projection of a 3-dimensional cube on a plane), for $N = 4$ a torus (as a projection of a 4-cube), for $N = 5$ a concentrical double torus (or 5-cube) etc. We also indicate to what extent the transitions can selectively be addressed in

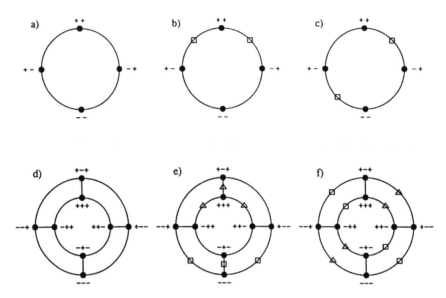

Figure 5 Network topologies in state space for N two-level systems. The edges ("optical wiring") are decorated to identify equal transition energies. (a) $N = 2$, no interactions, identical nodes: 1×4 (fold) transition energies. (b) $N = 2$, with interaction, identical nodes: 2×2 (fold) transition energies. (c) $N = 2$, no interactions, different nodes: 2×2 (fold) transition energies. (d) same as (a), but $N = 3$: 1×12 (fold) transition energies. (e) same as (b), but $N = 3$: 2×3 and 1×6 (fold) transition energies. (f) same as (c), but $N = 3$: 3×4 (fold) transition energies.

frequency space only, depending on local parameter variations ("disorder") and/or interactions. In Fig. 5c,f the transitions from any given node are distinct in frequency space, while in Fig. 5b,e this is not the case. The remaining degeneracies could be lifted if other types of interactions according to Eqs. (6) or (7) were included.

F. Adaptive Random Walk

Optical switching of a charge-transfer node will be connected with a change of the local dipole moment. This change is felt by other charge transfer centers reacting with a shift of their respective resonance frequencies. In this way the nodes become coupled in their driven dynamics (electron transfer between the nodes is excluded here). This coupling defines local stochastic rules of the type of Eq (3), where P denotes here the optical environment and can thus be manipulated: the rules specify the probability for a specific transition to occur in a given node as it depends on the states of the neighboring nodes. We note in passing that charge transfer plays a crucial role also in biological systems.

On a completely ordered lattice of nodes as introduced in IV.E, the local rules implemented by the optical environment will be the same everywhere. This significantly limits but also simplifies control. A given (constant) electromagnetic field will—depending on its spectral density—induce a characteristic stochastic dynamics in the network. This dynamics in state space maps onto a time-dependent pattern in real space. It has been shown (11) that the resulting pattern in real space can resemble, e.g., the dynamics of an Ising model at a specific temperature and with ferro or antiferro coupling. For a spatially varying electromagnetic field even some kind of image processing is possible: the type of processing depends on the type of observation (pump and probe) (24).

More complicated lattices offer additional possibilities. In lattices with alternating nodes (ABABAB . . .) one can alternatingly address A and then the B sublattice (11). This pulse sequence allows us to realize a kind of clocked dynamics, where one sublattice is updated in reference to the other sublattice as a fixed neighborhood. It is essential then that the charge transfer states have a long lifetime (at least large compared to the switching time and the operation time of the network). Figure 6 shows a single local transition frequency in an ABAB chain as it depends on the state of the neighbors: it is possible to comprise the influence of all but the next nearest neighbors into nonoverlapping bands; broad-band light fields can thus implement rules for nearest neighbor interaction only (even though the physical interaction does not have this property). The

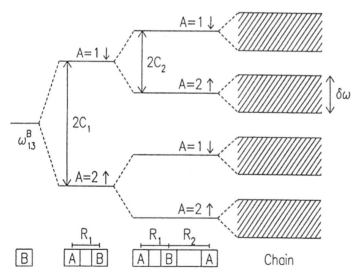

Figure 6 Infinite chain of alternating quantum dots: excitation frequency ω_{13}^{B}; from left to right: isolated node B, influence of left neighbor of type A, influence of left and right nearest neighbors, broadening by the rest of neighbors of the chain.

resulting dissipative pattern is, approximately, equivalent to a cellular automaton (cf. Fig. 7 and Ref. 25). In this case the finite lifetime of the respective metastable states and the finite errors induced by the unavoidable stochasticity of the switching is neglected. Without further pulse the present state would be preserved.

Finally, in the case of a random net, the response to a homogeneous light field will be inhomogeneous: the rules depend on space. Nevertheless, the various subsystems do interact (via dipole-dipole interaction). If only very specific frequencies are contained in the driving field, only a few localized sections of the network will be stimulated. With white light, on the other hand, reaction paths would be selected only according to their built-in probabilities. This type of adaption can be specified in more detail: Network states randomly visited for given parameters define a specific attractor basin. These basins, in turn, define also a local repertoire, which thus multiplies the local repertoire of built-in neigborhoods.

V. QUANTUM NETWORKS

A. Coherence and Entanglement

In the preceding section we have discussed networks states, which clearly are based on quasi-molecular nodes and thus certainly require a quantum description. Strong damping, however, has been assumed to prevent the system from developing the central feature of quantum behavior: the superposition of states (26).

The superposition principle is a formal consequence of the linearity of the Schrödinger equation. It means that if $|\psi_1 >$ and $|\psi_2 >$ are solutions, then also

$$|\phi > \, = \, \alpha_1|\psi_1 > \, + \alpha_2|\psi_2 > \tag{14}$$

where α_i are complex amplitudes. For a composite quantum system, any pure state can be written as a superposition of product states. For example, for two 2-level systems one may obtain

$$|\phi(1,2) > \, = \, \alpha_1|\psi_1(1) > |\psi_2(2) > \, + \beta|\psi_2(1) > |\psi_1(2) > \tag{15}$$

where $\psi_i(\mu)$ are the basis states for subsystem μ. An important criterion for so-called entangled states is that they cannot be written as simple products; they do not factor. One easily generalizes the concept of entanglement to more than two nodes and more than two states per node.

Entangled states have very strange properties, if exposed to local measurements (i.e., measurements affecting only one subsystem). Such measurements can be modelled as projections on the respective local measurement basis giving (for a two-dimensional state space) either "up" or "down." Such repeatedly occurring projections underlie the rate equation; here, they are supposed to be "rare" events (possibly even under external control) so that coherence has time to develop.

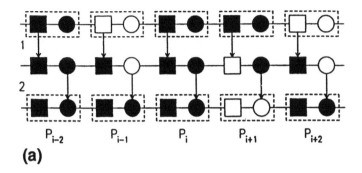

(a)

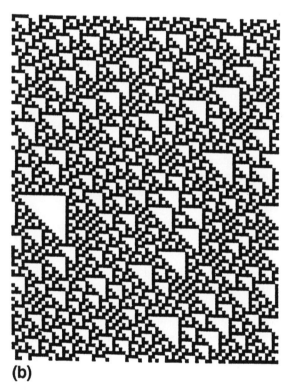

(b)

Figure 7 Controlled dynamics in a one-dimensional ABAB superlattice. (a) Chain and pulse sequence, corresponding to one "time step" of the automaton. (b) Clocked simulation of the unidirectional rule $P_i(t+1) = P_{i-1}(t)P_i(t)$ with $P = 0$ for state $-$, $P = 1$ for state $+$. Shown are the states $-$ (black) and $+$ (white) for the A cells only after each complete double pulse updating the A and the B cells. Time runs from top to bottom.

A measurement on subsystem 1 resulting in the state $\psi_1(1) >$, say, would thus imply by Eq. (15) that subsystem 2 is in state $\psi_2(2) >$, irrespective of its spatial position. After measurement the entanglement is completely destroyed; the system is in a product state.

These "spooky actions at a distance" have been confirmed by experiment. It has further been shown that the entangled states can be considered eigenstates of operators characterizing the *total system* such that the operators of any part are incompatible: local properties are thus not even defined! This is easily demonstrated by the fact that the local entropy (i.e., the missing information) of either subsystem is just 1 bit (for $\alpha = \pm\beta$) even though the state of the total system has zero entropy (it is completely specified by $|\phi(1,2) >$).

Are these states useful? A first explicit application has been proposed not in computation but in communication (29): entangled polarization states of photon pairs travelling away from a common source may be used as a communication channel for which eavesdropping would always be detected by the legitimate users. This idea is based on the fact that eavesdropping requires a local measurement, and any local measurement would destroy entanglement and thus distort the correlations found by the communicating parties.

As far as computation is concerned, one should note that the state space including all types of entanglement is huge compared to that without (which is the "classical" state space of Ising spin models). Consider a network with N two-level subsystems: the density matrix of each local subsystem is specified, in general, by three independent real parameters, so that any product state requires $3N$ numbers. However, allowing for general superpositions we need to consider a space of $2^{2N} - 1$ parameters. The ratio

$$\gamma = \frac{3N}{2^{2n} - 1} \tag{16}$$

is a measure for that "classical" fraction of the total state space, and amounts to 10^{-58} already for $N = 100$! Without decoherence effects (induced by the environment) even small networks would most likely live in highly entangled sections of Hilbert space. This exponential growth of state space makes it virtually impossible to study completely coherent networks under fairly general conditions.

This inherent difficulty in simulating quantum systems proper by classical computers has prompted R. Feynman (30) to suggest that also the reverse might be true: that quantum systems could, in principle, extremely efficiently simulate classical systems (and problems).

B. Unitary Quantum Computation

In order to exploit entanglement, one will have to concentrate on unitary dynamics, by which, in principle, any superposition can be generated. General unitary transformations are, of course, extremely involved and likely to be beyond any external control. It has therefore been of considerable interest that D. Deutsch (6,27,28) could show that any such transformation can be composed from elementary types, called quantum gates (in analogy to the gates constituting a classical computer). The resulting (highly formal) architecture has been termed the quantum Turing machine. It is a variant of reversible computation.

Up to now neither the implementation of such a machine nor its control appear to be within reach of available technology. One should bear in mind that this type of dynamics would be restricted to time scales small compared to the pertinent coherence time, which, for typical mesoscopic systems, will hardly exceed a few msec, often much less. Even if this limitation could be relaxed, it remains doubtful whether all the necessary "point-measurements" for preparation, checking "counter-bits", reading out of final data, etc, can actually be made efficient enough.

Nevertheless, such a (still distant) quantum technology should be seen as the real challenge for molecular electronics, as such a machine could hardly be built by other means; its simulation on a classical computer would make no sense. While present-day computer architectures are hard to compete with in terms of rather "exotic" molecular material, here is a goal worth going a long way for.

C. Clues from Experiments

Extended networks showing explicit quantum behavior have not yet been realized. Many aspects, though, have already been seen in small networks, often accidentally. Examples from nanotechnology and nanobiology exhibit striking analogies, which are not yet understood in detail. The following list is not intended to be complete.

1. Nanophysics

Stochastic control of individual atoms in a Paul trap (31,32). This shows that the stochastic dynamics of optically driven atoms can be changed by choosing an appropriate light-field environment.

Single defects in matrixes dynamically controlled by their environment (33, 34). These show that the stochastic dynamics of single quantum objects can also be changed by the *microscopic* environment. In the present case the resonant frequency of the respective defect is changed by the neighborhood being in different charge distribution states; the dynamics of the neighborhood is not controlled in the present case.

Spectral hole burning in electric fields.   This shows that optical transition frequencies of local defects are shifted externally due to the field-induced change of charge distributions (dipole-moments, etc.).

Laser-induced unimolecular dissociation (35).   This shows that jumps in total reaction rates occur as the number of accessible transient states is varied by vibrational excitation.

Coherent control (36).   This shows that molecular excitations, etc., can selectively be controlled by phase-sensitive laser pulses. The resulting dynamics can be interpreted to result from (interfering) paths in a coherent quantum network.

Electrical current through a point contact controlled by the cooperative action of a few nearby centers (37).   This shows that a macroscopic variable (current) can stochastically be varied by the correlated action of some quantum objects. Their dynamics is, in the present case, again uncontrolled (spontaneous).

STM study of switching atoms between tip and substrate (38).   This shows that by means of an external electrical field individual atoms can change places, which can be detected either by the STM image or by the dependence of the tunnelling current on the atomic position.

Dependence of molecular adsorption on geometry and charge transfer within a bimetallic surface (39).   This shows that the properties of surfaces are not necessarily fixed by their chemical nature. This gives space for dynamic manipulations.

Charge separation in halide–viologen complexes (40).   This shows that optically induced charge separation can be made long-lived also in synthetic molecular structures.

Quantized circulation of superfluid He (41).   This shows that single quantum events (here: phase-slips) can be observed also in macroscopic variables.

Superparamagnetism (42).   Arrays of nanometer-scale magnetic particles show quantum effects (tunnelling) associated with the mesoscopic magnetic degrees of freedom.

2. Nanobiology

DNA electrophoresis on synthetic surface structures (43).   This shows that also man-made structures can induce specific paths for organic macromolecules; the experiment uses optical techniques.

Chemically driven single motor cells (44).   The stochastic strike sequence shows that also biology goes quantum. The energy of one chemical reaction cycle is transformed into one quantum of mechanical energy. The sequence of events is stochastic and depends on the boundary conditions imposed by the respective experiment.

Current through individual ion channels (in living cell membranes) (45). This shows that the ion current controlled by an individual macromolecule within a living system exhibits two-level current fluctuations (on–off). Interestingly enough, the open- and shut state is induced chemically by the binding and unbinding of a specific molecule.

Stochastic signals of single neurons (46). These indicate that the dynamics and information coding of neurons might be similar to the behavior of optically driven atoms.

Influence of surface topology on molecular recognition (47). This shows details of how "pocket geometry" is used in nature for molecular recognition. Different pockets may have different functions and different selective powers.

Optical monitoring of brain function (48). This shows that dynamical features of a biological network can be recorded by locally detecting the reflection of specific light frequencies.

Neurobiology of olfactory receptors (49). This shows that the adsorption of odor molecules changes the dynamics of the substrate molecules for detection purposes.

Biological motors (50). These shows that chromosomes might be self-propelled, i.e., chemical–mechanical energy conversion appears possible on the molecular level.

Quantum effects thus also control biological systems on a certain level of description. Fröhlich (51) and others (52) have considered coherent and/or collective states like Bose-condensed or superconducting states. Experimental evidence, though, is still rather inconclusive. The more recent experiments cited above seem to indicate intriguing analogies between nanotechnological and nanobiological systems in terms of stochastic behavior. We take this as an indication that similar functions on similar scales are likely to rely on similar physcis.

VI. SUMMARY AND CONCLUSIONS

For a quantum system in a classical environment, additional information emerges at the interface via continuous measurement according to stochastic rules. These rules, in turn, which derive from quantum mechanics, can be manipulated by selecting an appropriate environment; it seems that, due to the sensitivity of the quantum system to its surrounding, the control by this environment has to be much more detailed than for classical systems. Coherence and entanglement effects further underline this observation.

Machines require control and are thus open systems. Control of nanostructures on the quantum-mechanical level appears to be possible. This possibility appears to depend on a hierarchical structure (network). The realization of such nanostructures can be based on different classes of materials; the optimum choice is not clear yet. However, if there will be any useful quantum machinery at all in the near future, it will emerge from materials research combined with theoretical simulations and modelling.

It appears likely that the functions of quantum systems will be restricted to computation, sensorics, and catalysis. The function of computation has been reserved to physical systems capable of producing many different elementary processes from which a large variety of dissipative patterns can emerge. It is expected that quantum dynamics may extremely efficiently give rise to complex behavioral patterns (an efficiency typical for analogue simulations), but hardly anything that could not, in principle, be simulated by a classical computer. Otherwise, quantum dynamics would have to involve something noncomputable, for which there is presently no indication. However, classical systems cannot simulate quantum systems in an analogue way. This is so because classical objects have just one complete measurement basis, while quantum objects have to be forced into one reference frame, while others then become incompatible. In present computers this potential is suppressed. Control includes the selection of rules from a repertoire and the measurement of the state evolution. The latter is severely constrained by temporal and spatial resolution limits. The detection on coarse-grained scales might, however, be combined with different types of observational levels. The flexibility is likely to be limited, though; this is why we did not require universal computation. There tends to be a trade-off between universality and efficiency. Information processing systems with largely different capabilities have been considered, starting from simple (molecular) storage devices to more sophisticated network dynamics. Some of these employ ensembles; however, the basic function does not depend on that, contrary to devices defined on the hydrodynamic level. Challenging applications might include hybrid scenarios in which a conventional computer controls a virtual environment for a molecular network. Besides the promise of technical applications there are strong indications that the dynamics of synthetic nanostructures might present useful analogies to the dynamics of nanobiological systems: they eventually operate on similar scales.

ACKNOWLEDGMENTS

Part of this work has been carried out while on sabbatical leave from the University of Stuttgart. I thank Carlton Caves, Stuart Kauffmann, Steen Rasmussen, and Hartmut Körner for valuable discussions. A travel grant from the Deutsche Forschungsgemeinschaft is gratefully acknowledged.

REFERENCES

1. Cf., e.g., Landauer, R., *Int. J. Theoret. Phys.*, *21*, 283 (1982).
2. Fredkin, E., and Toffoli, T., *Int. J. Theor. Phys.*, *21*, 129 (1982).
3. Meixner, A. J., Renn, A., Bucher, S. E., and Wild, U. P., *J. Phys. Chem.*, *90*, 6777 (1986).
4. Albert, J., Culik, K., II, *Complex Systems*, *1*, 1 (1987).
5. Zurek, W. H., *Nature*, *341*, 119 (1989).
6. Deutsch, D., *Proc. Royal Soc. London*, *A400*, 97 (1985).
7. Schneider, T. D., *J. Theor. Biol.*, *148*, 83 (1991).
8. Hong, F. T., *Nanobiology*, *1*, 39 (1992).
9. Anderson, P. W., *Physics Today*, July 1991, p.9.
10. H. Mühlenbein, this volume.
11. Mahler, G., Körner, H., and Teich, W., in *Festkörperprobleme/Advances in Solid State Physics* (U. Rössler, ed.), Vieweg, Braunschweig, 1991, Vol. 31, p.357.
12. Baer, E. N., Hiltner, A., and Morgan, R. J., *Phyics Today*, Oct. 1992, p.60.
13. Tredgold, R. H., *Rep. Progr. Phys.*, *50*, 1609 (1987).
14. Reed, M. A., *Sci. Am.*, Jan. 1993, p. 98.
15. Ekimov, A. I., Evros, Al. L., Onushchenko, A. A., *Solid State Commun.*, *56*, 921 (1985).
16. Schulz-Ekloff, G., in *Zeolite Chemistry and Catalysis* (P. A. Jacobs et al., eds.), Elsevier, Amsterdam, 1991, p.65.
17. Ross, M. M., and Callahan, J. H., *J. Phys. Chem.*, *95*, 5720 (1991).
18. Cram, D. J., *Nature*, *356*, 29 (1992).
19. Thoma, R., Hampp, N., Bräuchle, C., Oesterhelt, D., *Optics Lett.*, *16*, 651 (1991).
20. Nagayama, K., *Nanobiology*, *1*, 25 (1992).
21. Fano, U., *Rev. Mod. Phys.*, *29*, 74 (1957).
22. Körner, H., and Mahler, G., *Phys. Rev. Lett.*, *65*, 984 (1990).
23. K. K. Rebane, this volume.
24. Körner, H., and Mahler, G., this volume.
25. Teich, W. G., Obermaier, K., and Mahler, G., *Phys. Rev.*, *B37*, 8096 (1988).
26. Zurek, W. H., *Physics Today*, Oct. 1991, p.36.
27. Deutsch, D., *Proc. Roy. Soc. London*, *A425*, 73 (1989).
28. Deutsch, D., and Josza, R., *Proc. Roy. Soc. London*, *A439*, 553 (1992).
29. Ekert, A., *Phys. Rev. Lett.*, *67*, 661 (1991).
30. Feynman, R., *Int. J. Theoret. Phys.*, *21*, 467 (1982).
31. Nagourney, W., Sandberg, J., and Dehmelt, H., *Phys. Rev. Lett.*, *56*, 2797 (1986).
32. Sauter, Th., Blatt, R., Neuhauser, W., and Toschek, P. E., *Opt. Commun.*, *60*, 287 (1986).
33. Ambrose, W. P., and Moerner, W. E., *Nature*, *349*, 225 (1991).
34. Orrit, M., and Bernard, J., this volume.
35. Lovejoy, E. R., Kim, S. K., and Moore, Moore, C. B., *Science*, *256*, 1541 (1992) 1592) .
36. Rice, S. A., *Science*, *258*, 412 (1992).
37. Farmer, K. R., Rogers, C. T., and Buhrman, R. A., *Phys. Rev. Lett.*, *58*, 2255 (1987).

38. Eigler, D. M., *Nature*, *352*, 600 (1991).
39. Rodriguez J. A., and Goodman D. W., *Science*, *257*, 897 (1992).
40. Vermeulen, L. A., and Thompson, M. E., *Nature*, *358*, 656 (1992).
41. Avenel O., and Varoquaux, E., *Phys. Rev. Lett*, *60*, 416 (1988).
42. Awschalom, D. D., Vincenzo, D. P., and Smith, J. F., *Science*, *258*, 414 (1992).
43. Volkmuth, W. D., and Austin, R. H., *Nature*, *358*, 600 (1992).
44. Uyeda, T. P. Q., et al., *Nature*, *352*, 307 (1991).
45. Neher E., and Sakmann, B., *Sci. Am.*, March 1992, p.44.
46. Levitan, I. B., and Kaczmarek, L. K., *The Neuron*, Oxford U Press, Oxford, 1991.
47. Matsumura, M., Fremont, D. H., Peterson, P. A., and Wilson, I. A., *Science*, *257*, 227 (1992).
48. Haglund, M. M., Ojemann, G. A., and Hochman, D. W., *Nature*, *358*, 668 (1992).
49. Shepard, G. M., *Nature*, *358*, 457 (1992).
50. Mitchison, T., Evans, L., Schultze, E., and Kirschner, M., *Cell*, *45*, 515 (1986).
51. Fröhlich H., *IEEE Transactions MTT*, *26*, 613 (1978).
52. Dawydow, A. S., *Solitons in Molecular Systems*, Reidel, Dordrecht, 1985.

4

Langmuir-Blodgett Techniques

I. R. Peterson

Coventry University
Coventry, England

I. RELEVANCE TO THE CRITICAL AREAS

A. Architectures and Quantum Effects

Molecular electronics is a vision of the future, with very real promise. In some senses, it is already here. Transistors are available in which the active layer has a thickness of the order of 10 nm, comparable to the wavelength of thermal charge carriers and to the linear dimensions of many biomolecules (1,2). Input and output transducers are available in which molecular materials and processes play a vital role (3). However, molecular electronics, in the sense of data processing networks whose functional elements are of molecular size in all dimensions, has not yet arrived.

What, then, are the problem areas that must be addressed before a true molecular electronics technology can be implemented? In this chapter, component and systems requirements will be discussed in the light of the problem of manufacture. It will be shown that there are still problems to be solved in the fabrication of a molecular electronic system, and that the Langmuir–Blodgett technique is one of the more promising methods for assembling functional molecules.

Has perhaps a molecular electronics technology not yet been demonstrated because existing system architectures are inadequate? There is no doubt that a molecular electronics technology should lead to a cornucopia of extremely inexpensive processing elements, whose potential computing power would be wasted by connecting them in a simple von Neumann architecture. Of course, this situation is already being faced in conventional microelectronics. The challenge—scarcely a problem—of how most effectively to integrate extremely large numbers of components has already led to a number of imaginative new architectures consisting of arrays of von Neumann processors (4). Another proposed architecture is that of the Hopfield associative memory (5–7), which has also been implemented using conventional technology, although not yet on the mega-gate scale. Hence there is a choice of approaches to this challenge, and while it is fascinating it is not critical for further development.

Has a molecular electronics technology not been demonstrated because the existing principles of operation are inapplicable at the molecular level? Certainly, in a nanometer-sized device it will not be possible to ignore nonclassical quantum effects. Quantum uncertainty will play a large part in its noise analysis and may place an important lower bound on its size. However, in addition to imposing restrictions, quantum effects may also liberate. New principles of device operation should become possible and may provide a way around any limitations they impose on conventional types of device. Many chapters of the present book deal with just such alternative, inherently quantum, principles.

B. The Scope for New Principles

How different may a novel technology be? Present microelectronic technology involves charge carriers interacting with baseband electrical potentials, i.e., signals whose frequency spectrum extends down to zero, and the circuitry is powered from a low-impedance voltage source of zero frequency. Optical switching using nonlinear optical (NLO) effects is often considered as an alternative (8). In such a technology, both the signal and the power source are narrowband and are conveyed by waveguide structures.

However, it is possible to exclude this range of possibilities from consideration in the context of the present book. The volume per function of elements that process waveguided photons is inherently greater than λ^3, where λ, the optical wavelength, is of the order of 1 μm. In fact, functional units fabricated from presently known NLO materials are many times bulkier than this. While NLO technologies are of great interest for applications requiring maximum bandwidth and only low functional densities and levels of integration, e.g., telecommunications, they cannot be described as having functional elements of molecular size.

In conventional microelectronics, there is a trend towards the use of optical communications between subassemblies to minimize cross talk and dispersion, and these reasons will no doubt be just as cogent for molecular electronics. Clearly though, this is only an option if the subassemblies are significantly larger than λ. Waveguided communications are effectively ruled out at the component level when the latter have dimensions of the order of 10 nm.

Notwithstanding these considerations, optical communications will probably have a limited, but important, role to play in the development of a molecular electronics technology. For communication between the molecular system and the external world, it is still extremely difficult to make more than two electrical connections to a functional molecular system of a size within the capabilities of current synthetic techniques. It is most likely that optical signals will have to be used for input to or output from such systems (9).

With this one exception, the most promising option is to use baseband signal representation at device level, and to power the devices using a baseband (in fact DC) voltage supply. Interconnections between components will still have to use a conducting ''wire'' that conveys information by means of changes of electrostatic potential, and this potential will still have to be maintained, or changed rapidly when required, by the controlled flow of charge carriers, even if the control mechanism involves inherently quantum phenomena.

It is possible to imagine novel quantum baseband components. A recently proposed ''quantum transistor'' requires charge carriers to tunnel, which is only possible in a device of dimensions comparable to the wavelength of thermal carriers (10). In other respects, its principle of operation is conventional. The tunnelling occurs through a propagation region under the influence of a control electrode, in a manner exactly analogous to the operation of a vacuum tube, field effect transistor, or bipolar transistor.

In fact, ''quantum'' components are no newcomers to the field of data processing: the bipolar transistor is a very successful example. Not all of them have prospered. Just like the above-mentioned ''quantum transistor'', Esaki tunnel diodes also rely on the tunnelling of carriers (11). By the standards of the 1960s when they were first demonstrated, they were extremely high speed devices, but the attempts made to construct computers from them were unsuccessful. A more recent example is the Josephson junction (12), whose principle of operation involves interference between coherent Cooper pairs of electrons. In both cases the components were individually perfectly capable of performing the required operations, but insuperable difficulties were encountered in integrating them into a large system. There is now a consensus that a designable data-processing system can only be built from three-terminal devices showing gain (13).

It is therefore necessary to take a pragmatic approach. Novel components will almost certainly enrich the possibilities of design at the molecular level, although there is no fundamental reason why, given the appropriate materials, conventional principles cannot also provide competitive alternatives. As yet, no general agreement has emerged. It is clear that conventional technology using conventional materials cannot be pushed as far as the molecular level, because of problems with high-level doping and tunnelling between adjacent circuit elements. However, both of these problems are in principle solvable using new materials.

C. The Case for Organic Materials

Among the possible new materials, organics have a number of significant advantages. They are inherently designable: apart from a few very simple combining rules at very short length scales, there are very few limitations on the structures that can be synthesized. Unlike typical inorganic semiconductors, they need not represent a thermodynamic ground state: in chemically mild environments, covalent bonds are indefinitely metastable. They are usually anisotropic, restricting the directions in which carriers can move, thus overcoming a major disadvantage of conventional materials, where carriers can tunnel between adjacent devices at high packing densities. The most easily synthesized organization of an organic semiconductor is in fact linear, and highly anisotropic conductivity has been demonstrated in a number of systems (14,15).

For macroscopic systems, one-dimensional conductors are a decided disadvantage, because there is no way for charge carriers to avoid the traps that are inherently present in all charge transport media. However when the transport is only over molecular separations, this disadvantage should disappear. Another possible problem with 1D conductors is a marked tendency towards polaron formation, in which the presence of a charge carrier perturbs the nuclear positions. The movement of the carrier thus involves the correlated movement of many heavy nuclei, so that the former acquires a rather large mass. It is possible that with the appropriate choice of medium, this mechanism can be turned into a feature rather than a disadvantage. For example, in spite of their larger mass, acoustic polarons can display very high mobilities (16).

For many years, interest in polymeric semiconductors was largely limited to polyacetylene. While significant new processing methods have been developed to improve its characteristics, and the unusual soliton phenomena associated with its inherent Peierls distortion are of great scientific interest, the latter is also associated with very large electron–phonon coupling and essentially eliminates the possibility of significant optoelectronic effects. An extremely promising new development in this field is the demonstration of the injection and propagation of both electrons and holes and their radiative recombination in processable polyphenylenevinylene (PPV) derivatives (17,18). The backbone structure of PPV is shown in Fig. 1.

D. Fabrication

The only way to achieve consensus about the best principle of operation and the best material for its implementation will be to compare actual performances of

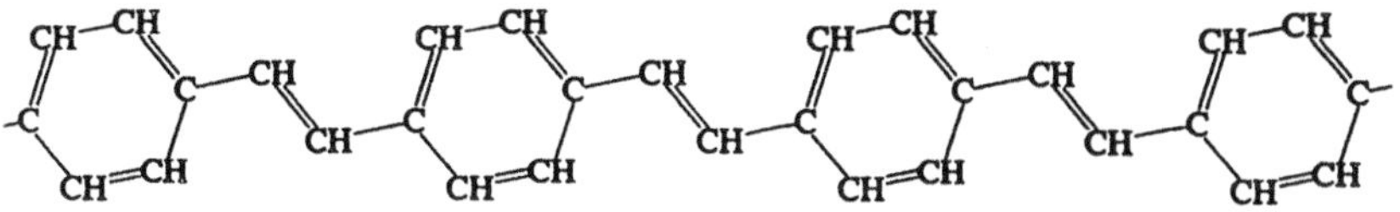

Figure 1 The molecular structure of unsubstituted polyphenylenevinylene (PPV). (Adapted from Ref. 9.)

demonstration devices and systems. Unfortunately the author is not aware of even an unsuccessful attempt to demonstrate a three-terminal device of nanometer dimensions. There have been demonstrations of a two-terminal device, the molecular rectifier (19,20). The fact that a two-terminal device can be attempted, while a three-terminal device has not, highlights a third problem that really must be solved before a molecular electronic device can be demonstrated. The required molecular structure must first be fabricated and connected to an external circuit.

There is no reason to think that the control of fabrication parameters will be any less important at the molecular level than it is in conventional technology. Indeed, there is every likelihood that it will be even more demanding, requiring extremely high levels of material purity and extremely low levels of particulate contamination, in addition to the obvious requirement of nanometer dimensional control. It is of course necessary to think of the future and to plan for arrays with billions of processing elements, but it would nevertheless be extremely exciting to demonstrate the correct operation of just a single gate, or even a single active device, of molecular size. In spite of a number of very promising new methods of molecular manipulation, this demonstration still lies in the future.

II. RELEVANCE TO DEMONSTRATORS

A. The Aviram–Ratner Molecular Rectifier

The technique of Langmuir–Blodgett (LB) deposition is indisputably important for molecular electronics in this capacity, as it is one of the few ways of fabricating low-defect organic structures patterned on a molecular scale. It is well adapted to the fabrication of molecular rectifiers. The first proposal of this sort was put forward two decades ago by Aviram and Ratner (AR) (21) and involves an oriented monolayer of molecules of the specific A–Σ–D structure shown in Fig. 2.

There have in fact been a couple of reports from the groups of Ashwell (20) and Sambles (19,20) of partial implementation of this concept using the LB technique. It should be noted that the molecules used bore only a faint resemblance to the original A–Σ–D concept, because the latter type of material is

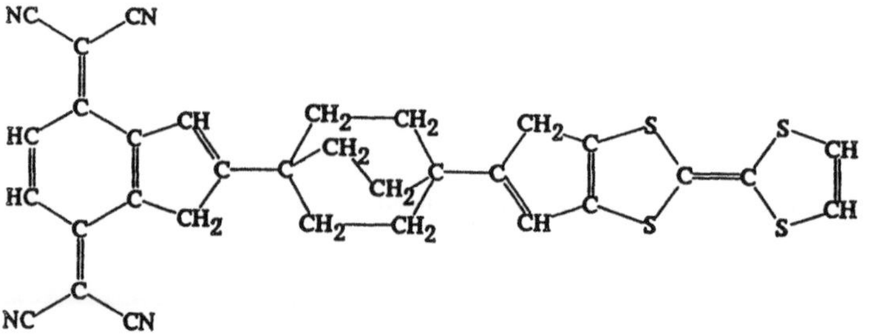

Figure 2 The A–Σ–D molecule proposed by Aviram and Ratner. (From Ref. 21.)

very difficult to deposit as an LB monolayer. It should also be noted that the reported characteristics of the device were extremely poor. They showed a very low rectification ratio, very low switching speed, and very high levels of forward-current noise. Their significance lies in that they are actual demonstrations involving the internal electronic transport properties of molecules.

Of course, since the lateral dimensions of the Ashwell–Sambles demonstrators were measured in millimeters, they were not much more "molecular electronic" than is an Esaki tunnel diode. However there is a difference, in that they were composed of individual molecules functioning independently. There is no reason in principle why a device of molecular dimensions would not work in the same way, and the measurement of the device characteristics allows the immediate deduction of the electrical characteristics of a single molecule in an appropriate environment. At the moment it is known that the electrical properties of bulk organic materials are dominated by extrinsic, defect effects. There is a basic need for more data on high-field intramolecular charge transfer on which to base a rational design.

It is true that there are models for the rates of charge transport in biomolecules and redox reactions in small organic molecules that are in order-of-magnitude agreement with some experimental results (22). However there is no detailed understanding on the electronic level (23). Best agreement is obtained when the driving force is of the order of kT. Although there is no method for determining the distribution of electric fields in these systems, those in the rate-limiting regions must be small, well under bulk breakdown levels, leading to essentially linear current–voltage response. An electronic data processing system must be based on nonohmic devices (24), and this can only be achieved (24) by operating them at voltages significantly higher than $kT/e \approx 25mV$.

In high-field experiments, there are disagreements of many orders of magnitude between theory and experiment. A major structural component of many organic compounds, the aliphatic chain, is normally considered to be chemically and electrically inert, with no electronic levels accessible to thermal charge carriers. However in a study of high-field conduction in LB films of the fatty acid soaps, Sugi et al. were forced to conclude that such levels exist on each terminal methyl group (25). It is also necessary to assume their existence to explain the slow dielectric breakdown of polyethylene (26). Under suitable circumstances polyethylene (27) and other long-chain compounds can metastably support extremely high current densities in a switching phenomenon that must involve accessible electronic levels. So far there is no understanding of their nature.

B. Comparison with Alternative Techniques

The Langmuir–Blodgett (LB) technique is not the only method for fabricating an oriented monolayer. There are alternative techniques of physisorption (28) and chemisorption (29) that have attracted much recent attention and that could in principle also be used. All of them rely on specific chemical interactions between an interface and a part of the molecule to be incorporated. In the case of the LB technique, the molecules are first assembled at an air–water interface before transfer to their ultimate substrate. In the case of physi- and chemisorption, they are assembled in situ.

All of these techniques provide a way in which to make two electrical contacts to a single molecule. In this way it is possible to apply potential differences of up to a few volts to its opposite end. While this corresponds to electric fields of GV/m, well above the threshold for dielectric breakdown of bulk materials, breakdown does not occur in the monolayer because charge carriers cannot acquire the energy of several eV required for an ionizing (Auger) collision. As mentioned above, a power supply voltage of much greater than 25mV to each functional element is essential to obtain a nonlinear response. Information theory indicates that this is also an essential condition for reliable computation (30).

A possible alternative to the monolayer is to assemble the functional elements into a very thin crystal of controlled thickness, greater than a monolayer. Each element of the crystal will then receive, on average, a defined fraction of the voltage applied across the whole layer. However in this case, the impedance of the supply to each molecule will not be low, and the independent operation of adjacent elements will cause unpredictable voltage fluctuations. Moreover the total voltage across the crystal must then be considerably higher than the threshold for Auger processes, so that catastrophic dielectric breakdown is much more likely. Hence a monolayer array of the basic functional elements is optimal, not just for a demonstrator, but also for the eventual molecular computer.

Even in a monolayer, the required electric fields will impose important constraints on other aspects of the molecular computer, and in particular on the allowed level of monolayer defects. A short circuit caused by a defect would result in an intolerably high fault current and catastrophic damage. It is likely that the criterion for the choice of fabrication technique will be the degree of perfection of the resulting monolayer. The great excitement that greeted the reports (31,32) from Kuhn's group in the 1970s on their LB film manipulation techniques was due in large part to their demonstration that it was possible to deposit a single monolayer with a reasonably low defect density. A large fraction of their metal–monolayer–metal cells were free of conductive defects. While it has since been suggested that this result may have been due to the protective role of the aluminium oxide layer in their cells (33), there have been more recent reports of defect-free metal–LB–metal cells on noble-metal substrates (34).

Very few studies of defect conduction are yet available for the alternative monolayer fabrication techniques, and it is possible that chemisorption or physisorption will provide monolayers with an acceptably low defect level. However it should be noted that the LB technique is unique in assembling the monolayer on a fluid surface prior to deposition on the substrate. Hence in this procedure, defects produced during the initial assembly can be eliminated by long-range cooperative interactions. Since the two adsorption techniques are random processes, there will be an inevitable level of local defects, which may not anneal out at a significant rate.

III. MONOLAYER FORMATION AND DEPOSITION

A. Formation and Stability of Langmuir Monolayers

The LB technique is a way of fabricating ultrathin anisotropic layers from molecules of an organic substance. The first basic step is the formation of an oriented monomolecular layer, or Langmuir monolayer, of the active substance on a water surface (35). This process is shown in Fig. 3(a–c). Not all molecules are suitable for the technique: most molecules forming a satisfactory monolayer have an asymmetric structure. At one end, they have a chemical group showing a high affinity for water, for example —OH, —COOH, or —NH_2, all of which are said to be hydrophilic. The rest of the molecule interacts only weakly with water and is described as hydrophobic. Other things being equal, hydrophilic materials are miscible with each other, as are hydrophobic substances, but materials from opposite categories, for example oil and water, do not mix. Molecules containing a moiety of each type are called amphiphilic.

Amphiphiles are quite common. For example, all detergents are built on the basic hydrophilic–hydrophobic scheme. At moderate concentrations in water, detergent molecules aggregate into spherical micelles with the hydrophilic 'headgroups' pointing outwards and the hydrophobic 'tails' shielded from contact with water in the interior (36). Hydrophobic dirt is readily incorporated in the micelle interiors and thus solubilized. At higher concentrations of the amphiphile in water, more extended structures are formed. The micelles can deform into ellip-

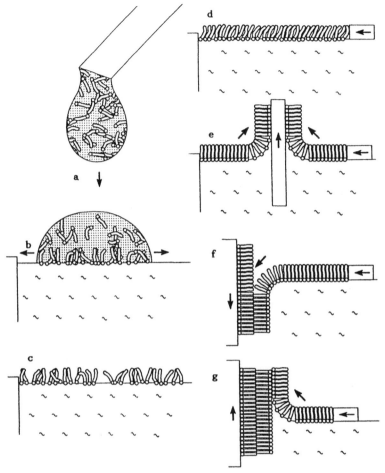

Figure 3 Elements of the Langmuir–Blodgett technique. (a) Dispensing the spreading solution; (b) spreading; (c) evaporation of solvent giving an expanded monolayer; (d) compression; (e) deposition of the first layer on a hydrophilic substrate; (f), (g) deposition of subsequent layers. (Not to scale.)

soids, then extend to form long rods, and further deform in a second direction to form planar membranes. These membranes have headgroups on both sides and can be considered to be formed from two monolayers placed back-to-back (or tail-to-tail). The bilayer membrane, usually formed from mixtures of phospholipids and cholesterol, is the basic scaffolding of all biological cell membranes (37).

A Langmuir monolayer is an individual monolayer of an amphiphile, assembled by allowing the substance to come into contact with an air–water interface. The molecule orients itself so that the hydrophilic headgroup is immersed in the water, while its hydrophobic tail points away. In all of these structures, the apparent tendency of the hydrophobic group to avoid water does not mean that it is actively repelled by the water surface; in fact, there is a weak attraction. However because the interaction of water with itself is so much stronger, the hydrophobic group is expelled, much as a cork floats to the top of a bucket of water in spite of its gravitational attraction to the earth.

The thermodynamic stability of the above structures depends critically on the nature of the interactions between the headgroups and water. Although these are of great biological and industrial importance, and the concepts of hydrophilicity and hydrophobicity in condensed phases are well known, it is interesting to realize that their physical basis was misunderstood and it was not possible to account for the stability of micelles and bilayers even as little as a decade ago.

Previous ideas were based on London's 1922 analysis of the molecular interactions in gases, based on measurements of their second virial coefficients (38). In the gas phase at low density, molecules normally interact two at a time, and rarely come into close contact. London divided the operative interactions into polar forces, associated with the average molecular charge distribution, and dispersion forces, associated with fluctuations about it. For molecular separations much greater than their hard-core van der Waals diameters, it is adequate to consider only the leading term in the spherical harmonic expansion of the electrostatic interaction, which corresponds to an interaction between dipoles μ considered to be located at the center of gravity of each molecule as a whole. Apart from this weak interaction, the molecules are free to rotate. On these assumptions, London showed that the total interaction energy between two different molecules 1 and 2 separated by a distance r has the form

$$U_{12} = \frac{-\kappa \mu^2_1 \mu^2_2}{r^6} - \frac{A_1 A_2}{r^6}$$

where μ_1 and μ_2 are their dipole moments and A_1 and A_2 their Hamaker coefficients; κ is a temperature-dependent coefficient. He showed that, in steam, 70% of the cohesive energy is associated with polar forces and 30% with dispersive forces.

Each of these two classes of interaction is represented by a product term combining corresponding parameters for the interacting molecules. It is readily shown that the geometric mean of two positive numbers is always smaller than their arithmetic mean. Hence if two type 1 molecules and two type 2 molecules interacting via the above type of potential are allowed to choose their partners, the total system energy will be minimized when each chooses its own kind.

If the interfacial free energy between two condensed phases had the above form, then bilayers should not exist. The interfacial free energy between the headgroups and water should always be higher than the average of the head-group–headgroup and water–water values. Like prefers like, and a bilayer membrane dispersed in water should always be unstable with respect to demixing into bulk water and bulk amphiphile. The clear contradiction with experiment has only recently been resolved.

The problem does not lie with London's analysis of the dispersion interaction. It was shown (39) in the 1960s that it is permissible to extrapolate his ideas to condensed matter: with the exception of perfluorinated substances, which remain anomalous (40,41), London's formalism applies to nonpolar organic liquids with good accuracy and even gives the correct value for the interfacial tension between water and mercury, which can interact only via dispersion.

The contradiction is resolved by the realization that the polar interactions in condensed phases cannot be handled in the same way as they are in gases. In contrast to gases, molecules in liquids and solids may typically have 12 neighbors in close contact at all times. The dipole moment μ of the total molecule is subject to conflicting fields from each of its neighbors, so that a summation of pairwise-calculated molecular interaction energies is no longer appropriate. At the same time, higher order terms in the spherical harmonic expansion of the electrostatic field cannot be neglected, and relative molecular rotation is restricted by hard-core repulsions. An alternative approach to the spherical harmonic expansion is needed. It turns out to be much more useful to analyze the molecules into distinct functional groups, and to combine the effects of local charge distributions and steric hindrance into a number of overall interaction parameters.

The hydrogen bond is the most important of these interactions. As was known even in the 1960s, there is no correlation between the strength of a hydrogen bond and the total dipole moment μ of the molecule (42): the charge distribution much closer to the center of the bond is more important (43). It was pointed out by Small that the two sides of a hydrogen bond are not equivalent (44). The Lewis acid side of the bond, which provides the proton, has a local dipole moment with the positive end pointing outwards across the van der Waals surface; while on the Lewis base side it points inwards, as shown in Fig. 4. Steric hindrance prevents the local dipoles of two Lewis acid sites, or two Lewis base sites, from interacting significantly with one another, so that only acid-base interactions need be considered. Hence this interaction is of the required type in which opposites attract.

This approach has been pursued vigorously in the last few years, so that there is now a significant literature and library of data from which interfacial tensions and solubilities of polar organic substances can be estimated. On the scheme of van Oss and Good (45), three parameters are required to characterize each surface: γ^{LW}, which defines the dispersion interactions together with the residual

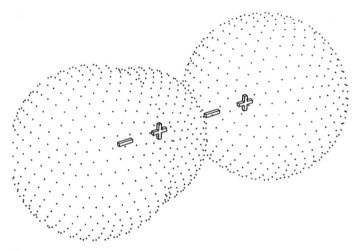

Figure 4 Orientation of the local dipoles with respect to the van der Waals surfaces of the Lewis acid and base partners (left and right hand sides, respectively) in a hydrogen bond

London-type dipole–dipole effects; γ^+, the Lewis acid strength; and γ^-, the Lewis base strength. They have had considerable success in explaining, among other things, the self-assembly properties of uncharged amphiphiles (46).

The formation of a hydrogen bond does not greatly affect the preexisting charge distribution of the molecules (43). Based on the work of Drago, Fowkes has extended the acid–base formalism to include a further important class of intermolecular interactions, the charge-transfer complexes. Following the nomenclature of Pearson, hydrogen bonds are called "hard" and charge-transfer complexes "soft," and it is found that there is a continuum of behavior between the two extremes. Fowkes' group (47,48) use a distinctly different notation to that of van Oss and Good, and the correspondence between the two methodologies is by no means obvious, although it is clear that they are very closely related.

With an appropriate headgroup, the interfacial energy of formation of an amphiphilic monolayer is negative, and the substance spreads spontaneously when brought into contact with an air–water interface. Because of the strong interaction between water and headgroup, the molecule remains captive at the air–water interface, with a preferred molecular orientation in which the hydrophobic group avoids immersion in the water.

Liquid amphiphiles, for example vegetable oils, spread rapidly on water, and the resulting monolayer is responsible for the well-known calming action of oil on stormy seas. However to form a stable LB film, a substance must be solid at room temperature, and the spontaneous spreading of solids is usually very slow. Figure 3 (a–c) shows a much faster and more convenient method, in which the substance is effectively converted to a liquid by dissolving it in a volatile organic solvent, immiscible in water. The solution spreads rapidly over the water surface, and afterwards the solvent evaporates, leaving a pure monolayer.

B. Elements of Langmuir–Blodgett (LB) Deposition

The second basic step of the Langmuir–Blodgett (LB) technique is to deposit the floating monolayer onto a solid substrate. To do this, the substrate is merely moved through the monolayer-covered interface. Under many conditions a monolayer is transferred to the substrate on each passage. Figure 3(d) to 3(g) illustrate the basic steps of LB transfer (49).

The apparatus used for this purpose is called a preparative film balance or LB trough. The trough holds the liquid, or subphase, on which the monolayer is to be spread, and the area of liquid surface available to it is controllable by means of a movable barrier. The barrier is operated by a servomechanism connected in a feedback loop so as to control the surface tension of the liquid within the barrier. A second servomechanism moves the substrate through the water surface.

Almost all modern troughs are of the rigid barrier type illustrated in Fig. 5, and almost all are made of polytetrafluoroethylene or a related fluoropolymer. These materials are both hydrophobic and oleophobic, that is to say, not only do they not participate in hydrogen bonds to any appreciable extent, but their van der Waals interactions with nonpolar hydrocarbons are anomalously weak.

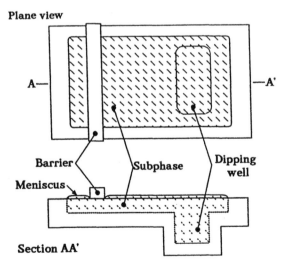

Figure 5 A rigid barrier trough. (Adapted from Ref. 49.)

This facilitates cleaning, and also allows a very simple barrier principle. Both the underside of the barrier and the mating part of the trough edges are machined flat. Because the polymer is not wetted by water, it is possible to fill the trough up to a meniscus height at least 3 mm above the level of the trough edges. The barrier then divides the trough surface into two noncommunicating halves.

During deposition, material is constantly being removed from the water surface onto the substrate, and it is necessary to advance the barrier as required to maintain constant conditions for monolayer transfer. The control signal is obtained by measuring the surface tension of the monolayer-covered surface, or equivalently, its surface pressure, defined as the difference between the prevailing surface tension and that of a clean water surface.

The most popular principle for measuring the surface pressure is that of the Wilhelmy balance, illustrated in Fig. 6. Since the Wilhelmy plate penetrates the monolayer, the force acting on it varies with the surface tension. The force on the plate also has components due to its weight and buoyancy. However the latter are constant and hence can be corrected for.

The component of the force on the Wilhelmy plate due to the surface tension γ is proportional not only to γ but also to $\cos \vartheta$, where ϑ is the equilibrium contact angle between the plate and the water meniscus. It is important to choose a plate material for which ϑ is not only as close to zero as possible, but remains close to zero in the presence of contaminants for as long as possible. Filter paper is a very popular choice because of its good characteristics in this regard.

Due to the extremely small amount of material present in the monolayer, even small traces of impurity can drastically modify its properties. All aspects

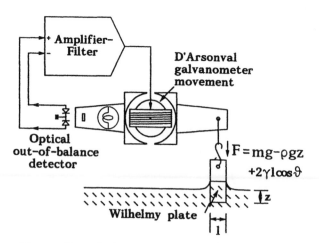

Figure 6 A Wilhelmy balance implementation in which the force on the Wilhelmy plate is converted to electrical form using an electrobalance.

of the LB laboratory must be planned to minimize accidental and systematic contamination. The most important is the purification of the water, required for cleaning and as the subphase. In a typical trough, the spread monolayer of thickness $\sim$ 2 nm floats on water of average depth $\sim$ 20 mm, a ratio of 10^7. For reproducible results the concentration of critical contaminants in the water must therefore be measured in ppb (parts per 10^9). Luckily, water purification units are now readily available that provide such electronic-grade water in large quantities, but even so the supply of water is one of the major items of expense in an LB laboratory. Almost as important are the solvents used for preparing the spreading solutions and also for cleaning purposes. The purest specified available grade should always be used.

Finger grease is a mixture of biological lipids and is highly surface-active. Nothing that is to come into contact with the monolayer or the water surface should be touched by human hands. One should avoid handling all these items, and an assortment of clean tweezers should be on hand in any LB laboratory. Another way is to wear disposable plastic gloves, preferably in embossed polyethylene, which overcomes the problem of sticking due to perspiration.

The trough must be cleaned regularly. The easiest procedure is a change of water with solvent clean. "Solvent clean" means that before adding the new water, the trough is wiped down with perfume-free paper tissues soaked first in 2-propanol to remove hydrophilic impurities, then in 1,1,1-trichloroethane to remove hydrophobic ones.

The solvent-clean procedure is unfortunately not effective for removing polymeric material (often of biological origin). Because of their high molecular weight, polymers are not appreciably soluble in organic solvents, yet they can diffuse slowly to the water surface in amounts significant enough to change the properties of the desired monolayer. Although it appears paradoxical, the best way of removing such surface active material is with another surfactant. Potassium octanoate is a good choice, because it has a very high critical micelle concentration, meaning that local concentrations of the detergent disperse rapidly into pure water. After cleaning with 2% detergent solution at 50°C, all traces must be removed by five rinses with dilute KOH solution followed by five with pure water, all at 50°C.

A material to be spread to form a monolayer must first be dissolved in a volatile organic solvent that is immiscible in water. Chloroform (trichloromethane, $CHCl_3$) is a common choice. Being volatile, it evaporates quite rapidly after spreading. The slight, but definite, Lewis acidity of the proton facilitates dissolution of many otherwise difficult-to-dissolve amphiphiles and aids in the spreading process without going so far as to make the solvent water miscible.

Unfortunately, chloroform has a number of aspects that are less than ideal. Especially in the presence of light, the hydrogen atom splits off to give two chemically reactive free radicals. This explains why chloroform is usually sold with a few percent of ethanol as free-radical scavenger. However this cannot protect LB film molecules, which are more reactive than ethanol. Even when the scavenger effectively protects the LB molecules, the end products of free radical attack tend to be nonvolatile and remain as surface-active contaminants in the spread monolayer. Because of its versatility, it is impossible to do without chloroform in an LB laboratory. It should however be bought in small quantities so that it is always fresh, and stored in a refrigerator in the dark.

To form the monolayer, the spreading solution is dispensed drop by drop onto the water surface, while monitoring the surface pressure on the Wilhelmy balance. Except for unusually stable materials, spreading is stopped at the first sign of a rise in surface pressure. The monolayer is allowed to settle for a while to allow the spreading solvent to dissipate before compression is started. As the barrier restricts the area available to the molecules, the surface pressure rises monotonically. When the final deposition surface pressure is reached, the transfer process can be initiated.

The curve of the surface pressure developed by a monolayer at a constant temperature plotted against the area per molecule is called its isotherm. The isotherm at a given temperature is characteristic for the substance. All modern LB troughs have a facility for plotting or otherwise displaying isotherms, which are a useful check of cleanliness and the correct operation of the trough.

Because problems with contamination are very commonly encountered, it is useful to have a number of reference substances whose isotherm shape is sensitive to contamination and can be used to narrow down the search for its source. One such substance, whose isotherm has been reported in many publications, is octadecanoic acid (which is also known by its nonsystematic name of stearic acid). The isotherm at 25°C on pure water is shown in Fig. 7. Note that it is composed of three regimes, with significantly different gradients and with sharply defined transitions between them. Note the straightness of the isotherm over most of the L_2 and LS regimes, with the G–L_2 transition extending over at most a 1 mN/m range and the L_2–LS transition even sharper. A smoothing-out of the transitions is indicative of organic contamination of the monolayer. Ionic contamination of the subphase is typically indicated by a reduction in the L_2–LS transition pressure.

IV. FILM STRUCTURE AND PROPERTIES

A. Molecular Structure of Langmuir Films

For many undemanding applications of LB films, it is not necessary to know the local arrangement of molecules, either before deposition on the water surface, or afterwards on the solid substrate. However it is now quite clear that the molecular packing in both cases can have a significant impact on film morphology

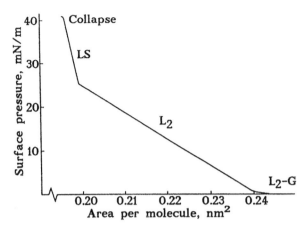

Figure 7 The isotherm of octadecanoic (stearic) acid at 25°C on pure water. (Adapted from Ref. 35.)

and quality. The particular packing adopted by the molecules depends on the energetics of the interaction between the headgroups and subphase, as well as on the lateral interactions between neighboring molecules in the monolayer. In multilayer films of the type shown in Figs. 3(f) and 3(g), interactions between adjacent monolayers must also be taken into account.

The structure of Langmuir monolayers on the water surface was, until very recently, known neither on visible length scales nor at the molecular level. The improvement in the situation in the last decade was due to the development of new measurement techniques capable of detecting the infinitesimally small quantity of monolayer material in the presence of 10^7 times as much water. Chronologically the first of these was fluorescence microscopy (50), or epifluorescence (51), in which the spatial distribution of the fluorescence of a dye added to a monolayer is imaged using the extremely sensitive TV cameras now available. The second was grazing incidence x-ray diffraction (52), which has now become practicable using the extremely bright and highly-collimated x-radiation now available from synchrotrons. Most recently, Brewster-angle microscopy (53,54) has provided a method for optically imaging a monolayer without adding any fluorescent impurity.

The new techniques have led to a revolution in the mainstream conception of monolayers. A decade ago, monolayers were believed to have at most four phases, only one of which was expected to show diffraction phenomena. They were thought to be homogeneous if prepared with reasonable care. In fact, many of the monolayer phases display diffraction spots (52,55–58), and x-ray diffraction has confirmed a very few previous reports (59,60) that the number of distinct phases is at least seven. The two optical techniques have revealed that monolayers display a rich variety of textures.

The most surprising result was that the phases from which multilayer films can be deposited do not have structures analogous to bulk crystalline solids, as had been thought, or even isotropic liquids, but instead all have an intermediate structure, analogous to bulk liquid crystals (also known as mesophases). In fact most of the monolayer phases can be assigned to bulk lamellar mesophase categories, such that the molecular order in the monolayer is the same as that found in each of the lamellae of the bulk phase (61).

The phase diagrams of amphiphilic substances that are emerging from recent investigations display an almost incredible richness of behavior. Figure 8 shows a composite phase diagram in which all the phases so far reported are plotted against surface pressure π and temperature T. Any particular substance displays only a restricted subset of those shown, because the experimentally accessible range is restricted in temperature by the freezing and boiling points of water,

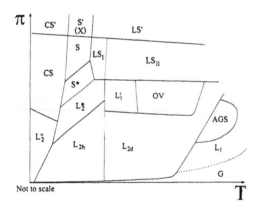

Figure 8 A generalized temperature–pressure phase diagram for aliphatic-chain amphiphiles, which attempts to show, in their correct relative positions, all phases for which experimental evidence now exists. In a number of cases the evidence has been gathered using different techniques on different monolayer systems, so that the correspondence and relative positions are not clear. For example, the X phase certainly belongs at high pressure, and could possibly be either CS' or S', while the Overbeck phase OV may or may not be distinct from the L_1' phase. Not all authors recognize the distinction between the pairs $(S^*, L_2^*) = L_2'$, (L_{2h}, L_{2d}), (LSI, LSII), or (AGS, L_1), and the phase nomenclature is still in a state of flux. References for each of the phases are as follows, in ASCII order:
AGS (62), CS (58,63,67), CS' (60), G (65,66,68), L1 (66,68), L1' (61), L_{2h} (58,63,67,68), L2' (58,63,67), L2'' (58,60), L2* (61,64), L_{2d} (69), LSI (59,60,67,71), LS' (60,64), OV (70), LSII (71), S (58,63,67), S' (60,64), S* (61,64), X (72,73). In the shaded region the monolayer collapses so rapidly that it is not possible to characterize its structure.

and in surface pressure by its surface tension. The regular progression of the phase diagrams with change of chain length justifies the idea that the one generic diagram applies to all amphiphiles with long aliphatic chains (74,75). Present progress in the field is extremely rapid.

In general, the monolayer structure does not remain the same after deposition onto a solid substrate. However enough data is now available to be able to say that all the molecular packings found on solid substrates can also be obtained on the water surface under suitable conditions. Based on this fact and on the observed correspondences between the phase diagrams of monolayers with different head groups and chain lengths, the present author and coworkers have proposed a way of comparing and classifying the molecular packings (76).

The CS and L_2'' phases shown at the low-temperature end of the phase diagram are both crystalline solids. The monolayer does not flow in response to surface pressure gradients and hence cannot be transferred onto a substrate using the LB technique. A modification, the Langmuir–Schaeffer technique, is available for rigid monolayers, but due to the film rigidity there is no tendency for holes in it to disappear. Grain boundaries are also long-lived metastable defects in monolayers in these two phases.

At the high-temperature end, the G, L_1, and AGS phases are fluids. G and L_1 are isotropic, and the AGS phase may be anisotropic (62), although this has not been positively confirmed. While a monolayer in any of these states flows readily, so that holes disappear, and it is transferred to a substrate as the latter is withdrawn from the subphase, its adhesion is poor, and the monolayer returns to the water surface on substrate re-immersion.

All the intermediate phases flow in response to surface pressure gradients (liquids) while giving rise to diffraction spots (crystals). Unlike the diffraction spots from perfect crystals, those from the intermediate phases have a finite width that does not decrease on annealing, and that changes reversibly on cycling the monolayer conditions of temperature or surface pressure. It is the lattice imperfections that allow the phase to show fluid behavior, by drastically reducing its threshold for shear deformation below the megapascals typical for molecular crystals.

The high-surface-pressure phase LS is optically isotropic in the plane, so that it is impossible to image its textures. However for the other phases it has proved possible to photograph the textures. Figure 9 shows computer-generated images of one of the domain textures observed in monolayer phases with tilted molecules using fluorescence (77) and Brewster-angle (78) microscopy. The contrast is due to the tilt-related optical anisotropy. The continuous change of tilt axis throughout the domain interior and its constant orientation to the boundary are characteristic for mesophases. This particular texture is also known in bulk liquid crystals (79) and is called a ''boojum'' because its morphology is identical to that of the defect of the same name in superfluid helium-3.

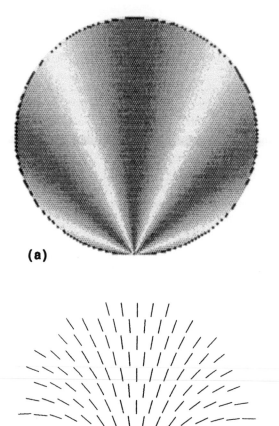

(a)

(b)

Figure 9 Computer-generated images of a "boojum" domain of a monolayer in a tilted hexatic phase (a) as visualized between crossed polarizers or in a Brewster-angle microscope (b) tilt director field. Note that the director orientation changes continuously within the domain, and that it is always perpendicular to the boundary.

Depending on the headgroup, the monolayer adhesion to the substrate in these phases, or others of its properties, still may not be sufficient for the formation of high-quality LB films. However all suitable phases are mesophases. Since the adhesion of the monolayer to the substrate can vary markedly from one phase to another, it is useful to know the phase diagram to ensure correct deposition. Another important factor is the monolayer texture. As will be discussed in greater detail in Sec. B, there exist specific textures in the Langmuir monolayer that are partially preserved on deposition and that lead to conducting defects of the deposited monolayer. It is important to know in which phases these textures are present in thermodynamic equilibrium, in which phases they do not exist at all, and in which phases they can exist but spontaneously anneal out. In the latter case it is of interest to know the annealing kinetics.

The phases suitable for LB film deposition are not only mesophases, but belong to a specific subset that has only been widely recognized in the last decade. Since they give rise to discrete diffraction spots, they may be classified as hexatics, which possess long-range lattice orientational order. A hexatic is a particular sort of smectic mesophase, or liquid crystal, formed from rodlike molecules. "Smectic" means that the molecules organize themselves spontaneously into layers, just like the layers of a multilayer LB film. There are four known categories of bulk hexatic (80): smectic BH, smectic I, smectic F, and smectic L, which differ in the details of their molecular tilt. Bulk hexatics clearly have a large number of layers, although it is possible to form very thin free-standing films of them in which the number of layers is very small. In the present case of monolayer hexatics, there is just one molecular layer, which however shows the same molecular organization as is found in each of the layers of a bulk hexatic.

The research interest accorded to liquid crystals in recent decades is a direct consequence of their application to electronic displays. In order to understand their behavior in this application it is not necessary to understand how each individual molecule behaves. Since the wavelength of visible light is less than 1 μm, and the resolution of the unaided eye no better than 100 μm, the material characteristics can be understood adequately in terms of averages over billions of molecules, in which unusual behavior by a small fraction of them makes no observable difference. As a consequence, most textbooks on liquid crystals limit themselves to phenomenological theories of mesophase behavior. The "artist's impressions" of local molecular order presented in them are often physically impossible, involving large voids or implying the overlap of molecular van der Waals surfaces, and are to be interpreted as a stylized way of expressing the long-range statistics of various order parameters.

In smectics, there are four main order parameters, whose asymptotic correlational decay law within each lamella is shown in Table 1. PO, the position order, is always short-range, symbolized by S, meaning that positional correlations decay exponentially with increasing separation. There are three orientational order parameters: LO (lattice orientational order), TO (tilt orientational order) and AO (axial order, of the molecular zigzag plane). The hexatic phases are those in which LO has quasi-long-range order, symbolized by Q, and meaning

Table 1 Statistics in the Different Smectic Phases of the Four Major Order Parameters PO, TO, LO, and AO[a]

Smectic	PO	TO	LO	T–L	AO	Fig. 8
A	S	S	S	—	S	L_1
BC	Q	S	Q	—	S	
BH	S	S	Q	—	S	RII
C	S	Q	S	—	S	(AGS)
E	Q	S	Q	—	Q	(S)
F	S	Q	Q	NNN	S	L_2*
G	Q	Q	Q	NNN	S	
H	Q	Q	Q	NNN	Q	S*
I	S	Q	Q	NN	S	L_3
J	Q	Q	Q	NN	S	L_2
K	Q	Q	Q	NN	Q	L_2''
L	S	Q	Q	I	S	(OV, L_1')

[a]PO, in-plane position correlations between different planes. TO, molecular tilt azimuth. LO, lattice or bond orientation. AO, axial order of molecular zigzag planes, normally occurring as herringbone order. Q, quasi-long-range order, indicates that the correlation of fluctuations decays algebraically with increasing separation. S, short-range order, indicates that the correlation of fluctuations decays exponentially with increasing separation. When both TO and LO are quasi-long-range ordered, there exists the possibility of a definite relationship between the two resulting macroscopic orientation fields, shown in the column labelled T–L. NN, tilt locked towards nearest neighbor in hexagonal lattice. NNN, tilt locked towards next-nearest neighbor. I, tilt locked in a direction intermediate between NN and NNN.
Source: Adapted from Ref. 61.

that the correlations decay with increasing separation according to a power law. Cartoons representing these asymptotic decay laws in coded form are shown in Fig. 10.

These decay laws for the various order parameters are of great experimental interest because they are directly measurable. Optical techniques are sensitive to the optical anisotropy, and hence TO, averaged over linear dimensions of the order of 1 μm. X-ray and electron diffraction techniques provide evidence for PO and LO. Up till now, the mobility of the molecules in a mesophase has prevented directly imaging them. However for the present application it is no longer true that only macroscopic averages are important. As already discussed in Sec. II.B, the enormous electric fields that must be present across the monolayer mean that local deviations from the ideal molecular arrangement may lead to catastrophic damage. Even if they occur with very low probability, they are important for molecular electronics. The concepts required to handle these local deviations are under current development (80,81).

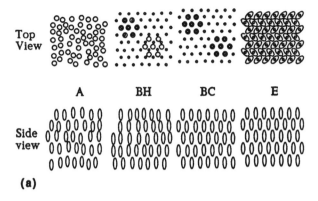

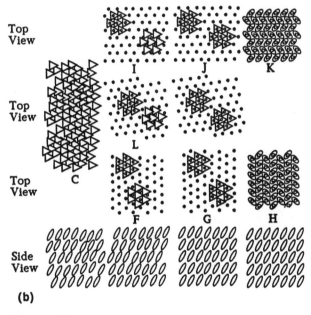

Figure 10 (a) Cartoons representing the asymptotic decay laws of order parameter correlations in the orthogonal smectic phases. (b) Cartoons representing the asymptotic decay laws of order parameter correlations in the tilted smectic phases. (From Ref. 80.)

B. Specific Defects of Langmuir–Blodgett Films

The lateral interactions between molecules in a deposited monolayer are the same as those in the Langmuir film on the water surface, but the interaction between the headgroups and its environment may be quite different. It has now been demonstrated conclusively in a number of different systems (72,82,83) that the local packing of the molecules can change considerably on deposition, depending on the nature of the substrate. At the same time, their macroscopic textures are sometimes preserved to a considerable degree.

All other factors being equal, changes of local molecular packing brought about on deposition are probably not relevant to molecular electronics. The prime requirement is for a uniform distribution of molecules. Changes in the particular orientation of the molecule with respect to its neighbors or the angle of molecular tilt will probably not seriously affect its molecular electronic function.

However, all other factors are not equal, because the local packing can influence many other important monolayer parameters. For example, long-range order affects molecular mobility. As already noted, the monolayer on the water surface must be in a mesomorphous phase on the water surface, to allow both the rapid annealing of defects and the shear deformation of LB deposition. However if the organization remains mesomorphous on the substrate during subsequent processing steps, the monolayer is soft and readily damaged. This may, for example, lead to the formation of short circuits during the deposition of a top metal electrode. The molecules may also diffuse later to unwanted positions as the device ages. Hence it is clearly advantageous for the deposited monolayer to be in a crystalline phase. The desirable combination of a mesophase on the water surface and a crystalline phase after deposition occurs with some monolayer/substrate combinations. Its failure to occur in others may explain, for example, why noble metal–monolayer–noble metal cells are often short-circuited whenever the monolayer is a pure fatty acid, with no metallic counterions.

Another factor influenced by the local packing is the monolayer texture on a mesoscopic (1 μm) scale. When texture is preserved on passage through a phase transition, the transition is said to be paramorphotic. It is normally a rare phenomenon. For example, in the freezing of an isotropic liquid to a crystalline solid, the new domains are nucleated with random positions and orientations, so that the final polycrystalline texture is unrelated to any structure in the previous phase. However it is quite common to encounter paramorphotic transitions between mesophases (84), particularly of the more ordered variety, and evidence has been provided for their occurrence in monolayers, both on the water surface and on deposition (82,85).

If local packing is important, it is also important to know how it changes in a monolayer on compression, and on changing the temperature, i.e., its phase behavior. For a long time this was thought not to be particularly important for the deposition of LB films or for molecular electronics. Mann and Kuhn had demonstrated in 1971 that individual monolayers could be prepared with adequately low defect density (31,32). The first strong suggestion that all was not so simple came from Tredgold and Winter (86), who in 1982 reported that they were unable to repeat Mann and Kuhn's results. Their metal–monolayer–metal cells, fabricated using apparently the same procedures and conditions as those used by Kuhn's group, had a conductivity orders of magnitude greater, indicating the presence of conductive defects. On storage of the cells at low temperature in a helium atmosphere, the conductivity of their cells decreased, reaching values comparable to Kuhn's after a month. If there were defects in the monolayers deposited by Tredgold's group, but not those of Kuhn's, some important aspect of the procedure was clearly not being reproduced. In fact, of those groups interested in fabricating devices in which monolayers play a role as ultrathin insulators, none has reported current densities in as-fabricated cells as low as those of Kuhn's group.

In a subsequent paper, Tredgold, Vickers, and Allen (87) reported that the conductivity of a 15-layer film of cadmium stearate could be changed by a factor of 30 by changing the conditions of deposition of the first layer only. The fact that defects in the first layer were not immediately covered over on deposition of the second and subsequent layers showed that they were not just holes but possessed a specific structure capable of propagating through subsequent layers. This is consistent with the discovery of the author that deposition onto an existing LB layer shows orientational epitaxy (85,88) and suggests that the defects are associated with some aspect of orientational order. The fact that the conductivity of the cells was strongly influenced by annealing of the monolayer on the water surface prior to deposition indicates that the initial deposition was paramorphotic, preserving the defects that have been present in some form in the Langmuir layer.

These implications of Tredgold, Vickers, and Allen's report (87) are very exciting. Firstly, if conducting defects have a specific structure, then it is in principle possible to eliminate them: in fact, the report demonstrated that annealing was one such method. Secondly, if functional defects are identical to structural ones, it is possible to detect them rapidly using noninvasive and nondestructive methods, and thus readily to evaluate new elimination schemes. It was this possibility that motivated the investigations of monolayer structure discussed in the previous sections. To detect areas of film conduction, it is necessary to apply electrodes to both sides and to have some method of determining where current has flowed. One such method, consisting in decorating conductive defects by electroplating them with copper, has recently been used by several groups (89–91). However it is possible that the copper sulfate electrolyte that must be brought into contact with the LB film causes changes in the film structure other than those caused by a metal top electrode, and the deposited copper cannot subsequently be removed. In contrast, orientational defects can be detected by polarized optical microscopy. In the first such demonstrations, the images were viewed with the naked eye or recorded on photographic film, and required layer thicknesses of the order of 1 μm to achieve adequate image brightness. It is now possible to detect the anisotropy in a single monolayer using an ultrasensitive CCD TV camera (92).

A few studies have linked conductive defects with structural defects, although there is a lack of agreement between them. The author has demonstrated conditions under which conducting and orientational defects are connected (91), while Matsuda et al. (90) found no correlation in their eicosanoate monolayers. The difference may be due to the different experimental procedures used to decorate the conducting defects.

Irrespective of this structure–property relationship, there is much more agreement on the effectiveness of annealing for removing conductive defects. The initial report of Tredgold et al. (87) has been confirmed by Kato and Ohshima (93) and Bibo (94) under a number of different conditions. Clearly this rules out any process occurring during or after transfer as the cause of defect conduction.

The program of research discussed here has not yet been completed. There is as yet no universally agreed method for producing monolayers free of conducting defects, using either the Langmuir–Blodgett technique or any other. The author believes strongly that defects will not disappear of their own accord, and that high-performance devices of whatever sort cannot be fabricated reproducibly until a procedure to control them has been developed.

V. PERSPECTIVES

There are many problems to be solved before a working molecular electronics technology is achieved, and the fabrication of the system is one of them. Of the available methods, the Langmuir–Blodgett technique is a very promising way of manipulating functional molecules and assembling them into ordered structures. It has deficiencies and is incompletely understood, but these areas are the subject of ongoing investigations that have already yielded fundamental new knowledge about the behavior of molecular systems.

Apart from some general considerations at the beginning, this chapter has been mainly concerned with the basic techniques of Langmuir–Blodgett deposition. Except to illustrate more vividly some of its aspects, it has been inappropriate to discuss in detail proposals for its application to molecular electronics, to which the reader has either been referred or which are the subject of other chapters. Most of the illustrations of the LB technique, and indeed most of the specific knowledge about it, concerns molecules that are clearly incapable of performing any active data transformation function. It is likely that a molecule capable of functioning as a logic gate with an acceptable error rate will have ten times the molecular weight of most present LB molecules, while the smallest feasible von Neumann processor must consist of approximately 1000 logic gates. At this length scale techniques of microlithography will most probably be available for connecting the individual molecules into arrays of even greater complexity.

While at mesoscopic length scales other techniques are available for the manipulation of matter, the Langmuir–Blodgett technique will always represent a method of choice for assembling functional molecules into a layer in order to make electrical connection to them. It will probably be necessary to develop new methods for the organic synthesis of large molecules with repetitive structure, as well as new concepts to predict the phase diagram of monolayers of such large molecules. Using the new ideas about molecular interactions now being developed, it may prove possible to use self-assembly in mixed monolayers to lighten the synthetic task. While it is as yet impossible to fix the date for the birth of molecular electronics, it is highly likely that the Langmuir–Blodgett technique will have been present at the delivery.

ACKNOWLEDGMENTS

The author would like to thank Prof. C. M. Knobler for critically reading the draft of this paper, Dr. F. Rondelez for providing challenging ideas, Dr. R. Bruinsma and Dr. V. M. Kaganer for stimulating discussions, and the Deutsche Forschungsgemeinschaft, the Bundesministerium für Forschung und Technologie, and the Fonds Henri de Rothschild for financial assistance.

REFERENCES

1. Morkoc, H., and Solomon, P. M., The HEMT: a superfast transistor, *IEEE Spectrum* (Feb. 1984).
2. Fritzsche, D., Heterostructures in MODFETs, *Sol. State Electron.*, *30* (1987).
3. Ashwell, G. J., ed., *Molecular Electronics*, Research Studies Press, Taunton, UK, 1992.
4. Arthur, C., *New Scientist*, *1885*, 5 (1993).
5. Hopfield, J. J., Neural networks and physical systems with emergent collective computational abilities, *Proc. Natl. Acad. Sci. USA*, *79*, 2554 (1982).
6. Ripley, B. D., Statistical aspects of neural networks, *Statistical and Probabilistic Aspects of Neural Networks* (O. E. Barndorff-Nielsen, D. R. Cox, J. L. Jensen, and W. S. Kendall, eds.), Chapman and Hall, London, 1993.
7. Domany, E., Neural networks: a biased overview, *J. Stat. Phys.*, *51*, 743. (1988).
8. Stegeman, G. I., and Seaton, C. T., Nonlinear integrated optics, *J. Appl. Phys.*, *58*, R57. (1985).
9. Peterson, I. R., Langmuir–Blodgett films: a route to molecular electronics, *Nanostructures Based on Molecular Materials* (W. Göpel and C. Ziegler, eds.), VCH, Weinheim, Germany, 1992.

10. Sols, F., Macucci, M., Ravaioli, U., and Hess, K., On the possibility of transistor action based on quantum interference phenomena, *Appl. Phys. Lett.*, *54*, 350 (1989).

11. Gentile, S. P., *Basic Theory and Application of Tunnel Diodes*, Van Nostrand Reinhold, Princeton, NJ, 1962.

12. Zappe, H. H., *Advances in Superconductivity* (B. Deaver and J. Ruvalds, eds.), Plenum Press, New York, 1983.

13. Keyes, R. W., What makes a good computer device? *Science*, *230*, 138 (1985).

14. Kuzmany, H., Mehring, M., and Roth, S., eds., Electronic properties of conjugated polymers, *Springer Series in Solid State Sciences*, Vol. 76, Springer-Verlag, Berlin, 1987.

15. Skotheim, T. A., ed., *Handbook of Conducting Polymers*, Vol. I., Marcel Dekker, New York, 1986.

16. Wilson, E. G., Polarons and excition-polarons in one dimension: the case of polydiacetylene, *J. Phys. C*, *16*, 6739 (1983).

17. Braun, D., and Heeger, A. J., Visible light emission from semiconducting polymer diodes, *Appl. Phys. Lett.*, *58*, 1982 (1991).

18. Burroughes, J. H., Bradley, D. D. C., Brown, A. R., Marks, R. N., Mackay, K., Friend, R. H., Burns, P. L., and Holmes, A. B., *Nature*, *347*, 539 (1990).

19. Geddes, N. J., Sambles, J. R., Jarvis, D. J., Parker, W. G., and Sandman, D. J., Fabrication and investigation of asymmetric current–voltage characteristics of a metal/ Langmuir–Blodgett monolayer/metal structure, *Appl. Phys. Lett.*, *56*, 1916 (1990).

20. Ashwell, G. J., Sambles, J. R., Martin, A. S., Parker, W. G., and Szablewski, M., Rectifying characteristics of $Mg|$ ($C_{16}H_{33}$-Q3C LB Film) | Pt structures, *J. Chem. Soc. Chem. Commun.*, 1374 (1990).

21. Aviram, A., and Ratner, M. A., Molecular rectifiers, *Chem. Phys. Lett.*, *29*, 277 (1974).

22. Marcus, R. A., and Sutin, N., Electron transfers in chemistry and biology, *Biochim. Biophys. Acta*, *811*, 265 (1985).

23. Larsson, S., Electron transfer in chemical and biological systems. Orbital rules for nonadiabatic transfer, *J. Am. Chem. Soc.*, *103*, 4034 (1981).

24. Keyes, R., Physical limits in digital electronics, *Proc. IEEE*, *63*, 740 (1975).

25. Sugi, M., Fukui, T., and Iizima, S., *Mol. Cryst. Liq. Cryst.*, *50*, 183 (1979).

26. Markewicz, A., and Fleming, R. J., Simultaneous thermally stimulated luminescence and conductivity in low-density polyethylene, *J. Phys. D*, *21*, 349 (1988).

27. van Roggen, A., Electronic conduction of polymer single crystals, *Phys. Rev. Lett.*, *9*, 368 (1962).

28. Lvov, Y., Decher, G., and Möhwald, H., Layer-by-layer deposited ultrathin films, *Langmuir*, *9*, 481 (1993).

29. Ulman, A., and Tillman, N., Self-assembling double layers on gold surfaces: the merging of two chemistries, *Langmuir*, *5*, 1418 (1989).

30. Wallmark, J. T., Fundamental physical limitations in integrated circuits, *Electron. Eng.*, *47*, 52 (1975).

31. Mann, B., and Kuhn, H., Tunneling through fatty acid salt monolayers, *J. Appl. Phys.*, *42*, 4398 (1971).

32. Mann, B., Kuhn, H., and v. Szentpaly, L., Tunneling through fatty acid salt

monolayers and its relevance to photographic sensitization, *Chem. Phys. Lett.*, *8*, 82 (1971).

33. Peterson, I. R., Defect density in a metal–monolayer–metal cell, *Aust. J. Chem.*, *33*, 1713 (1980).

34. Geddes, N. J., Sambles, J. R., Parker, W. G., Couch, N. R., and Jarvis, D. J., Electrical characterisation of M/I/M structures incorporating thin layers of 22-tricosenoic acid deposited on noble metal base electrodes, *J. Phys. D*, *23*, 95 (1990).

35. Gaines, G. L., Jr., *Insoluble Monolayers at the Gas-Water Interface*, Wiley Interscience, New York, 1966.

36. Mukerjee, P., Micelles, *Encyclopedic Dictionary of Physics*, Pergamon Press, 1961.

37. Sackmann, E., Molecular and global structure and dynamics of membranes and lipid bilayers, *Can. J. Phys.*, *68*, 999 (1990).

38. Eisenschitz R., and London, F., Über das Verhältnis der van der Waalsschen Kräfte zu den homöpolaren Bindungskräften, *Z. Physik*, *60*, 491 (1930).

39. Fowkes, F. M., Intermolecular and interatomic forces at interfaces, *Surfaces and Interfaces. I.* (J. J. Burke, N. L. Reed, and V. Weiss, eds.), Syracuse Univ. Press, Syracuse, N.Y., 1967, p. 197.

40. Croll, I. M., and Scott, R. L., Fluorocarbon solutions at low temperatures. III. Phase equilibria and volume changes in the CH_4–CF_4 system, *J. Am. Chem. Soc.*, *62*, 954 (1958).

41. Mukerjee, P., Fluorocarbon–hydrocarbon interactions in interfacial and micellar systems, *J. Am. Oil Chem. Soc.*, *59*, 573 (1982).

42. Pimental, G. C., and McClellan, A. L., *The Hydrogen Bond*, Freeman, San Francisco, CA, 1960.

43. Bernstein, J., Etter, M. C., and Leiserowitz, L., *The Role of Hydrogen Bonding in Molecular Assemblies*, in press.

44. Small, P. A., Some factors affecting the solubility of polymers, *J. Appl. Chem.*, *3*, 71 (1953).

45. van Oss, C. J., and Good, R. J., Prediction of the solubility of polar polymers by means of interfacial tension combining rules, *Langmuir*, *8*, 2877 (1992).

46. van Oss, C. J., and Good, R. J., Relation between the apolar and polar components of the interaction energy between the chains of nonionic surfactants and their CMC in water, *J. Dispersion Sci. Technol.*, *12*, 45 (1991).

47. Fowkes, F. M., Kaczinski, M. B., and Dwight, D. W., Characterization of polymer surface sites with contact angles of test solutions. I. Phenol and iodine adsorption from CH_2I_2 onto PMMA films, *Langmuir*, *7*, 2464 (1991).

48. Fowkes, F. M., Dwight, D. W., Cole, D. A., and Huang, T. C., Acid–base properties of glass surfaces, *J. Non-Crystalline Solids*, *120*, 47 (1990).

49. Peterson, I. R., Langmuir–Blodgett films, *Molecular Electronics* (G. J. Ashwell, ed.), Research Studies Press, Taunton, UK, 1992.

50. Lösche, M., and Möhwald, H., Fluorescence microscopy on monomolecular films at an air/water interface, *Colloids and Surfaces*, *10*, 217 (1984).

51. McConnell, H. M., Tamm, L. K., and Weis, R. M., Periodic structures in lipid monolayer phase transitions, *Proc. Natl. Acad. Sci. USA*, *81*, 3249 (1984).

52. Wolf, S. G., Deutsch, M., Landau, E. M., Lahav, M., Leiserowitz, L., Kjaer,

K., and Als-Nielsen, J., A synchrotron x-ray study of a solid–solid phase transition in a two-dimensional crystal, *Science*, 242, 1286 (1988).

53. Hénon, S., and Meunier, J., Microscope at the Brewster angle: direct observation of first-order phase transitions in monolayers, *Rev. Sci. Instr.*, 62, 936 (1991).

54. Hönig, D., and Möbius, D., Direct visualization of monolayers at the air–water interface by Brewster angle microscopy, *J. Phys. Chem.*, 95, 4590 (1991).

55. Barton, S. W., Goudot, A., Bouloussa, O., Rondelez, F. R., Lin, B., Novak, F., Acero, A., and Rice, S. A., Structural transitions in a monolayer of fluorinated amphiphile molecules, *J. Chem. Phys.*, 96, 1343 (1992).

56. Helm, C. A., Tippman-Krayer, P., Möhwald, H., Als-Nielsen, J., and Kjaer, K., Phases of phosphatidyl ethanolamine monolayers studied by synchrotron x-ray scattering, *Biophys. J.*, 60, 1457 (1991).

57. Schwartz, D. K., Schlossman, M. L., and Pershan, P. S., Re-entrant appearance of phases in a relaxed Langmuir monolayer of tetracosanoic acid as determined by x-ray scattering, *J. Chem. Phys.*, 96, 2356 (1992).

58. Lin, B., Shih, M. C., Bohanon, T. M., Ice, G. E., and Dutta, P., The phase diagram of a lipid monolayer on the surface of water, *Phys. Rev. Lett.*, 65, 191 (1990).

59. Stenhagen, E., Surface films, *Determination of Organic Structures by Physical Methods* (E. A. Braude and F. C. Nachod, eds.), Academic Press, New York, 1955, pp. 325–371.

60. Lundquist, M., The relation between polymorphism in "two-dimensional" monomolecular films on water to polymorphism in the three-dimensional state, and the formation of multimolecular films on water, Parts I and II, *Chem. Scripta*, 1, 5, 197 (1971).

61. Bibo, A. M., Knobler, C. M., and Peterson, I. R., A monolayer phase miscibility comparison of the long chain fatty acids and their ethyl esters, *J. Phys. Chem.*, 95, 5591 (1991).

62. Albrecht, O., Gruler, H., and Sackmann, E., Polymorphism of phospholipid monolayers, *J. Phys. France*, 39, 301 (1978).

63. Ställberg-Stenhagen, S., and Stenhagen, E., Phase transitions in condensed monolayers of normal chain carboxylic acids, *Nature*, 156, 239 (1945).

64. Lawrie, G. A., and Barnes, G. T., Octadecanol monolayers: the phase diagram, *J. Colloid Interface Sci.*, 162, 36 (1994).

65. Adam, N. K., The structure of thin films. Part VII. Critical evaporation phenomena at low compressions, *Proc. Roy. Soc. London*, A110, 423 (1926).

66. Kim, M. W., and Cannell, D. S., Experimental study of a two-dimensional gas–liquid phase transition, *Phys. Rev.*, A13, 411 (1976).

67. Kenn, R. M., Böhm, C., Bibo, A. M., Peterson, I. R., Möhwald, M., Kjaer, K., and Als-Nielsen, J., Mesophases and crystalline phases in fatty acid monolayers, *J. Phys. Chem.*, 95, 2092 (1991).

68. Harkins, W. D., Young, T. F., and Boyd, E., The thermodynamics of films: energy and entropy of extension and spreading of insoluble monolayers, *J. Chem. Phys.*, 8, 954 (1940).

69. Kaganer, V. K. and Loginov, E. B., Phys. Rev. E, 51, 2237 (1995).

70. Overbeck, G. A., and Möbius, D., A new phase in the generalized phase diagram of monolayer films of long chain fatty acids, *J. Phys. Chem.*, 97, 7999 (1993).

71. Shih, M. C., Bohanon, T. M., Mikrut, J. M., Zschack, P., and Dutta, P., X-ray diffraction study of the superliquid region of the phase diagram of a Langmuir monolayer, *Phys. Rev. A*, *45*, 5734 (1992).

72. Engel, M., Merle, H. J., Peterson, I. R., Riegler, H., and Steitz, R., Structural relationships between floating and deposited monolayers of docosanoic acid, *Ber. Bunges. Phys. Chem.*, *95*, 1514 (1991).

73. Shih, M. C., Bohanon, T. M., Mikrut, J. M., Zschack, P. and Dutta, P., Pressure and pH dependence of the structure of a fatty acid monolayer with calcium ions in the subphase, *J. Chem. Phys.*, *96*, 1556 (1992).

74. Bibo, A. M., and Peterson, I. R., Monolayer phase diagrams of the long chain fatty acids, *Advanced Materials*, *2*, 309 (1990).

75. Peterson, I. R., Phase diagrams and chain order in monolayers of aliphatic chains, *Organic Thin Films and Surfaces*. Vol. 1 (A. Ulman, ed.), Academic Press, Boston, 1993.

76. Peterson, I. R., Brzezinski, V., Kenn, R. M., and Steitz, R., Equivalent states of amphiphilic lamellae, *Langmuir*, *8*, 2995 (1992).

77. Qui, X., Ruiz-Garcia, J., Stine, K. J., Knobler, C. M., and Selinger, J. V., Direct observation of domain structure in condensed monolayer phases, *Phys. Rev. Lett.*, *67*, 703 (1991).

78. Hönig, D., Overbeck, G. A., and Möbius, D., Morphology of pentadecanoic acid monolayers at the air/water interfaces studied by BAM, *Adv. Mater.*, *4*, 419 (1992).

79. Langer, S. A., and Sethna, J. P., Textures in a chiral smectic liquid-crystal film, *Phys. Rev, A A34*, 5035 (1986).

80. Brock, J. D., Birgeneau, R. J., Litster, J. D., and Aharony, A., Liquids, crystals, and liquid crystals, *Physics Today*, *42*, 52 (1989).

81. Glaser, M. A., and Clark, N. A., Melting and liquid structure in two dimensions, *Advances in Chemical Physics, Vol. 83*, (I. Prigogine and S. A. Rice, eds.), John Wiley, New York, 1993.

82. Steitz, R., Mitchell, E., and Peterson, I. R., Relationships between fatty acid monolayer structure on the subphase and on solid substrates, *Thin Solid Films*, *205*, 124 (1991).

83. Shih, M. C., Peng, J. B., Huang, K. G., and Dutta, P. Structures of fatty acid monolayers transferred to glass substrates from various Langmuir monolayer phases, *Langmuir*, *9*, 776 (1993).

84. Gray, G. W., and Goodby, J. W., *Smectic Liquid Crystals: Textures and Structures*, Leonard Hill, Glasgow, 1984.

85. Peterson, I. R., and Russell, G. J., The deposition and structure of LB films of long-chain acids, *Thin Solid Films*, *134*, 143 (1985).

86. Tredgold, R. H., and Winter, C. S., Tunnelling currents in Langmuir–Blodgett monolayers of stearic acid, *J. Phys.*, *D14*, L185 (1981).

87. Tredgold, R. H., Vickers, A. J., and Allen, R. A., Structural effects on the electrical conduction of Langmuir–Blodgett multilayers of cadmium stearate, *J. Phys.*, *D, 17*, L5 (1984).

88. Peterson, I. R., Optical observation of monomer Langmuir–Blodgett film structure, *Thin Solid Films*, *116*, 357 (1984).

89. Bourgoin, J. P., Palacin, S., Vandevyver, M., and Barraud, A., Electrical defect visualization in insulating Langmuir–Blodgett films, *Thin Solid Films*, *178*, 499 (1989).

90. Matsuda, H., Kawada, H., Takimoto, K., Morikawa, Y., Eguchi, K., and Nakagiri, T., Conducting defects in Langmuir–Blodgett films, *Thin Solid Films, 178*, 505 (1989).

91. Peterson, I. R., A structural study of the conducting defects in fatty-acid Langmuir–Blodgett films, *J. Mol. Electron., 2*, 95 (1986).

92. Overbeck, G. A., Hönig, D., Wolthaus, L., Gnade, M., and Möbius, D., Observation of bond orientation order in floating and transferred monolayers with Brewster angle microscopy, in press.

93. Kato, T., and Ohshima, K., Defect-free LB films by the isobaric thermal treatment of barium arachidate monolayers, *Jpn. J. Appl. Phys. II, 29*, L2102 (1990).

94. Bibo, A. M. and Peterson, I.R., Defect annealing rabe in monolayers displaying a smectic-L phase. *Thin Solid Films*, 210/211, 515 (1992).

5

Spectroscopy of Single Molecules in Solids

M. Orrit and J. Bernard

CNRS and University of Bordeaux
Talence, France

I. INTRODUCTION

The need for ever smaller, ever more intricately connected electronic components has led researchers to envision devices working at the nanometer scale. The power and richness of organic chemistry, together with the wonderful success of living things in storing and processing information at the molecular scale, suggest that molecular interactions and cooperativity could be exploited in practical devices. If this dream of a molecular electronics can be realized, it could be on operation principles that are quite different from those of today's silicon technology. Hereafter, we give our personal view of the main conditions necessary to realize such a concept. At present, none of these conditions can be satisfactorily fulfilled, but identifying them could help progress.

First, to realize such a device practically, we need control of the structure at the molecular (nanoscopic) level. The classical way to achieve this control is to apply the powerful but heavy methods of chemical synthesis. The way used by living things is recognition and self-assembly of complex molecules. As described in Chap. 4, new methods like the Langmuir–Blodgett technique or self-assembly are starting to be investigated to the effect of controlling molecular assemblies. Alternatively, directed intervention at the molecular scale becomes feasible now with near-field microscopes (STM or AFM). Manipulation of atoms and molecules on solid surfaces by fine tips has already been demonstrated (see Chap. 4).

Second, we require operating principles and concepts for molecular devices. Several possible concepts have been proposed, for example, cellular automata and neural networks (Chap. 7) and highly parallel optical computers (Chap. 6).

Third, in order to take advantage of the small size of its constituent building blocks (so-called molecules), the device will have to perform tasks at the molecular level, but in a highly parallel way.

Near-field microscopies are today one of the few techniques allowing local addressing of single atoms or molecules. Therefore, they are of paramount importance for fundamental structural studies at the nanometer scale. However, for applications in working molecular electronic devices, these techniques have two drawbacks. First, the type of addressing can only be serial, because a single tip must address a large area with many systems. Second, only the surface of the sample is accessible to this technique. Hereafter, we propose a scheme that we think could overcome these two drawbacks.

Let us think of a sample as a bank of local oscillators with high quality factors, each oscillator having a different resonance frequency. Each oscillator can be retrieved by its frequency, so that the bank is ordered along a one-dimensional axis according to frequency. We call "mapping" the correspondence between the position of the oscillator and its frequency. In the case of disordered solid solutions the mapping would be determined essentially by the particular disordered configuration of our sample. In principle, however, the mapping could be fixed at will if we had full control of the sample's structure. An important example of a simple and well-defined mapping is that of a nonuniform magnetic field by the resonance frequencies of nuclear spins used in NMR imaging. The spatial resolution achieved by such a mapping has little to do with the wavelength of the radiation used to reach the resonance. Like optical near-field microscopies, this technique can attain subwavelength resolution.

The scheme of an oscillator bank together with its mapping would present several advantages: (1) since the addressing is obtained by the resonance of the incoming wave with the frequency of a local oscillator, it will work for a three-dimensional sample as well as for a surface; (2) different points can in principle be addressed simultaneously, in parallel, by a superposition of waves with different frequencies, (3) the addressing of a single oscillator allows in principle to funnel energy or information into a well-defined local spot, without cross talk to others.

Whatever solutions will eventually prevail for molecular electronics, we need a better fundamental understanding of structure and dynamics at the nanometer scale, in order to plan efficiently the working of a molecular device. All techniques enabling us to address and probe nanoscopic systems will deepen our understanding of molecular dynamics and interactions. The selection of a single oscillator eliminates the usual averaging over the whole sample, which cannot be avoided by conventional methods. The molecular mechanisms will then be much simpler, and a more direct comparison with theory will follow. This will be the case of the technique discussed in the present article, the isolation of the optical signal from a single molecule in a solid matrix. The main drawback of this method is that optical resonances broaden very fast when the temperature increases. A good frequency selectivity therefore requires liquid helium temperatures, at least for the systems studied so far.

The oscillators we are dealing with in this article are single organic molecules dissolved at low concentration in a solid matrix. Other systems with allowed electronic transitions, like single ions or quantum dots, could in principle be studied in the same way. The system consisting of the ground and excited states of the optical electron of a single molecule in a solid at low temperature is a good example of the local oscillator discussed above. The local disorder of the matrix shifts the transition frequency at random. If the number of molecules in the addressed sample is small enough, the resonances of single molecules will be resolved from each other in the spectrum. Tuning the laser to one of these resonances will allow us effectively to select a single molecule in the sample.

The study of single atoms and ions started several years ago in the gas phase. There, single atoms or ions can be isolated and studied in atomic beams (1) or in electromagnetic traps (2,3) respectively. The interest and the feasibility of the study of single atoms or molecules in solids at the nanometer scale became appearent in the early 1980s with the scanning tunnelling microscope (4,5). While the possibility of attaining subwavelength resolution with optical probing was soon realized (6), experimental demonstrations were restricted to volumes barely smaller than the wavelength. Early attempts to detect statistical fluctuations of the number of molecules in the small probing volume were unsuccesful because of the low absorption cross-section of molecules at room temperature. The cross-section can be enhanced dramatically by the resonance of a single-frequency laser with the narrow zero-phonon line of an impurity in a solid solution at low temperature. In this way, the oscillator strength of the transition is concentrated on a narrow but very intense line. The statistical fluctuations of the absorption of a solid solution at low temperature (statistical fine structure) were demonstrated experimentally by Moerner and colleagues (7,8). The same group later used a sophisticated modulated absorption technique to demonstrate the first single molecule detection in a solid (9,10). The signal/noise ratio was, however, too small for more than a mere detection of the molecule's signal. As we showed later (11,12), the fluorescence excitation technique provides a very good signal/noise ratio, opening the way for a genuine spectroscopy of the single molecule, at least on well-chosen systems. These experiments for the first time showed the feasibility of addressing and probing a single center on nanometer scales within a transparent solid, a potentially useful ingredient for molecular electronics. Since then, several experiments have been performed on single molecules (13–30), showing the potential scope of this technique. The present article illustrates the main results of these experiments by taking as examples results obtained by our group. From these results, the reader can imagine what could be achieved by combining very sensitive spectroscopy with other techniques described in this book, like self-assembly and near-field microscopies.

The principles and experimental methods of single molecule spectroscopy (SMS) are discussed in Sec. II, together with the general appearance of single molecule lines and their dependence on exciting intensity, temperature, and external perturbations. Section III is concerned with time-resolved measurements using the intensity autocorrelation method; these experiments are unique to SMS and give access to intra- and intermolecular dynamics over a wide range of time scales. The last section presents the main conclusions of this study and ideas as to possible directions to extend SMS.

II. SINGLE MOLECULE SPECTROSCOPY: STATIC PROPERTIES

A. Principles of SMS

Solutions at room temperature have extremely broad optical spectra. This large width is caused by two fundamentally different processes, a static one and a dynamic one. First, the instantaneous disorder in the neighborhood of different molecules gives rise to inhomogeneous broadening, accounting for the main part of the room temperature width (about 300 to 1000 cm^{-1}). Second, the linewidth of each molecule is determined by the very short correlation time (of the order of 0.1 ps at room temperature) of the electronic oscillation responsible for absorption. Fast thermal fluctuations are responsible for this coherence loss (for fluctuation rates faster than the optical frequency change, they give rise to dephasing). In addition to these, slower fluctuations cause the molecule's frequency to drift quickly within the inhomogeneous band, as microscopic configurations evolve (one uses the phrase ''spectral diffusion'' to describe the spread of a frequency packet). It is now possible, using femtosecond laser spectroscopy, to find the timescales of the processes leading to broadening of optical transitions at room temperature (31).

Here, we are concerned with high-resolution spectroscopy by means of cw lasers. The only possibility of getting narrow lines in solid solutions is to slow down the system's dynamics by reducing its temperature. The optical spectrum of a frozen solution, determined by the interaction of the external electron with its surroundings, will depend primarily on structure (32,33). For a quenched glass, the disorder is quite close to that of the liquid state, and so is the inhomogeneous width. In crystals, on the other hand, residual structural defects and impurities often lead to small inhomogenous widths of the order of a few cm^{-1}. The homogeneous linewidth of a single molecule, however, being determined by dynamic effects, changes by several orders of magnitude between room and helium temperature. The homogeneous width often reaches the natural width of the excited molecular state in liquid helium (the natural width is of the order of 30 MHz for an allowed transition). The inhomogeneous absorption band of a macroscopic sample is therefore a superposition of the many much narrower lines of single molecules (see Fig. 1). Spectral diffusion at low temperature is much slower and much more limited than at room temperature, due to the much lower number of accessible states, but it exists and will be discussed in the following.

Let us consider more closely the optical spectrum of a single molecule in a solution at low temperature, by taking into account the quantification of vibrational levels (33). In the simplest adiabatic and harmonic approximation, the shift of vibration coordinates between ground and excited electronic states brings about linear coupling between excitation and vibrations and leads to vibronic lines in the optical spectrum. Intramolecular vibration modes often form well-shaped progressions in the spectrum, while low-frequency matrix modes (pho-

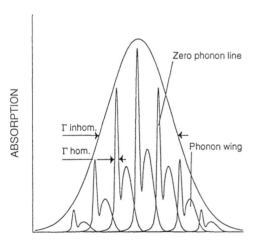

Figure 1   Schema of an inhomogeneously broadened absorption band. The broad Gaussian band, with width Γ_{inhom} results from the superposition of the complex profiles of single molecules, shifted at random by disorder in the matrix. These profiles are represented in the picture by packets of molecules, all with the same resonance frequency. They consist of a sharp Lorentzian zero-phonon line, with width Γ_{hom} and of a broad phonon wing at higher frequency.

nons) give continuous bands (phonon wings) on the high-frequency wing of the absorption transition. One transition is of special interest: the zero-phonon line, which corresponds to an optical excitation of the molecule without creation or annihilation of any phonon quanta (33,34). At zero temperature, the purely electronic zero-phonon line is the lowest-frequency transition, presenting the smallest width (Fig. 1). The ratio of this line's intensity to the total intensity of the purely electronic absorption profile is known as the Debye-Waller factor. It varies as the quantum overlap between fundamental vibration states in ground and excited electronic states. Therefore, its amplitude decreases exponentially with the number of quanta released on excitation. In particular, the Debye-Waller factor is very small for strong shifts of low-frequency modes upon excitation. Since this factor decreases sharply with temperature (33,34) owing to the activation of matrix modes, zero-phonon lines are seen only at low enough temperatures. Hereafter, because we investigate only small portions of the spectra at high resolution, we concentrate on the zero-phonon line and forget about phonon wings.

At low temperature, homogeneous broadening is caused by residual motion in the surrounding matrix. We present here a simple discussion of the effect of this motion on a single molecule's line. We adopt a semiclassical picture in which the resonance frequency of the molecule evolves as a classical function of time (35). This stochastic model is well suited to describe slow processes and homogeneous broadening, in which quasi-elastic interactions with the thermal bath take place (36). Depending on the magnitude of the molecule's frequency change $\Delta\omega$ with respect to the characteristic time of the motion τ_c, one can distinguish between slow and fast modulation (35,36). Fast modulation occurs for $\tau_c^{-1} > \Delta\omega$. It causes the phase vector to perform a random walk similar to Brownian motion. After some characteristic time T_2 such that

$$T_2^{-1} \approx \Delta\omega^2 \, \tau_c \qquad\qquad\qquad (1)$$

the phase memory is lost. T_2 is called the dephasing time of the optical electron. The dephasing rate is reduced by motional narrowing as compared to the frequency excursion $\Delta\omega$. Dephasing is especially important at temperatures above 10K, where activation of quasi-local modes broadens the zero-phonon line exponentially. At low temperature, T_2 is often in the range of 1 ps to 1 ns. For slow modulation, $\tau_c^{-1} < \Delta\omega$, the optical frequency follows the fluctuations of the surrounding matrix adiabatically. The frequency excursion of the optical transition is then $\Delta\omega$, and it is in principle possible to follow the spectral diffusion of the molecule's line in the spectrum on times longer than the dephasing time T_2. For a group of molecules with the same initial frequency, the resonances are shifted at random and diffuse in the spectrum. This process is therefore called spectral diffusion. It is unlikely that soft harmonic modes of the matrix are responsible for motion slower than a nanosecond. The period of the slowest localized modes is expected to lie in the picosecond domain, and the only harmonic modes having frequencies in the range of nanoseconds and longer will be acoustic phonons. However, in amorphous solids there exist the low-energy excitations of so-called two-level systems (37,38) (TLS). A TLS represents two nearly degenerate positions of a particular atom or group of atoms in the matrix. Jumping between the two wells will be possible at low temperature via tunnelling only. Tunnelling processes can be extremely slow because they depend exponentially on barrier parameters, which are thought to be widely spread in amorphous systems. TLS's dominate several properties of glasses at low temperature (39), including specific heat (40), ultrasound absorption (41) and microwave absorption (42,43), optical dephasing, and spectral diffusion (44,45).

The optical spectroscopy of molecular solutions at low temperature started some twenty years ago (46,47) after tunable lasers became widely available. The main technique for the investigation of homogeneous linewidths is persistent spectral hole-burning (32), which is described in Chap. 6 of this book. In this method, the sample is irradiated with a very narrow laser. Some absorption strength is removed from the spectrum around the irradiation frequency, because the excited molecules can undergo photochemical or photophysical processes changing their resonance frequency. The width of the persistent hole thus burnt in the spectrum is related (for low enough burning power and fluence (48)) to the homogeneous width, i.e., to dephasing (49) and spectral diffusion processes (50) on the time scale of the measurement. More recently, the time scale of the spectral diffusion processes was investigated by time-resolved spectral hole-burning (51,52) or stimulated photon echoes (53).

In the present work, the problem of isolating homogeneous structures in an inhomogeneous band is tackled in another way. The simple idea of SMS is to reduce the number of molecules in the sample under study, so as to achieve the spectral resolution of single molecule lines in the spectrum. Figure 2 shows how this regime is reached when the number of molecules decreases. Starting from the inhomogeneous spectrum for many molecules, statistical absorption fluctuations appear when the number of molecules decreases. These fluctuations scale as the inverse square root of the number of molecules absorbing at a given frequency (within their homogeneous width). They were first observed by Moerner and coworkers by modulated absorption (7,8). Finally, when the number of molecules is small enough, at most one, but most probably no molecule will absorb at any given frequency. To reach this limit, two conditions must be fulfilled. (1) The number of molecules in the sample must be smaller than the ratio of inhomogeneous to homogeneous widths. Even when this criterion is not met, it is possible to find isolated single molecule lines by searching far out in the red wing of the inhomogeneous distribution (10). (2) The sensitivity of the detection method must be high enough to detect the weak signal from a single molecule. This last condition is the most difficult to realize and is the main subject of the next paragraph.

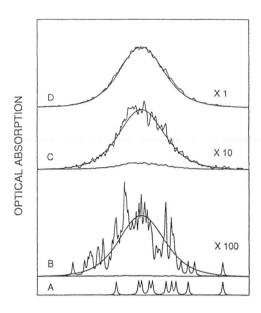

OPTICAL ABSORPTION

FREQUENCY

Figure 2   Numerical models of the absorption spectrum of a sample with small numbers of molecules. These were obtained by adding 10 (A), 100 (B), 1,000 (C), and 10,000 (D) Lorentzian lines of single molecules with resonance frequencies distributed at random according to a Gaussian probability density. For large numbers of molecules (D), the absorption profiles follow closely the probability density (smooth line), with some slight deviation. When the number decreases, the statistical fluctuations increase in relative strength (statistical fine structure), until the regime is reached where single molecule lines are completely resolved from each other (A).

B.   Experimental Methods

Let us first discuss experimental methods to isolate a small number of molecules in the excited part of the sample. The spatial extent of the exciting beam being limited by diffraction, a spot of a few μm^2 is excited. The maximum concentration of the solute molecules needed to obtain spectral resolution of single molecule lines in the band center is then determined by the thickness of the sample. This is typically 10^{-6} mole/L for a 10 μm thick sample but would be about 10^{-3} mole/L for a 100 Å thick Langmuir–Blodgett film.

The samples we used in our experiments were either a molecular single crystal or a polymer (polyethylene) doped with low concentrations of absorbing molecules. The structures of the two colored compounds we used, pentacene

(Pc) and terrylene (Tr), are shown in figures. (1) The thin crystal flakes of para-terphenyl (pTP) were grown by sublimation. Under suitable conditions, pTP crystals grow as thin flakes (a few μm to a few tens of μm thick). They incorporate very low concentrations (around 10^{-9} mole/mole) of pentacene during their growth, due to the much lower vapor pressure of this compound. pTP was extensively zone-refined to reject unwanted impurities. (2) Polyethylene (high density, 70% cristallinity) was obtained from Aldrich. Terrylene (Tr) was synthesized and kindly given to us by Prof. K. Müllen. The polyethylene (PE) pellets were doped with Tr by sublimation and diffusion of this compound into the polymer at about 200°C. To make the sample, a small piece of doped polymer (concentration about 10^{-6} mole/L) was flattened between two glass plates down to about 20 μm thickness.

In order to work with measurable concentrations of absorbing molecules in the sample and to decrease the background from matrix molecules and other impurities, it is advisable to reduce the excited volume, i.e., to tighten the focus of the exciting beam as much as possible. Up to now, three different possible schemes have been proposed. In our first work (12), we used a single-mode optical fiber, at the end of which a thin sample (a few tens of μm) is contacted. The sample is placed at the focus of a large-aperture parabolic mirror that collects the fluorescence light (Fig. 3). We still use this method in most of our experiments because of its simplicity and of the low level of stray light it allows. This sample mounting has some drawbacks, due to the fiber: it is impossible to scan the exciting beam across the sample, and both the light intensity and the polarization are difficult to control accurately. Other groups proposed different schemes. The IBM group (16) uses an aspheric lens to focus the exciting beam tightly onto the sample and a parabolic mirror similar to ours to collect fluorescence. This requires adjustments but gives much flexibility in the control of optical parameters. The Zürich group (21) limited the incoming beam by a pinhole on whose back the sample was fixed. Then, they collected the fluorescence outside the cryostat by means of a large-aperture lens, with a somewhat lesser collection angle than with the parabolic mirror. Here too, the exciting point is fixed, but the polarization can be controlled.

The next problem we must address is that of detectivity. The host-guest system must be carefully chosen in order to optimize the detection conditions of the weak signal from a single molecule. For any optical detection scheme, we need the molecule to interact strongly with a large number of photons. This leads to the following requirements. (1) The absorption cross-section must be as large as possible. The general expression for this cross section σ (54,55) is

$$\sigma = 6\pi \left(\frac{\lambda}{2\pi}\right)^2 \cos^2 \theta \; \alpha_{DW} \frac{\gamma_0}{\gamma_h} \tag{2}$$

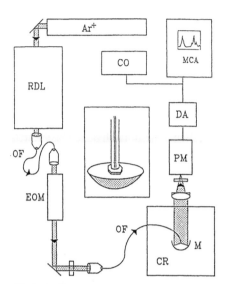

Figure 3 Diagram of the setup used for measuring fluorescence excitation lines of single molecules. The excitation source is a single mode ring dye laser (RDL) pumped by a cw argon laser (Ar +). A piece of optical fiber (OF) spatially filters out the pointing jitter of the laser beam. The beam power is stabilized by an electrooptic modulator (EOM). A second optical fiber is used to excite the sample in the cryostat (CR). The fiber's end carries the sample and is placed at the focus of a parabolic mirror that collects fluorescence (see inset). Fluorescence photons reach the detector (PM) through a filter that eliminates the exciting wavelength. The photoelectron pulses are sorted out and amplified (DA), then recorded as a function of exciting frequency in a multichannel analyzer (MCA) and sent to a correlator (CO) for analysis of the photon pair distribution.

λ is the wavelength of light, θ the angle between laser polarization and the transition moment of the molecule, and α_{DW} the Debye–Waller factor or relative strength of the zero-phonon line. γ_0/γ_h is the ratio of radiative to homogeneous widths. Equation (2) shows that the cross-section can in principle become about as large as the square of the wavelength of the exciting light, due to resonant enhancement. This large value is reduced by the Debye-Waller factor and by the ratio γ_0/γ_h, which is always smaller than unity. α_{DW} is largest for a rigid matrix and for a low coupling to the solute molecule. This will be easier to satisfy for aromatic hydrocarbons than for ionic dyes whose excitation usually involves significant charge redistribution and therefore strongly couples to the environment. In order to concentrate the oscillator strength on a narrow but intense peak, we must reduce γ_h, i.e., work at low temperature where dephasing and spectral diffusion are weak. (2) The weak molecular signal must be accumulated over long time intervals, the resonance frequency of the molecule remaining constant. This requires both molecule and matrix to be photochemically and photophysically very stable. Hole burning should be very difficult, which, at least for a first try, rules out many systems presenting hydrogen bonds (56,57). (3) Metastable molecular states will trap the molecule in a dark state, out of resonance with the exciting laser. They will therefore limit the rate of absorption of the molecule, making the signal difficult to detect against detector background and stray light and lengthening the experiment unacceptably. For organic molecules, there is usually one metastable state, the triplet T_1. The guest molecule should be chosen with a small intersystem crossing (ISC) rate to T_1, and its ISC rate from T_1 to the ground singlet state should be as high as possible.

Depending on the method used for detection of the resonance, other conditions will arise. Two methods have been tried so far, absorption and fluorescence excitation. In absorption, the weak molecular signal must be discriminated against strong shot noise from the unabsorbed photons (9,10). The signal/noise ratio thus achieved was barely sufficient to detect single molecules, but not for spectroscopic studies. The absorption method will work best for very tightly focused beams, when the absorption cross-section will be a significant part of the beam cross-section. In fluorescence excitation on the contrary, the signal originates from absorbed photons only, i.e., it appears on a practically dark background of Raman scattering from the matrix and luminescence by residual impurities. For the latter method, a good fluorescence yield of the molecule is necessary. A more quantitative evaluation of the signal/noise ratios for both methods can be found in Ref. (19).

Single molecules have been detected and studied in only three systems so far: pentacene in a molecular crystal (Pc/pTP) and perylene or terrylene in a polymer (Pr/PE and Tr/PE). The Pc/pTP system has been extensively studied in the literature (58–60) (hence the name organic ruby it is sometimes given). Pc has a very good emission yield (about 80%) and a low intersystem crossing rate in this matrix (58), so that the above conditions are satisfied almost ideally. Pr and Tr are known for having high fluorescence and low triplet yields. PE was chosen as a matrix because it is fairly stable (nonphotochemical hole burning is difficult) and it gives narrow zero-phonon lines at low temperature (49). Attempts to detect single molecules in Pc/PE were unsuccessful, probably because Pc molecules are highly distorted in the polymer and have a much higher triplet yield (61,62).

Until now, most energy was devoted in the different groups to trying new experiments on single molecules in systems where SMS was known to work. The probe molecules were chosen for their convenience for excitation with visible dye lasers (Pc/pTP absorbs at 592–593 nm for the two lowest-frequency sites, Tr/PE (22) between 570 and 585 nm, and Pr/PE (19) between 442 and 450 nm in their red absorption wings). The signal/noise ratio achieved in these three model systems is so high that we think single molecules can be observed, perhaps with somewhat more difficulty, in many other systems. There is little doubt that new probes and new matrices will be found in the future. One can look for possible matrices among other molecular crystals and polymers, but also among inorganic matrices, surfaces, Langmuir–Blodgett films, etc.

Our optical setup for the measurement of fluorescence excitation spectra (12,24) is shown in Fig. 3. The exciting source is a tunable single-mode cw dye laser (CR 699-21 operated with rhodamine 590) pumped by all lines of a 6 W argon laser. The laser's output beam is spatially filtered by a piece of single mode optical fiber and its intensity stabilized by an electrooptic modulator. An additional frequency stabilization can be used to improve stability and resolution of the laser (24). The exciting light is then sent to the optical fiber, whose end carries the sample in the helium cryostat. All experiments reported here except the temperature dependences were done in superfluid helium at 1.8K. The optical quality of the mirror is not very high, and the adjustment of the sample at the focus is done approximately. Therefore, the image of the sample is rather large and falls on the extended cathode (about 1 cm^2) of a photon counting photomultiplier tube (RCA C31034 A02). The direct exciting light from the fiber is eliminated by a small screen, while stray and scattered exciting light is blocked by a red colored glass (Schott RG), or a holographic filter (Kaiser notch) when the excitation frequency remains in a narrow range of 2–4 nm. The photocounts are discriminated, amplified, and recorded in a multichannel analyzer as a function of the excitation frequency. The autocorrelation function of the intensity can be recorded by a logarithmic correlator (ALV 5000) on time ranges between 200 ns and several minutes.

C. Excitation Lines of Single Molecules

Fluorescence excitation lines of single molecules were first observed in the Pc/pTP system (11,12). Here, we shall recall the main qualitative features of these spectra as an illustration of the method. At temperatures below 193K, the pTP crystal has a triclinic structure with four molecules in its unit cell (63–65). On substitution of each of these molecules by one Pc molecule, four absorption sites are obtained (58,60). The single molecule experiments were done in the neighborhood of the lowest-frequency O_1 and O_2 sites at about 16883 and 16887 cm^{-1}, usually in the red wing of the O_1 site. These sites are very convenient for the study of coherent transients and intersystem crossing (58,59). The inhomogeneous width in bulk samples (grown by the Bridgman method) is about 40 GHz (60). Depending on the strains induced in the crystal by dilation of the silica fiber on cooling, the inhomogeneous broadening could be comparable to that of bulk crystals (12) or much larger. In this last case, the position of lines in the red wing of O_1 did not ensure that these sites belonged to O_1 sites.

A typical line of a single Pc molecule in our small sample at 1.8K is displayed in Fig. 4 at different frequency resolutions. The isolated peak resolves into a Lorentzian line, as expected from the simulation of Fig. 2. The distribution, intensity, width, and shape of the lines are all compatible with their attribution to single molecules. The good signal/noise ratio suggests that many experiments can be done on these lines, and that it should be possible to obtain interesting results in less favorable host–guest systems. The linewidth of many molecules reaches the natural width of 7.8 MHz (determined by photon echoes on a bulk sample (66,67). In other cases, however, the linewidth was larger, up to 20 MHz (20), even when the laser drift and jitter were suppressed. We attribute this additional broadening to spectral diffusion of the molecule's line during the measurement. This broadening effect could be larger in our sample than in the bulk because of defects induced by the fiber.

Spectral diffusion manifests itself much more strikingly than by an additional

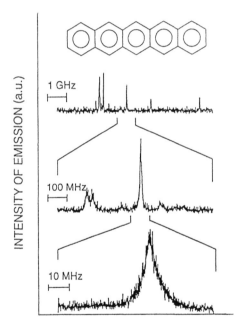

FREQUENCY

Figure 4 Excitation spectrum of a small sample of para-terphenyl crystal doped with absorbing pentacene molecules (see molecular structure) at the end of the fiber. The frequency scale is enlarged to show the resolution of the structures in Lorentzian lines. The width of the line in the bottom spectrum is about 8 MHz, as expected from photon echo decay in bulk samples. The weak structure in the middle spectrum is clearly much broader, which could be due to spectral diffusion during the measurement.

broadening. On one of our first spectra, we observed sudden jumps of the fluorescence intensity of a single Pc molecule (12), corresponding to jumps of the resonance frequency. This effect was later studied in detail by Moerner and his group (15,16), who showed it to be independent of light intensity. They observed for different molecules several different behaviors (16): some molecules were stable, others jumped into and out of well-defined positions, or drifted at random around a center wavelength, and still others crept slowly toward the inhomogeneous band's center. The rate of spectral diffusion was observed to increase with temperature. The modelization of the low-temperature dynamics by two-level systems (TLS's) and their time-resolved study will be discussed in Sec. III. In the spectral diffusion phenomenon, each single molecule acts as a very sensitive probe of the dynamics of its local surroundings. The large sensitivity is due to that of the optical electron to the environment and to the high quality-factor of the resonance.

The single molecule lines were studied as temperature was varied. The general trend is a broadening above some 4K, attributed to dephasing by a libration mode of Pc (16), as was demonstrated in the bulk (59). More surprising is the different behavior of different molecules (16,17): in some cases a narrower structure is observed after the temperature is raised. In Fig. 5 we present one of two such cases we found, where the two peaks of a single molecule close in and merge into a single one when the temperature is raised (20). This evolution was reproduced a few times on the same molecule. We interpret the observed changes as follows. The two peaks, stemming from only one molecule, would correspond to different configurations of the matrix around the molecule. At low temperature, the jump rate is low, and two distinct frequencies are obtained, depending on the actual configuration. As the temperature is raised, the jumping rate between the two positions increases. When it becomes larger than the frequency difference, a fusion of the two lines is expected. If this interpretation is correct, the jumping rate becomes higher than 100 MHz at 3K in the present case. More direct measurements of the rates of similar jumps will be presented in Sec. III.

The second system we studied is Tr/PE (22). The absorption spectrum of Tr in cyclohexane can be found in (68). Fluorescence excitation and fluorescence spectra of Tr in PE at room temperature are shown on Fig. 6. We note the good overlap of the spectra in the 0–0 region around 570 nm, and the marked vibronic progressions with vibrational frequency about 1500 cm^{-1}. The emission yield of Tr is good, 70% according to (69). The inhomogeneous width, about 500 cm^{-1}, is of course much larger in PE than that of Pc in pTP crystal because of the polymer's disordered structure. We found single molecule lines throughout the red wing of the broad excitation spectrum of a diluted sample, between 570 and 585 nm (the center and blue wing were not accessible to our laser). As compared with the single molecule lines in the crystal matrix, those in the polymer

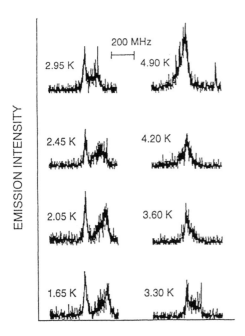

EMISSION INTENSITY

EXCITATION FREQUENCY

Figure 5 Dependence of the excitation structure of a particular single pentacene molecule on temperature. We attribute the two lines seen on the low-temperature spectra to the same molecule because their respective positions remained identical after several thermal cycles up to 7K. The general trend is an activated broadening above 3K. Here, the two lines close in between 1.7 and 3K, then merge into a single broad structure. We can interpret this behavior by fast jumping between the two configurations of a nearby two-level system. When the jumping rate is less than the frequency difference (here about 100 MHz), two lines appear. The fusion of the lines occurs for exchange faster than 100 MHz, which could happen on a temperature increase.

have a much broader distribution of linewidths. Fig. 7 presents a histogram of the measured widths for some 200 single molecule lines (30). The lines were recorded at low exciting power (about 1 mW/cm^2) to avoid power broadening and, especially, irreversible photoinduced jumps (see below). Lorentzians were fitted to the profiles to obtain the full half-maximum width. The cutoff of the distribution at about 50 MHz occurs approximately at the natural width corresponding to the fluorescence lifetime measured at room temperature (about 2.6 ns (30), corresponding to a FWHM of 60 MHz). Although only a very small fraction of the molecules has the natural width, this result shows that some

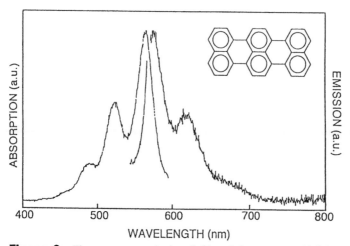

Figure 6 Fluorescence excitation (left) and fluorescence (right) spectra of terrylene (see molecular structure) in polyethylene at room temperature. At helium temperature, single molecule lines are found between 570 and 585 nm, in the overlapping 0–0 region between absorption and emission.

molecules can have negligible dephasing and spectral diffusion even in a disordered matrix. The histogram shows a most probable width of about 100 MHz, with possible widths as large as 400 MHz. We think that the broadening stems from spectral diffusion on time scales shorter than the measurement. More direct evidence for spectral diffusion on short time scales in this system will be discussed in Sec. III. The asymmetric shape of the width distribution is expected for a random spatial distribution of perturbers interacting with the molecule (30).

Our observations of spectral jumps in this system are very similar to those of the IBM group with Pr/PE (18, 19). In particular, spontaneous as well as photoinduced spectral jumps can be observed. Sometimes, the photoproduct line could be found at a different frequency within the same 30 GHz part of the spectrum. Then, the excitation of the photoproduct led, in a few cases, to reappearance of the original line. These are examples of processes leading to reversible nonphotochemical hole burning in bulk materials. Although the structures of Pr and Tr are akin, Tr is a much larger molecule. Yet, it seems that the nature of the colored molecule plays but a minor role in the spectral jumps of single molecules. The aromatic molecule can be considered a mere probe for the spontaneous thermal fluctuations of the polymer matrix around it. Even for some photoinduced jumps (induced by light absorbed by the molecule), the conformational change could occur far enough from the molecule that its actual structure does not critically affect the observed phenomena.

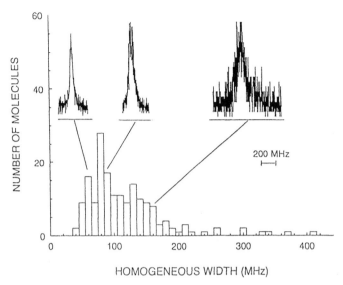

Figure 7 Histogram of measured linewidths for 177 single terrylene molecules in polyethylene at 1.8K. A few profiles of excitation lines are presented, corresponding to different parts of the histogram. We note the cutoff of the width distribution at about 50 MHz, which corresponds roughly to the natural width of terrylene. The width distribution is asymmetric, with a most probable width of about 100 MHz. We do not know yet whether the spikes in the distribution are real structures or statistical noise.

D. Effect of External Perturbations

Narrow homogeneous structures in the optical spectra of molecules can be used as probes for external perturbations. They are extremely sensitive probes for two different reasons. First, the optical electrons usually belong to delocalized π systems and are therefore very sensitive to their environment. Second, the quality factor of the resonance, i.e., the ratio of the transition frequency to the linewidth, is usually very high, 10^7 or higher, so that even small frequency shifts can be easily detected. The effect of external perturbations on persistent spectral holes has been studied extensively. In particular, shift and broadening of holes under electric field give information about the molecular dipole moments and the way they are affected by the matrix (70,71). Linear (34) and even weak quadratic (72,73) Zeeman effects can be measured, and the effect of hydrostatic pressure (74) or stress on a spectral hole gives information about the local compressibility around the molecule (75). The same type of experiments could now be done on single molecule lines. They will give similar information about a much smaller region of the sample, without averaging. For example, we observed that pressure variations from 0 to 1 atmosphere shift single molecule lines in Pc/pTP by about 1 GHz. More accurate measurements of this effect could be performed for different molecules using a pressure cell at constant temperature, instead of the combined pressure and temperature changes we apply.

In this section, we discuss the effect of a static electric field (Stark effect). The quadratic shift caused by an oscillating electric field (light shift or ac Stark effect (76) could also be measured using a nearly resonant pump light beam. The Stark effect on single molecules in Pc/pTP was investigated by the Zürich group (21). They applied the electric field between the metal pinhole, which limited the exciting light beam, and a transparent electrode (SnO_2-coated glass), 300 μm apart. The applied field was as high as 33 kV/cm, high enough for the quadratic Stark effect to appear clearly. In the symmetry of the perfect pTP crystal, the pentacene molecules should have an inversion center, and only a quadratic effect was expected (77). However, the molecules also present a slight linear Stark effect, with random sign and amplitude corresponding to very small dipole moment changes on excitation of about 10^{-4} D. These dipole moments are attributed to defects in the molecule's surroundings, also responsible for the inhomogeneous broadening of the absorption band.

The Stark effect on the Tr/PE system was measured by our group (22). The Stark cell consisted of two electrodes 600 μm apart, between which the fiber's end and the sample were placed. The fields achieved in this way were rather small (up to 1.5 kV/cm), but quite sufficient to measure a clear linear Stark effect. The quadratic effect, being much smaller, could not be detected in these experiments. The signs and amplitudes of the linear shifts were distributed at random, corresponding to dipole moment changes of 0.5 D on average. About 5 to 10% of the 60 molecules studied had much higher dipole moments of up to 2.5 D, a very large value for a nonpolar molecule in a nonpolar matrix. Since the Tr molecule is centrosymmetric, the observation of a linear effect shows that the molecular symmetry is broken by the local field of the polymer. The randomness in sign and amplitude is bound to the random disordered configurations and the random orientations of the molecules in the sample. The qualitative comparison of the induced dipole moment to that in pTP crystal shows that the disordered local field (or the short-range interactions equivalent to it) is much stronger in the disordered polymer than in the crystal. This conclusion is in qualitative agreement with that drawn from the inhomogeneous broadening, which is about three orders of magnitude larger in the polymer than in the crystal. The projection of the dipole moment change along the applied field provides a means to recognize single molecules by their local environment. This can be important after a temperature cycle below the glass transition of the polymer, whereby the frequencies of the molecules are redistributed, but the orientations and local surroundings should remain nearly the same. After a spectral jump, the induced dipole moment usually does not change (see the Stark shift before and after a jump in Fig. 8), as can be expected if the dipole moment is determined by the close surroundings of the molecule. The spectral jump, on the other hand, is probably caused by the local rearrangement of a group of atoms rather far from the molecule and therefore has little influence on the induced dipole moment.

The interaction of a single molecule with faraway defects is thought to be responsible for dephasing and spectral diffusion. The interactions having the longest range in dielectrics are probably of the dipole–dipole type. The associated perturbing field could be either an electric field or an elastic strain field. The study of externally applied electric fields and pressure is therefore of fundamental interest to determine which effect dominates for a given host–guest couple. The sensitivity of single molecule lines to an electric field is also potentially interesting for molecular electronics. By combining the electric field with optical probing of resonance frequencies, a variety of electrooptical experiments can be imagined, e.g., such operations as are now performed with spectral holes (78,79).

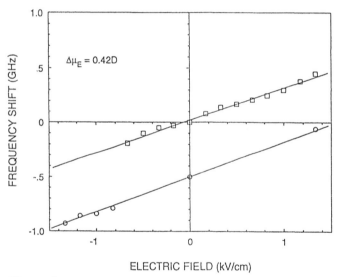

ELECTRIC FIELD (kV/cm)

Figure 8   Change of resonance frequency of a single terrylene molecule in polyethylene at 1.8K as a function of an applied electric field. The Stark effect is linear, although the terrylene molecule is centrosymmetric. The molecular symmetry is therefore broken by short-range forces from polymer molecules. The sign and magnitude of the slope of the plots depends on the particular molecule studied. Here, the squares were first recorded, then the molecule jumped to a new frequency, and the circles were recorded. The slope is not affected by the spectral jump. This observation agrees with a picture of the system where the slope is determined by the structure of the close surroundings of the molecule, whereas the comparatively small frequency jump is caused by a configuration change of the polymer at a larger distance.

If a group of atoms having two quasi-degenerate configurations (thus forming a two-level system) can be driven by an external field, information could be electrically stored in the sample and recovered by "reading" the resonance frequency of single molecules nearby.

III. TIME-CORRELATED STUDIES OF SINGLE-MOLECULE DYNAMICS

A. Introduction

Spectroscopists usually analyze the light emitted by a sample in the frequency space, i.e., by dispersing its spectrum and collecting the information contained in its components. A completely equivalent information is included in the auto-correlation function of the emitted field or emitting dipole, related to the spectrum by a Fourier transform. However, the optical frequency is much too high for our detectors to measure the dipole's correlation directly. The next possible thing is to measure a correlation function of higher order: the autocorrelation function of the emitted intensity, a quantity keeping track of the fluctuation patterns of the light intensity. In some cases, for instance when the observed intensity arises from the interference of many nearly coherent sources (as in scattering) (80,81), the intensity correlation function is related to the spectrum. In cases where no interference takes place (as in fluorescence), the total intensity is a sum of intensities emitted by N independent systems:

$$I(t) = \sum_{i=1}^{N} I_i(t) \tag{3}$$

The correlation function is defined as

$$g^{(2)}(\tau) = \frac{<I(t)\,I(t+\tau)>}{<I(t)>^2} \tag{4}$$

We see from Eqs. (3) and (4) that, for a large number N of independent systems, the $N(N-1)$ uncorrelated cross terms dominate the N autocorrelated terms (of the type $I_i(t)\,I_i(t+\tau)$), which hold potentially interesting information. This latter term is on the contrary the only one left for a single molecule. SMS is thus unique in that it gives direct access to this new source of information. More generally, "pauci-molecular" systems giving signals from a small number of particles (82) or molecules will present time-correlated properties that have no equivalent for large ensembles of independent quantum or classical systems. The correlation method can be applied to very small volumes of liquid solutions of a fluorescent dye at room temperature (83). The different processes affecting the emitted intensity appear on different time scales, from translational and rotational diffusion to intermolecular interactions and vibrational motion affecting the emission yield. The authors of (83) even used their confocal detection to isolate signals from single dye molecules with very good signal/noise ratio. Similar results were obtained by delaying the detection with respect to the exciting laser pulse (84). However, spectroscopy of single molecules at room temperature is much more difficult than detection because of the breadth of optical bands.

The correlation function is measured experimentally by counting the number of photon pairs separated by a given time interval. In experiments on single molecules, the shortest times appearing in the correlogram will be fixed by the optical Rabi frequency in the laser field and the dephasing time T_2, typically in the nanosecond range. The dynamical processes have a very broad distribution of characteristic times in amorphous solids. The shortest ones are on the order of $h/k_B T$ (in the picosecond range at helium temperature), but the longest ones can last for years or longer because tunnelling processes may occur through many kinds of potential barriers. Therefore, the correlation function should be measured on a very broad time scale. A logarithmic correlator fits this requirement ideally.

B. Intramolecular Dynamics

1. Photon Antibunching

For a classical function of time $I(t)$, one can easily see that the second order auto correlation function $g^{(2)}(\tau)$ must satisfy the condition $g^{(2)}(0)/g^{(2)}(\infty) > 1$. This follows from the fact that, in classical theory, $I(t)$ is a well-behaved random function of time. In quantum theory, $I(t)$ depends on the quantum state of the system. According to the principles of quantum mechanics, the wave packet suddenly collapses after a measurement. This implies that the quantum state of the system discontinuously jumps after a measurement and, as a consequence, that the "function" $I(t)$ is not well-behaved (one needs operators in order to correctly describe $I(t)$. It turns out that the behavior of the correlation function of the emission of a single quantum system can display its quantum nature directly by violating the above inequality: $g^{(2)}(0)/g^{(2)}(\infty) < 1$. The mechanism behind this inequality is easy to explain for a single two-level atom or molecule oscillating in a resonant laser field (85). After a photon has been detected, the measurement has caused the state to collapse into the ground state. Since emission is impossible from the ground state, the next photon cannot be detected immediately ($g^{(2)}(0) = 0$); we have to wait for the atom to be excited again by the laser. This needs some average time, which at high laser power is inversely proportional to the laser field, of the order of the Rabi period. The photons thus tend to be separated by some time interval, i.e., antibunched. The discontinuity introduced by measurement is unmistakably revealed in the antibunching phenomenon. We must remark that it plays an essential role in photon bunching too, although bunching does not violate the inequality $g^{(2)}(0)/g^{(2)}(\infty) > 1$ (see next section).

Photon antibunching was first demonstrated on single atoms in a low-density beam (1), then in electromagnetic traps on single ions (2,3). The same antibunching effect can be seen on a single molecule in a solid matrix, as predicted in (12) and recently demonstrated experimentally by Basché and Moerner (23) using Pc/pTP. At short times, it is simpler to measure the distribution of consecutive photon pairs instead of the correlation function, which is nearly the same. In order to avoid dead time and after-pulses of the detector, the signal was divided and directed to two photomultipliers, one giving the start, the other the stop signal. The correlation function was measured in the range 0–200 ns, covering the fluorescence lifetime of Pc in O_1 and O_2 sites (24 ns (58). At high laser power, the experiment was difficult because the molecule's signal saturates, but the background does not.

The results of these measurements clearly show the expected correlation hole at short times. At high laser power, the correlation function starts to oscillate, because the transition dipole undergoes Rabi nutation between the detections of two spontaneous photons. The fitting of measured data with the analytical solution of optical Bloch equations gives a very good agreement between measured and calculated Rabi frequencies (23). Finally, the comparison of correlation functions of the fluorescence of one and two molecules shows a contrast twice smaller for the latter, as is expected from Eqs. (3,4) for two independent emitters. The observation of these short transients demonstrates that the measurement of dephasing times and of Rabi frequencies is possible on single molecules, giving information on individual dipole moments and on the fast local dynamics responsible for dephasing.

2. Bunching Due to Intersystem Crossing

In the preceding discussion, we assimilated a single Pc molecule to a two-level system. In doing so, we forgot about the vibrational levels of electronic states and about the presence of a triplet subspace between the two main singlet levels (ground S_0 and excited S_1 singlet states), between which the optical transition occurs. The transitions between vibrational levels are very fast (picosecond), so that their relaxation is practically instantaneous on the time scales of nanoseconds and longer. The relaxation to and out of the triplet states is much slower (typically in the μs to ms range), and, as we discuss below, these states deeply affect the correlation function on this time scale. In this subsection, the triplet manifold will be represented by a single sublevel to simplify the discussion. A single triplet sublevel suffices to explain our results within experimental error, as long as no microwave photons are applied to the molecule (see next section).

After switching the exciting laser on, the molecule performs excitation-emission cycles between S_0 and S_1, giving intense fluorescence. The transition rate from S_1 to the underlying triplet T_1, k_{23}, is small, but the intersystem crossing transition finally takes place. Since there is no more resonance when the molecule is in the triplet state, the fluorescence stops for a time interval, whose average is the triplet lifetime. Then the metastable triplet state finally decays to the ground singlet and the whole cycle starts all over again. The distribution of the emitted photons is strongly nonrandom: the photons are grouped in packets or bunches, separated by "dark" intervals, when the molecule is in the triplet state. The theory of the three-level system (S_0, S_1, T_1) (24) predicts an exponential decay of the correlation function between $g^{(2)}(0)$ and $g^{(2)}(\infty)$ for times much longer than T_2, i.e., outside the region where antibunching is observed. The time constant of the correlation decay corresponds to the bunch duration at high intensity. Then, the bunches are much shorter than the dark intervals. The contrast is high, and the average bunch duration is approximately twice the inverse of the ISC rate k_{23}, because the population of S_1 is ½ at saturation. For low excitation power, on the other hand, the bunches are longer than the dark intervals. The contrast is weak and the decay rate is the ISC rate k_{31}, corresponding to the shortest time in the kinetics.

The experimental results for Pc/pTP (24) clearly confirm the theory as illustrated in Fig. 9. As the laser power increases, the single molecule line (inset) gathers intensity and starts to broaden due to saturation of the three-level system. Meanwhile, the correlation function acquires a stronger contrast and its decay becomes steeper. The measured contrast $g^{(2)}(0)/g^{(2)}(\infty)$ reaches 3 (which corresponds to a contrast of 8 when the background's contribution is taken into proper account). Such high contrast values can only be obtained with one or a very small number of independent systems. The correlation data are well fitted by single exponentials at all powers.

We have performed such correlation measurements for eight different molecules. Plots of the decay rate of the correlation as a function of the exciting intensity (on a logarithmic scale) are given in Fig. 10 for some molecules. The theory leads us to expect S-shaped curves starting from k_{31} at low power (right end of the plots) to reach $k_{31} + k_{23}/2$ at high power. The ISC rates can thus be read directly on the plots.

The ISC rates were found to be different for different molecules. We attribute these differences to the influence of neighboring defects on the geometry of Pc molecules. This result is surprising because ISC normally is an intramolecular process and should depend but little on environment. However, Pc is probably an atypical case in this respect. The ISC rate k_{23} of Pc is known to change by two orders of magnitude between the longwave sites O_1, O_2 and the shortwave sites O_3, O_4 (58). Distortions of the O_3, O_4 sites are thought to favor intersystem crossing by increasing spin-orbit coupling (61,62) (the transition would be forbidden for a perfectly flat Pc molecule). Similarly, even small distortions of the Pc

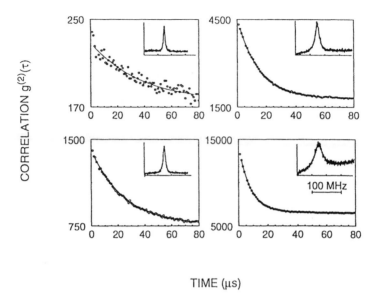

TIME (μs)

Figure 9 Excitation spectra (insets) and correlation functions of the fluorescence of single pentacene molecules in a para-terphenyl crystal for different exciting intensities. The correlation functions were recorded while the laser was tuned to the maximum of the excitation line. All correlation data are well fitted by single exponentials (solid lines). As the intensity increases, the fluorescence and correlation signals increase too, and the lines broaden due to saturation of the three-level system (ground singlet, excited singlet, triplet). The saturation means that short photon bunches are emitted and separated by comparatively long dark intervals, lasting for the triplet lifetime. Indeed, the correlation function decays much faster at high power, which allows us to determine the intersystem crossing rates.

molecule by crystal defects in nearly undistorted O_1 and O_2 sites could lead to significant changes in the ISC rates. This study illustrates the power of SMS to provide the distributions of molecular parameters, where conventional methods only give average values. In the present case, intramolecular processes are probed in a reliable way by their effect on intensity fluctuations.

3. Optically Detected Magnetic Resonance in a Single Molecule

At high excitation power, the molecule spends most of its time in the triplet subspace, and the fluorescence intensity is limited by the dwell time in this subspace. Usually, the triplet sublevels are populated differently by ISC from the excited singlet S_1 and have different ISC rates to the ground singlet S_0 (see (86) for the case of Pc in naphthalene crystal). Therefore, by applying a resonant microwave field between the zero-field split magnetic sublevels, it is possible

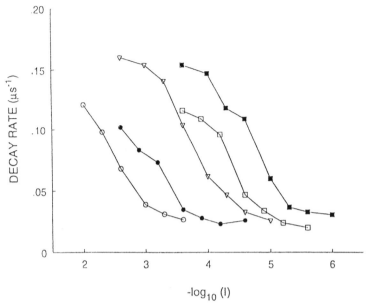

Figure 10 Plots of the correlation decay rate (see Fig. 9) as a function of the logarithm of exciting intensity for different pentacene single molecules. The plots have been shifted along the intensity axis for clarity. The S-shape of the curves is expected from saturation theory. The limit of the rate for low intensity is k_{31}, the rate of intersystem crossing (ISC) from triplet to ground singlet. The limit for high intensity is the sum $k_{31} + k_{23}/2$, k_{23} being the ISC rate from excited singlet to triplet. These limits are seen to depend on the particular molecule. ISC rates are therefore modified by defects in the local environment.

to change the average dwell time in the triplet, and thus the average fluorescence intensity. This double resonance experiment is called "optically detected magnetic resonance" (ODMR).

The ODMR experiment has been performed recently and independently by two groups (28,29). Two microwave transitions were detected at about 1.48 GHz and 1.36 GHz. The amplitude of the ODMR effect can reach 30%. The strongly asymmetric shapes of the microwave transitions show that the nuclear (proton) spins of the molecule are relaxing during the long time scale of the measurement, probably during passage in the singlet states, when nuclear spins are no longer blocked by the electrons' magnetic field. Work is now in progress to interpret the correlation functions with and without microwaves and to investigate various coherent effects with microwave pulses. Single molecule ODMR extends the sensitivity and selectivity of SMS to magnetic resonance, a powerful tool of physical chemistry. The selection of a single molecule suppresses the inhomogeneous broadening and orientation disorder. It will allow us in principle to perform experiments with disordered samples, which are now possible only with single crystals, an appealing perspective for biological molecules that are difficult to crystallize.

C. Dynamics of the Environment

Being very narrow, single molecule lines are extremely sensitive to motion in their environment. As discussed in Sec. II, fast fluctuations are responsible for dephasing. The molecular frequency change they cause is reduced by motional narrowing and is therefore not directly accessible to experiment. Slow fluctuations, on the other hand, with characteristic times longer than the inverse frequency excursion they cause, allow one in principle to follow the optical frequency of the molecule as a function of time. Accordingly, the molecular frequency diffuses within the inhomogeneous spectrum. In glasses, such spectral diffusion will result from many possible modes with very different characteristic times.

The different possibilities for thermal motion in the molecule's neighborhood must be considered according to their characteristic times. In crystals at low temperature, acoustic and localized phonons are the only excitations. Since they have large frequencies they will contribute to dephasing only. Amorphous systems (glasses) present excitations that have no equivalent in crystals (39). This is because glasses are metastable solids that can evolve between different configurations even at low temperature. The glass passes from one configuration to the other by tunnelling through a barrier. Because of the disorder, barrier parameters are distributed, which leads to a very broad distribution of jumping rates between configurations. Particular pairs of configurations between which jumping is most frequent are modeled by two-level systems (37,38) (TLSs). Many of the anomalous low-temperature properties of glasses can be explained very well with a suitable distribution of TLS's. Because TLS's can jump slowly, they will play the leading role in the optical spectral diffusion of impurities dissolved in the glass. If the exciting laser frequency is tuned to the resonance of one molecule, the spectral jumps due to a neighboring TLS will manifest themselves by fluorescence intensity fluctuations, as the molecule jumps into and out of resonance with the laser. The schema of the tunneling TLS is presented in Fig. 11. As it tunnels between configurations 1 and 2, the molecular frequency accompanies it in its slow motion, giving rise to two distinct lines in the excitation spectrum. If the exciting laser is tuned to the resonance of one of these lines, intensity fluctuations will produce photon bunching and become apparent on the correlation function.

We measured the correlation functions of many single molecules of the system Tr/PE discussed in section II.C. For this study, the broad time range given by our correlator was extremely important (it extends from 200 ns to 100 s). Several

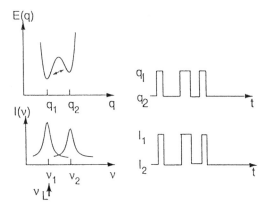

Figure 11 Schematic diagram of the influence of a two-level system (TLS, upper left) on the fluorescence of a single molecule in its neighborhood. As the group of atoms described by coordinate q tunnels between the two positions q_1 and q_2, the resonance frequency of the molecule is shifted from ν_1 to ν_2 (transitions between the configurations are sudden because the barrier crossing time is very short as compared to the inverse tunnelling rate). If the exciting laser is tuned to one of the resonances, say ν_1, signal interruptions arise due to switching of the TLS. The corresponding intensity fluctuations are directly visible as an exponential decay of the intensity correlation function.

different correlation patterns were observed (26,30). The most common occurrence was a flat correlation within the experimental window, but we could find one, two, or more well-defined exponential decays, depending on the molecule under study. Exponential decays correspond to well-defined jumping rates, which we observed to be distributed throughout the accessible time range. We attribute them to single TLS's in the molecule's environment. Figure 12 gives an example of correlation functions of a single molecule's fluorescence presenting a single exponential decay. That the two lines in the spectrum (inset of Fig.12) stem from the same molecule is indicated by their identical correlation time and confirmed by frequency changes where both peaks jump together to a new frequency. Such observations, which neatly confirm the TLS model, could be reproduced on some molecules. In most cases, however, only one peak could be recorded. The other must have lain outside the tuning range of the laser (30 GHz).

Once a particular TLS has been identified in the correlation function of a single molecule, its jumping rate can be studied as a function of external parameters. The study of the rate as a function of exciting power shows that TLS jumps can be spontaneous as well as photoinduced. This study, however, is difficult because the lines of single Tr molecules in PE tend to undergo large irreversible jumps under high-power irradiation.

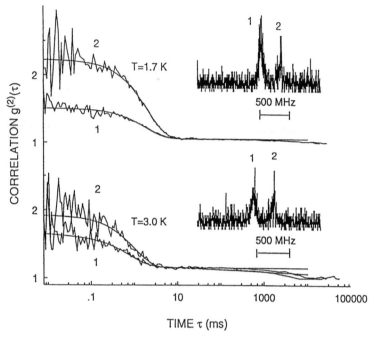

Figure 12 Example of a single terrylene molecule in polyethylene presenting two excitation lines. We interpret the two lines as the resonances of the molecule for two different configurations of a nearby TLS. The correlation rates of single molecules are usually spread throughout the accessible range of the correlator (note the logarithmic time scale). Therefore, the identity of the correlation times strongly supports the assignment of the two lines to only one molecule. Spectral jumps of the two lines together further confirm this assignment. When the temperature is raised, the intensities of the two lines tend to equalize (as expected for Boltzmann equilibrium), and so do the contrasts of the correlation functions. All these observations support the model of the coupling of the molecule to a single TLS.

By limiting ourselves to low power, we could measure the jumping rate of single TLS's as functions of temperature, in the range from 1.4 to 4.5K (26,30). For some molecules, the decay rates of the correlation showed irreproducible variations, probably because jumps of other TLS's were affecting the jump rate of the TLS under study. In other cases, the temperature dependence was reproducible. Figure 13 shows the variations of the jump rate for three single TLS's in the vicinity of three different single molecules. For two of them, the dependence is well fitted by a power law with exponent 1 or 3. We found two other cases of a linear dependence and one other case of a cubic one. These dependences are expected for a tunnelling system coupled to acoustic phonons,

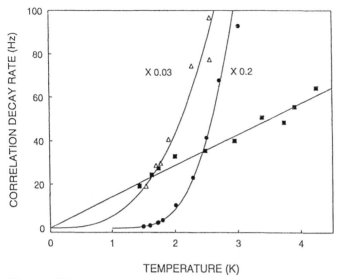

Figure 13 Temperature dependence of the correlation decay rate for three single terrylene molecules. Note the wide spread of rates and their strong dependence on temperature. We chose molecules with well-defined exponential decays in their correlation functions, which could be attributed to single TLS's. One TLS has a linear dependence of the jump rate within experimental error (squares). The other two present much stronger variations. One has a cubic dependence within experimental error (triangles, fitted by a T^3 law). The cubic behavior is expected for Raman processes in which one acoustic phonon is absorbed and another emitted during a TLS jump. Finally, the last TLS has a variation of 500 of its jump rate between 1.5 and 3K. The solid line is a fit using an Arrhenius activation law, which represents the data very well.

as in the case of unsymmetric energy transfer (87). Power laws of temperature are obtained only when the energy difference of the states in the two wells is much smaller than temperature. Emission or absorption of one acoustic phonon leads to a linear dependence. The cubic temperature dependence is expected for Raman processes, in which one acoustic phonon is emitted and another one with nearly equal energy is absorbed.

Finally, one of the measured TLS's presented a change of a factor of more than 500 between 1.5 and 3.5K (Fig.13). The variation could be very well fitted by an Arrhenius law. Large though it is, this variation of the jump rate is probably too small to be explained by classical activation. The probability of barrier crossing would be on the order of the ratio of the jump rate (about 10^3 s^{-1}) to the oscillation frequency in the well (about 10^{12} s^{-1} from general arguments). Such a large factor would depend even more strongly on temperature than the observed rate does. We therefore think that this TLS jumps by tunnelling. Additional stabilization in either well by slow matrix modes could explain the activated behavior (30,88).

These results illustrate the sensitivity of single molecule lines to subtle tunnelling in systems that are probably several nm away from the molecule. They also show the variety and complexity of molecular systems already on the nanometer scale and at temperatures of a few K. However, many of the observed behaviors could be rationalized using very simple models, which gives hope that this complexity can be understood and controlled. Generally, the observation of single systems, by eliminating all averages, should lead to a much more reliable comparison with theoretical models.

IV. CONCLUSION AND OUTLOOK

In this article, we have presented the main results obtained by isolating the lines of single molecules in fluorescence excitation spectra of small and dilute samples. In less than three years since this new spectroscopy arose, new and important results were obtained. We stressed our own results and mentioned the main results of others. These data show that, at least in the few systems investigated so far, the signal/noise ratio of single molecule fluorescence is large enough for a true spectroscopy at the nanometer scale. We have shown that this technique can be used for molecular spectroscopy, since photophysical processes within the molecule can be studied separately on individuals and then compared. We have also used single molecules as sensitive probes for external perturbations and dynamical processes in their environment on intermediate time scales (from one microsecond to seconds).

The difficulty of obtaining a good signal/noise ratio should be easily offset by the specific advantages of single molecule spectroscopy (SMS). (1) Since it involves optical electrons, SMS is sensitive and fast and can address molecules in the bulk of transparent media. (2) SMS gives direct access to the distributions of molecular quantities (e.g., fluorescence rate, ISC rates, vibrational frequencies), where most conventional methods only give average values. (3) SMS can probe and address single localized spots in a solid. A general scheme using SMS would include the sample with single molecules acting as doorways for energy or information, together with its mapping, i.e., the correspondence between geometrical coordinates of each point and frequency. This scheme could be exploited to address molecular electronics devices in a highly parallel fashion, without need for a physical contact with the spots addressed. (4) SMS is of great interest for the fundamental study of molecular processes. It suppresses all averagings entailed by other methods, even selective methods like spectral hole burning. The comparison of experimental results to theoretical models is therefore much simpler and more reliable.

The field of SMS is very new, and it is likely that it will develop quickly in the next few years. The direction now explored is the systematic application of conventional spectroscopic techniques to single molecules in the well-known systems Pc/pTP, Tr/PE, and Pr/PE. High-resolution spectroscopy, the study of spontaneous and photoinduced spectral jumps, transient regimes due to Rabi nutation and intersystem crossing, external field effects, and optical detection of magnetic resonance have all been demonstrated on one of these systems. Eventually, experiments in the broad and important field of nonlinear spectroscopy will also be performed on single molecules.

In the future development of SMS, it will be very important to improve both the time and the space resolution of the method. The time resolution given by the correlation method is currently 1 μs, but it reached 10 ns in the case of Pc/pTP. Even higher time resolutions could be achieved in systems with good fluorescence efficiencies where no triplet bottleneck exists, as in some radicals, for example. The space resolution could be greatly improved by coupling the SMS experiment to near-field microscopy techniques such as STM or AFM. In this way, resolutions of a few tens of nanometers should not be difficult to obtain, but would nevertheless improve by several orders of magnitude the current resolution of a few μm.

So far, all results were obtained with only three host–guest couples giving high signal/noise ratios. It is likely that other such systems will be discovered in the future, but a lower signal/noise ratio would still provide important microscopic insight into dynamics and structure of many less favorable host–guest couples. A better control of the nanoscopic structure of the matrix is highly desirable both for a better knowledge of molecular processes and for applications in molecular electronics. It would therefore be very promising to generalize SMS to Langmuir–Blodgett films and other autoassembling matrices.

In these more complex systems, we can imagine that single molecules can be brought into interaction, for example by tuning their resonance energies by externally applied fields. Single molecules could probe the motion of charge carriers through the system, or they could be used as doorways efficiently and selectively to funnel energy or information into local nanoscopic devices. Of course, all these ideas are only dreams at this point. The experiments described here show the complexity of soft matter at the molecular scale. Despite the relative simplicity of the systems studied and despite the low temperature, many different behaviors are observed, depending on the local configuration of the matrix around the molecule. It is clear that the ultimate aim of molecular electronics, i.e., controlling and harnessing this complexity for planned purposes, is a formidable task that cannot be achieved in a few years. Yet SMS is but one of several new techniques providing direct control and knowledge of matter at the nanoscopic scale. Our hope is that this technique will start to lend some credibility to the dream of molecular electronics, if only by lifting the last psychological barriers to studying and manipulating single molecules and their microscopic surroundings.

ACKNOWLEDGMENTS

The work presented here was done in part by Prof. R. I. Personov, Prof. Ch. von Borczyskowski, Dr. R. Brown, Dr. H. Talon, L. Fleury, A. Zumbusch, and J. Wrachtrup. We thank Prof. K. Müllen for the gracious gift of terrylene. We thank Dr. W. E. Moerner for communicating results of his group prior to publication.

REFERENCES

1. Kimble, H. J., Dagenais, M., and Mandel, L., *Phys. Rev. Lett.*, *39*, 691 (1977).
2. Bergquist, J. C., Hulet, R. G., Itano, W. M., and Wineland, D. J., *Phys. Rev. Lett.*, *57*, 1699 (1986).
3. Diedrich, F., and Walther, H., *Phys. Rev. Lett.*, 58, 203 (1987).
4. Binnig, G., Rohrer, H., Gerber, Ch., and Weibel, E., *Phys. Rev. Lett.*, *50*, 120 (1983).
5. Meyer, E., Howald, L., Overney, R. M., Heinzelmann, H., Frommer, J., Güntherodt, H.-J., Wagner, Schier, T., H., and Roth, S., *Nature*, *349*, 398 (1991).
6. Fischer, U. Ch., *J. Vac. Sci. Techn.*, B3(1), 386. (1985).
7. Moerner, W. E., and Carter, T. P., *Phys. Rev. Lett.*, *59*, 2705 (1987).
8. Carter, T. P., Manavi, M., and Moerner, W. E., *J. Chem. Phys.*, *89*, 1768 (1988).
9. Moerner, W. E., and Kador, L., *Phys. Rev. Lett.*, *62*, 2535 (1989).
10. Kador, L., Horne, D. E., and Moerner, W. E., *J. Phys. Chem.*, *94*, 1237 (1990).
11. Bernard, J., and Orrit, M., *C. R. Acad. Sci. Paris, t. 311*, Série II, 923 (1990).
12. Orrit, M., and Bernard, J., *Phys. Rev. Lett.*, *65*, 2716 (1990).
13. Orrit, M., and Bernard, J., *Mod. Phys. Lett.*, *B5*, n°11, 747 (1991).
14. Moerner, W. E., and Ambrose, W. P., *Phys. Rev. Lett.*, *66*, 1376 (1991).
15. Ambrose, W. P., and Moerner, W. E., *Nature*, *349*, 225 (1991).
16. Ambrose, W. P., Basché, T., and Moerner, W. E., *J. Chem. Phys.*, *95*, 7150 (1991).
17. Orrit, M., and Bernard, J., *J. of Luminescence*, *53*, 165 (1992).
18. Basché, Th., and Moerner, W. E., *Nature*, *355*, 235 (1992).
19. Basché, Th., Ambrose, W. P., and Moerner, W. E., *J. Opt. Soc. Am.*, *B9*, 829 (1992).
20. Talon, H., Fleury, L., Bernard, J., and Orrit, M., *J . Opt. Soc. Am.*, *B9*, 825 (1992).
21. Wild, U. P., Güttler, F., Pirotta, M., and Renn, A., *Chem. Phys. Lett.*, *193*, 451 (1992).
22. Orrit, M., Bernard, J., Zumbusch, A., and Personov, R. I., *Chem. Phys. Lett.*, *196*, 595 (1992); erratum *199*, 408 (1992).

23. Basché, Th., Moerner, W. E., Orrit, M., and Talon, H. *Phys. Rev. Lett.*, *69*, 1516 (1992).
24. Bernard, J., Fleury, L., Talon, H., and Orrit, M., *J. Chem. Phys.*, *98*, 850 (1993).
25. Moerner, W. E., and Basché, T., to appear in *Angew. Chem. Int. Ed.Engl.*, *32*, 457 (1993).
26. Zumbusch, A., Fleury, L., Brown, R., Bernard, J., and Orrit, M., *Phys. Rev. Lett.*, *70*, 3584 (1993).
27. Tchenio, P., Myers, A. B., and Moerner, W. E., *J. Phys. Chem.*, *97*, 2491 (1993).
28. Köehler, J., Disselhorst, J. A. J. M., Donckers, M. C. J. M., Groenen, E. J. J., Schmidt, J., and Moerner, W. E., *Nature*, *363*, 242 (1993).
29. Wrachtrup, J., von Borczyskowski, C., Bernard, J., Orrit, M., and Brown, R., *Nature*, *363*, 244 (1993).
30. Fleury, L., Zumbusch, A., Orrit, M., Brown, R., and Bernard, J., *J. Luminescence*, *56*, 15 (1993).
31. Brito Cruz, C. H., Fork, R. L., Knox, W. H., and Shank, C. V., *Chem. Phys. Lett.*, *132*, 341 (1986).
32. Moerner, W. E., ed., *Persistent Spectral Hole-Burning: Science and Applications*, Springer-Verlag, Berlin, 1988.
33. Sild, O., and Haller, eds., K., *Zero Phonon Lines*, Springer-Verlag, Berlin, 1988.
34. Personov, R. I., in *Spectroscopy and Excitation Dynamics of Condensed Molecular Systems* (V. M. Agranovich and R. M. Hochstrasser, eds.), North Holland, Amsterdam, 1983.
35. Abragam, A., *The Principles of Nuclear Magnetism*, Oxford Univ. Press, London, 1961, p. 450.
36. Molenkamp, L. M., and Wiersma, D. A., *J. Chem. Phys.*, *83*, 1(1985).
37. Anderson, P. W., Halperin, B. I., and Varma, C. M., *Phil. Mag.*, *25*, 1 (1972).
38. Phillips, W. A., *J. Low Temp. Phys.*, *7*, 351 (1972).
39. W. A. Phillips, ed., *Amorphous Solids; Low Temperature Properties*, Springer-Verlag, Berlin, 1981.
40. Pohl, R. O., in *Amorphous Solids; Low Temperature Properties* (W. A. Phillips, ed.), Springer-Verlag, Berlin, 1981, p. 27.
41. Hunklinger, S., and v. Schickfus, M., in *Amorphous Solids; Low Temperature Properties* (W. A. Phillips, ed.), Springer-Verlag, Berlin, 1981, p. 81.
42. Golding, B., and Graebner J. E., in *Amorphous Solids; Low Temperature Properties* (W. A. Phillips, ed.), Springer-Verlag, Berlin, 1981, p. 107.
43. Golding, B., Graebner, J. E., and Haemmerle, W. H., *Phys. Rev. Lett.*, 44, 899 (1980).
44. *Optical Linewidths in Glasses*, special issue of *J. Luminescence*, *36* (1987).
45. Breinl, W., Friedrich, J., and Haarer, D., *J. Chem. Phys.*, *80*, 3496 (1984).
46. Kharlamov, B. M., Personov, R. I. and Bykovskaia, L. A., *Optics Commun.*, *12*, 191 (1974).
47. Gorokhovskii, A. A., Kaarli, R. K., and Rebane, L. A., *Optics Commun.*, *16*, 282 (1976).
48. Völker, S., *Chem. Phys. Lett.*, *120*, 496 (1985).
49. Thijssen, H. P. H., and Völker, S., *J. Chem. Phys.*, *85*, 785 (1986).

50. Walsh, C. A., Berg, M., Narasimhan, L. R., and Fayer, M. D., *J. Chem. Phys.*, *86*, 77 (1987).
51. Littau, K. A., and Fayer, M. D., *Chem. Phys. Lett.*, *176*, 551 (1991).
52. Müller, K. P., and Haarer, D., *Phys. Rev. Lett.*, *66*, 2344 (1991).
53. Meijers, H. C., and Wiersma, D. A., *Phys. Rev. Lett.*, *68*, 381 (1992).
54. Jackson, J. D., *Classical Electrodynamics*, John Wiley, New York, 1975, p. 803.
55. Loudon, R., *The Quantum Theory of Light*, Oxford Uni. Press, 1973, p. 287, Chaps. 5, 9.
56. Fearey, B. L., Carter, T. P., and Small, G. J., *J. Phys. Chem.*, *87*, 3590 (1983).
57. Kokai, F., Tanaka, H., Brauman, J. I., and Fayer, M. D., *Chem. Phys. Lett.*, *143*, 1 (1988).
58. de Vries, H., and Wiersma, D. A., *J. Chem. Phys.*, *70*, 5807 (1978).
59. Orlowski, T. E., and Zewail, A. H., *J. Chem. Phys.*, *70*, 1390 (1979).
60. Patterson, F. G., Lee, H. W. H., Wilson, W. L., and Fayer, M. D., *Chem. Phys.*, *84*, 51 (1984).
61. Kryschi, C., Wagner, B., Gorgas, W., and Schmid, D., *J. Luminescence*, *53*, 468 (1992).
62. Kryschi, C., Fleischmann, H.-C., and Wagner, B., *Chem. Phys.*, *161*, 485 (1992).
63. Baudour, J.-L., Delugeard, Y., and Cailleau, H., *Acta Cryst.*, *B32*, 150 (1976).
64. Baudour, J.-L., Cailleau, H., and Yelon, W. B., *Acta Cryst.*, *B33*, 1773 (1976).
65. Baudour, J.-L., *Acta Cryst.*, *B47*, 935 (1991).
66. Morsink, B. W., Aartsma, T. J., and Wiersma, D. A., *Chem. Phys. Lett.*, *49*, 34 (1977).
67. de Vries, H., de Bree, P., and Wiersma, D. A., *Chem. Phys. Lett.*, *52*, 399 (1977).
68. Clar, E., *Polycyclic Hydrocarbons*, Academic Press/Springer, New York/Berlin, 1964, p. 226.
69. Bohnen, A., Koch, K.-H., Lüttke, W., and Müllen, K., *Angew. Chem. Int. Ed. Eng.*, *29*, 525 (1990).
70. Kador, L., Haarer, D., and Personov, R. I., *J. Chem. Phys.*, *86*, 5300 (1987).
71. Meixner, A. J., Renn, A., and Wild, U. P., *Chem. Phys. Lett.*, *190*, 75 (1992).
72. Ulitsky, N. I., Kharlamov, B. M., and Personov, R. I., *Chem. Phys.*, *141*, 441 (1990).
73. van den Berg, R., van der Laan, H., and Völker, S., *Chem. Phys. Lett.*, *142*, 535 (1987).
74. Sesselmann, Th., Richter, W., Haarer, D., and Morawitz, H., *Phys. Rev.*, *B36*, 7601 (1987).
75. Zollfrank, J., and Friedrich, J., *J. Opt. Soc. Am.*, *B9*, 956 (1992).
76. Cohen-Tannoudji, C., and Kastler, A., *Progress in Optics*, Vol. V (E. Wolf, ed.), North Holland, Amsterdam, 1966, p. 1.
77. Meyling, J. H., and Wiersma, D. A., *Chem. Phys. Lett.*, *20*, 383 (1973).
78. Wild, U. P., de Caro, C., Bernet, S., Traber, M., and Renn, A., *J. Luminescence*, *48, 49*, 335 (1991).
79. Rebane, A., Bernet, S., Renn, A., and Wild, U. P., *Opt. Commun.*, *86*, 7 (1991).
80. Loudon, R., *The Quantum Theory of Light*, Oxford Univ. Press, 1973, Chaps. 5, 9.

81. Cummins, H. Z., and Pike, E. R., eds., *Photon Correlation and Light Beating Spectroscopy*, Plenum Press, New York, 1974. See also R. Pecora, ed., *Dynamic Light Scattering*, Plenum Press, New York, 1985.

82. Schaefer, D. W., and Berne, B. J., *Phys. Rev. Lett.*, *28*, 475 (1972).

83. Rigler, R., Widengren, J., and Mets, U., in *Fluorescence Spectroscopy* (O. Wolfbeis, ed.), Springer-Verlag, 1992.

84. Brooks Shera, E., Seitzinger, N. K., Davis, L. M., Keller, R. A., and Soper, S. A., *Chem. Phys. Lett.*, *174*, 553 (1990).

85. C., Cohen-Tannoudji, Ecole d'été des Houches 1975, in Frontiers in Laser Spectroscopy (R. Balian, S. Haroche, and S. Liberman, eds.), North Holland, Amsterdam, 1977.

86. van Strien, A. J., and Schmidt, J., *Chem. Phys. Lett.*, *70*, 513 (1980).

87. Holstein, T., Lyo, S. K., and Orbach, R., in *Laser Spectroscopy of Solids*, (W. M. Yen and P. M. Selzer, eds.), Springer-Verlag, Berlin, 1981, p. 39.

88. Kagan, Yu., *J. Low Temp. Phys.*, *87*, 525 (1992).

6

Transfer of Single Electrons and Single Cooper Pairs in Nanojunction Circuits

Michel H. Devoret, Daniel Esteve, and Cristian Urbina

CEA-Saclay
Gif-sur-Yvette, France

I. INTRODUCTION

One individual atom, a purely theoretical entity a hundred years ago, can now be imaged and manipulated at the surface of bulk matter (1) or, free-standing, in vacuum (2). Is the electron, the simplest and most thoroughly studied particle, amenable to such ultimate control? In vacuum, the detection of single electrons is now routine. A spectacular example of the control of individual electrons travelling in a vacuum chamber is the experiment in which Dehmelt et al. (3) were able to probe during three months a single electron kept in an electromagnetic trap, thereby measuring to unprecedented accuracy the anomalous part of its magnetic moment. In matter, the manipulation of individual electrons is a very different game, because the separation between electrons is of the same order as their quantum mechanical wavelength. Here we focus on the most basic type of such manipulation. We explain how it is possible to take, at a precise instant, exactly one electron from a first electrode and transfer it with certainty to a second electrode. By making these electrodes part of an electrical circuit and by continuously repeating this transfer process we can achieve a perfectly controlled current source. In particular, for a sequence of single-electron transfers clocked by a radio-frequency signal at frequency f, the current I will be given simply by $I = ef$, where e is the quantum of charge, a fundamental constant. We also explain how, when at least one of the electrodes is in the superconducting state, electron pairing favors charge transfer by units of $2e$.

II. BASIC PRINCIPLES OF SINGLE ELECTRON TRANSFER

Although the charge of the electron was measured as early as 1911 (4), the granularity of electricity does not usually show up in the macroscopic quantities such as current and voltage, which describe the state of an electric circuit. This is not just a matter of the number of electrons being very large in typical devices. Charge flow in a metal or a semiconductor is a continuous process because conduction electrons are not localized at specific positions. They form a quantum fluid that can be shifted by an arbitrarily small amount. The variations of the charge Q on a capacitor C and of the associated potential difference $U = Q/C$ illustrate this property. The charge Q can be any fraction ϵ of the charge quantum e: if ρ denotes the electron density in the metallic plates of the capacitor and S their surface area, it is easy to see that a bodily displacement $\delta = \epsilon/(\rho S)$ of the electronic fluid with respect to the ionic background, in the direction perpendicular to the plates, produces the charge $Q = \epsilon e$.

There exists, however, a solid-state device in which electric charge flows in a discrete manner. It consists of two metallic electrodes separated by an insulating layer so thin that electrons can traverse it by the tunnel effect (5) (Fig. 1). Tunnelling can be considered as an all-or-nothing process because electrons spend a negligible amount of time under the potential barrier corresponding to the insulating layer (6,7). If one applies a voltage V to such a tunnel junction, electrons will randomly tunnel across the insulator at a rate given by V/eR_t, where the tunnel resistance R_t is a macroscopic parameter of the junction that depends on the area and thickness of the insulating barrier. Apart from allowing the tunnel effect, the two facing electrodes behave as a capacitor whose capacitance C is the other macroscopic parameter of the junction. It is important to stress that the transport of electrons in a tunnel junction and in a metallic resistor are fundamentally different, even though the current–voltage characteristic is linear in both cases. Charge flows continuously along the resistor, whereas it flows across the junction in packets of e. Obviously, a tunnel junction provides the means to extract electrons one at a time from an electrode. With a single voltage-biased tunnel junction, however, it is not possible to control the instants at which electrons pass from the upstream electrode to the downstream electrode, because of the stochastic nature of tunnelling. A further ingredient is needed.

Suppose that instead of applying directly a voltage source to the junction one biases it with a voltage source U in series with a capacitor C_s (we reserve the letter symbol V for transport voltage sources that have to deliver a static current). A metallic electrode entirely surrounded by insulating material is formed between the junction and the capacitor (see Fig. 2a). We will call such an isolated electrode, which electrons can enter and leave only by tunnelling, an "island." The island is coupled electrostatically to the rest of the circuit by the capacitances C and C_s whose charges are denoted by Q and Q_s, respectively. Although, as

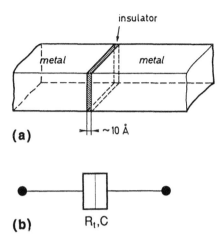

insulator

metal   metal

~10 Å

(a)

(b) R_t, C

Figure 1   (a) Tunnel junction traversed by a current $I(t)$ that consists, when a fixed voltage is imposed to the junction, of uncorrelated charge packets corresponding to individual electrons. Electrons tunnel through the thin layer of insulator sandwiched between the metal electrodes. The junction is represented in circuit schematics by a double box symbol (b) and is characterized by the tunnel resistance R_t and capacitance C. It is worth noting that although R_t is called a "resistance," it characterizes a purely elastic process. At the insulating barrier, the electron wave function is partially transmitted and reflected. Its energy does not change. The tunnel resistance is inversely proportional to the barrier transmission coefficient, which decreases exponentially with the thickness of the insulating layer. In practice, measurable tunnel resistances can be achieved only with insulating layers a few nanometers thick.

we have seen, Q and Q_s are both continuous variables, their difference is the total excess charge of the island. Because charge can enter the island only by tunnelling, this total charge is a multiple of the electron charge: $Q - Q_s = ne$. Suppose furthermore that the island dimensions are small enough that the electrostatic energy $E_c = e^2/2C_\Sigma$ of one excess electron on the island is much larger than the characteristic energy $k_B T$ of thermal fluctuations. Here, $C_\Sigma = C + C_s$, and k_B and T denote the total capacitance of the island, the Boltzmann constant, and the temperature, respectively. This Coulomb energy E_c is the other ingredient of controlled electron transfer.

When $U = 0$, n will stay identically zero because the entrance or exit of an electron would raise the electrostatic energy of the island to a level much higher than permitted by thermal fluctuations. As U increases from zero, however, the total energy difference between the $n = 0$ and $n = 1$ states of the whole circuit decreases, because when an electron tunnels to the island the potential drop $C_s U/C_\Sigma$ partly compensates the electrostatic energy of the island. In fact, a straightforward calculation of the total energy of the circuit yields $E = E_c(n - C_s U/e)^2 +$

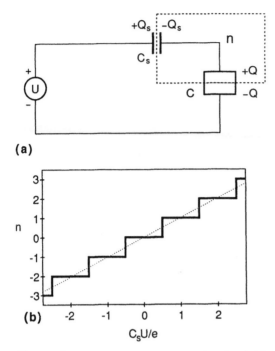

Figure 2 (a) Single electron box: a junction is biased by a voltage source U in series with a capacitance C_s. The metal electrode between the junction and the capacitance forms an isolated "island" (box in dashed line), which contains n excess electrons. (b) Variation of $\bar{n}$, the average of n as a function of U, when $k_B T \ll E_c$ (full line) and $k_B T \gg E_c$ (dashed line).

terms independent of n. Thus, when $U = e/2C_s$, the $n = 0$ and $n = 1$ states will have the same energy and an electron can tunnel in and out freely. As U is increased further, the $n = 1$ state becomes the lowest energy state. The maximum stability of the $n = 1$ state against fluctuations is reached at $U = e/C_s$ where, as in the case $U = 0$, and $n = 0$, the charge Q vanishes. It is now easy to see that each time the voltage U is increased by e/C_s, the number n of excess electrons of the island is increased by one. If one plots n, the average of n, as a function of U, one gets the staircase function shown in Fig. 2b. It is therefore possible to control exactly the number of excess electrons of the island by adjusting the voltage U.

As the temperature is increased, the staircase becomes rounded, and for temperatures $k_B T \gg E_c$ it approaches the straight dotted line of Fig. 2b. In practise, one can reliably cool tunnel junctions down to 50 mK but not much below. To satisfy $E_c \gg k_B T$, $C\Sigma$ must be of the order of or smaller than one femtofarad. This requires the fabrication of junctions with typical areas of 50

× 50 nm and hence the use of nanofabrication techniques. With such low values of capacitance, the typical voltage corresponding to the addition of an electron is of the order of 100 μV–1 m V, a value that can be easily controlled electronically. To summarize, tunnelling breaks the continuity of the electron fluid into charge packets corresponding to single electrons. The Coulomb energy of excess charges on an island provides a feedback mechanism that regulates the number of electrons tunnelling in and out the island. At sufficiently low temperature, the exact number of excess electrons on the island does not fluctuate and can be entirely determined by an externally applied voltage. The quenching of the island charge fluctuations for the "single electron box" (the circuit of Fig. 2a), has been demonstrated experimentally by Lafarge et al. (8).

We have considered so far only thermal fluctuations of the number n. This variable is also subject to quantum fluctuations. In our analysis of the circuit of Fig. 2a, we have neglected the delocalization energy associated with tunnelling. This energy is very small compared with the Coulomb energy. Perturbative calculations (9,10) show that the quantum fluctuations of n become negligible in the limit $R_t \gg R_K = h/e^2$, where h is Planck's constant. The constant $R_K \approx$ 26 kΩ is the resistance quantum. In the next three sections we shall consider tunnel barriers sufficiently opaque that this latter condition is fulfilled.

III. SINGLE-ELECTRON EFFECTS: A BRIEF HISTORY

A large class of phenomena exist that combine the partial localization of electrons due to tunnelling and the Coulomb charging energy and may be called "single-electron effects." Decades ago, it was proposed that the variation of the island potential due to the presence of only one excess electron could be large enough to react back on the probability of subsequent tunnelling events (11–15). At that time, the effect could only be observed in granular metallic materials. It was realized that the hopping of electrons from grain to grain could be inhibited at small voltages if the electrostatic energy of a single electron on a grain was much larger than the energy of thermal fluctuations. The interpretation of these pioneering experiments, in which there is an interplay between single-electron effects and random media properties, was complicated by the limited control over the structure of the sample. With modern nanofabrication techniques, it is possible to design metallic islands of known geometry separated by well-controlled tunnel barriers (16). This led Fulton and Dolan to perform the first unambiguous demonstration of single-electron effects in an island formed by two junctions (17). Meanwhile, Likharev and coworkers (18,19) had produced detailed predictions of single-electron effects in a nanoscale current-biased single junction (this system was also considered in Refs. 20 and 21, but only for junctions in the superconducting state) and proposed various applications of the new effects. This current-biased scheme is analogous in some ways to the circuit of Fig. 2a, but with the capacitance replaced by a large resistance. In that case there is no island enforcing charge quantization, because an arbitrarily small amount of charge can flow through the resistance. Only the charge on the junction capacitance would provide the feedback of Coulomb energy on tunnelling.

It was later understood that, in general, the quantum electromagnetic fluctuations due to the finite value of the resistance wash out single-electron effects in this single-junction no-island system. Only if the value of resistance is made much larger than the resistance quantum R_K up to frequencies of the order of $e^2/(hC)$ (22–24) can tunnelling be Coulomb blocked in the current-biased junction system. In spite of the experimental difficulties involved in fabricating the resistance with adequate characteristics, the competition between single-electron effects and quantum electromagnetic fluctuations has been observed (25,26). The single-junction no-island system is certainly of interest as an illustration of the foundations of the field, but it is not suited for practical applications because getting rid of quantum electromagnetic fluctuations is so difficult experimentally. In what follows, we resume the discussion of systems that contain at least one island and are thus immune to quantum electromagnetic fluctuations. We focus mainly on the controlled transfer of single electrons. For general introductions to single-electron effects in normal and superconducting junction systems, see Refs. 27–30. For recent snapshots of the state of current research, see Refs. 31 and 32.

IV. THE SINGLE-ELECTRON TRANSISTOR

The one-junction one-island circuit of Fig. 2a is the simplest in which single-electron transfer can occur. On the other hand, it cannot produce an externally measurable static current, as the island is a cul-de-sac for electrons. Let us consider the next order of complexity, the two-junction one-island circuit of Fig. 3a (17). The state of the circuit is now characterized by the two numbers N and N' of electrons having passed through the two junctions. (The sign of N and N' is positive if during tunnelling the electron flows in the direction of increasing voltage, and negative otherwise.) It is convenient to introduce the number $n = N - N'$ of excess electrons on the island and the charge flow index $p = (N + N')/2$. The state $(n, p + 1)$ only differs from the state (n, p) in that one electron has been transferred from one terminal of the transport voltage source V to the other. The electrostatic energies of the various capacitances of the circuit are the same. As the precise value of p does not matter here, we condense the notations (n, p) and $(n, p + 1)$ into (n) and $(n)^*$. With regard to the total energy of the circuit, which includes the work of the transport voltage, state $(n)^*$ is lower by eV than state (n), and hence the circuit has no absolutely stable states. In principle, a steady current I could flow around the loop formed by the two junctions and the transport voltage V. To go from state (n) to state $(n)^*$, however,

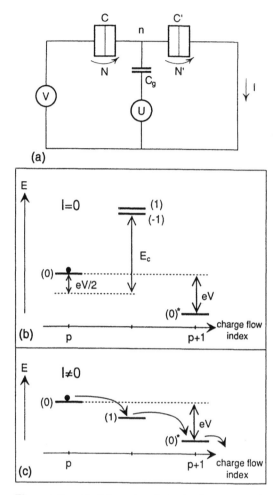

Figure 3 (a) Schematic of single-electron transistor (SET). Energy of the states of the circuit when (b) $U = 0$ and (c) $U = e/C_g$. The numbers in parentheses are the values of the number n of excess electrons on the SET island. The charge flow index is half the sum of the numbers N and N' of electrons that have traversed the junctions. In (b) no current can flow through the device: this is the Coulomb blockade.

the circuit must go through state $(n + 1)$ or state $(n - 1)$, because tunnel events occur one at a time. States $(n + 1)$ and $(n - 1)$ differ by state (n) by an electron having tunnelled through the first and second junction, respectively. This is where the single-electron Coulomb energy $E_c = e^2/2C_\Sigma$ comes into play (C_Σ is, as before, the island total capacitance given now by $C_\Sigma = C + C' + C_g$, where C and C' are the two junction capacitances). To simplify the discussion, suppose that $eV \ll E_c$.

When the control "gate" voltage is set at $U = 0$, the energy of states (-1) and (1) will be $E_c - eV/2 \approx E_c$ above the energy of state (0) (see Fig. 3c). At low temperature, this will provide a Coulomb barrier for the transport of electrons around the circuit. In this case the current I should be strictly zero. This situation is called Coulomb blockade. On the other hand, when the control voltage is such that $C_g U \approx e/2$, states (0) and (1) have nearly the same energy (Fig. 3c). As soon as the energy of the (1) state is lowered below that of the (0) state, the $(0) \rightarrow (1)$ transition becomes possible, and an electron enters the island through the first junction. If U is such that the energy of the (1) state, although below that of the (0) state, is still above the energy of the $(0)^*$ state, the transition $(1) \rightarrow (0)^*$ takes place and the electron leaves the island through the second junction. Apart from an electron having gone through the device, one is now back to the initial electrostatic state and the cycle can start over again. This cascade of transitions produces a current of order $V/(R_t + R'_t)$ through the device (R_t and R'_t are the tunnel resistances of the two junctions). When U is increased further, the energy of the (1) state goes below the energy of the $(0)^*$ state and one enters a new Coulomb blocked state with one excess electron on the island. The domains of the Coulomb blocked states in the set of U values are in a one-to-one correspondence with the flat portions of the staircase of the electron box (Fig. 2b), and it is easy to show that, at voltages low compared with the Coulomb voltage E_c/e, the current I is maximum when $C_g U = (n + \frac{1}{2})e$.

In practise, a current of the order of 10^9 electrons per second can be switched on and off by the presence or absence of half the electron charge on the gate capacitor, hence the name "single electron transistor" (SET) given to this device. The remarkable charge sensitivity of the SET is unrivalled by other devices: it is six orders of magnitude better than conventional field-effect transistor (FET) electrometers (28). A possible application is the detection of individual photoinduced electron–hole pairs in semiconductors (33). But the input capacitance of the SET is, by construction, so tiny that its voltage sensitivity is not high. In this respect, it does not compare favorably with the FET, the semiconductor device on which most of today's applications of solid-state electronics are based. Furthermore, in the SET the modulation of electron flow by the gate ceases as soon as the bias voltage becomes of the order of the Coulomb gap voltage E_c/e, whereas in the FETs used in digital circuits the modulation of the source–drain current by the gate only saturates at large bias voltages (34). It is this latter feature that ensures enough voltage gain to compensate for the dispersion in device parameters and that makes robust integrated digital circuit design possible with FETs.

An analogy (28) can be drawn between the SET and the d.c. SQUID, (superconducting quantum interference device) with an input coil (35) (see Fig. 4).

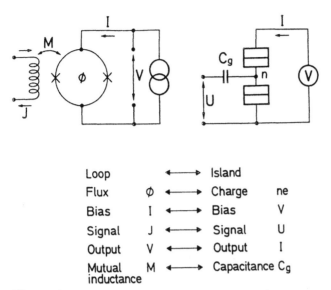

Loop		Island	
Flux	ϕ	Charge	ne
Bias	I	Bias	V
Signal	J	Signal	U
Output	V	Output	I
Mutual inductance	M	Capacitance	C_g

Figure 4   Comparison between the d.c. SQUID (left) and the SET electrometer (right).

The d.c. SQUID consists of two Josephson junctions in parallel biased by a static current. In the d.c. SQUID, the output voltage is a periodic function of the current in the input coil, whereas in the SET, the output current is a periodic function of the voltage on the input capacitor. For the d.c. SQUID the period is set by the flux quantum $h/2e$, whereas for the SET the period is set by the charge quantum e. It is tempting to speculate that the SET will play the same role for ultrasensitive electrometry that the d.c. SQUID plays for ultrasensitive magnetometry. However, the fundamental impossibility of building the charge analog of the superconducting flux transformer which is so crucial to the use of d.c. SQUIDs may severely limit the use of SETs.

The junctions that have been described so far consist in practise of two overlapping metallic films. It is also possible, instead of the three-dimensional gases that conduction electrons form in a metal, to use two-dimensional electron gases that are found in semiconductor heterostructures such as GaAs/GaAlAs. The detailed manifestations of Coulomb blockade have been thoroughly studied in these systems where single-electron effects may coexist with the quantum Hall effect (36).

Finally, Coulomb blockade has been observed with a scanning tunnelling microscope (STM) placed over a tiny metallic droplet (37). The role of the island is played by the droplet. Unfortunately, it has so far been impossible to modulate the gate voltage independently in the droplet–STM systems. On the other hand, very small island dimensions (a few nanometers) can be achieved in this manner, and Coulomb blockade at room temperature has been reported (38). In principle the island could even be reduced to a single molecule (39).

It is important to note at this point that we describe the state of a circuit such as the SET by discrete variables like n and p, and not by continuous variables like currents and voltages as in classical electronics. What makes nanojunction circuits of fundamental interest is that we must treat them as single atomlike quantum systems. Although we use macroscopic concepts like capacitance, we analyze charge flow in terms of quantum transitions between discrete energy levels of the whole circuit.

V. CONTROLLED TRANSFER OF CHARGE FLOWING IN AN EXTERNAL CIRCUIT

Although the principle of the SET involves the electrostatic energy of a single electron on the SET island, the charge flow through this device is not controlled at the single-electron level. The voltage U controls only the average value of the current. The instants at which electrons pass through the device are random, as in a single junction. A control of the charge flow electron by electron would mean that, using the control voltage U, one would make a single electron enter the island from the left junction, hold it in the island for an arbitrary time, and finally make it leave the island through the right junction. One could then go continuously from a Coulomb blocked state with $n = 0$ to a Coulomb blocked state with $n = 1$. This is not possible with only one island. When the energy of the (1) state dips below the energy of the (0) state, it is necessarily above the energy of the (0)* state to which it can decay (see Fig. 3c). An electron cannot be made to enter the island through one junction without setting the electrostatic energies so that it is energetically favorable for another electron to leave the island through the other junction.

The control of charge flow at the single-electron level requires at least three junctions (40). Let us consider the three-junction two-island circuit of Fig. 5a. As in the case of the SET, the state of the circuit can be described using the numbers n_1 and n_2 of excess electrons on each island and the charge flow index given by the third of the algebraic sum of the number of electrons having tunnelled through each junction. Using the condensed notation defined above, (n_1, n_2) and $(n_1, n_2)^*$ denote two states whose charge flow index differ by one, that is, states differing by an electron that has lost energy eV by passing through the entire device.

We suppose $V \ll \min(e/C_{\Sigma 1}, e/C_{\Sigma 2})$ where $C_{\Sigma 1}$ and $C_{\Sigma 2}$ denote the total capacitances of the two islands. The two control voltages U_1 and U_2, applied to the two gate capacitances C_1 and C_2, allow us to change the relative energy of the various states of this circuit. If we set U_1 and U_2 to $e/2C_1$ and $e/2C_2$, respectively, the energies of states (0, 0), (1, 0), (0, 1), and (0, 0)* form a

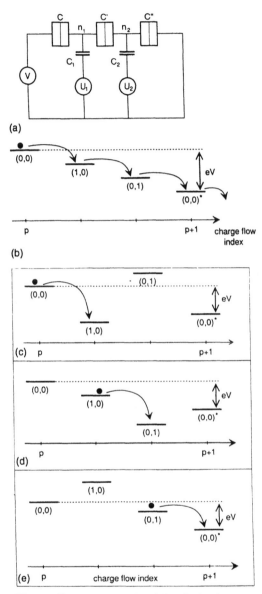

Figure 5 (a) Schematic of the single electron pump. (b) Energy states of the circuit when the control voltages U_1 and U_2 are set so that Coulomb blockade is suppressed. (c–e) Pumping cycle that transfers one electron around the circuit of (a). It is obtained by superposing two phase-shifted modulation signals on the values of U_1 and U_2 corresponding to (b).

cascade (Fig. 5b). We are in a situation equivalent to the suppression of Coulomb blockade depicted in Fig. 3c, and a stochastic current flows through the device. Because there are three junctions instead of two, there are now two intermediate states (0, 1) and (1, 0) in the cascade. Each one of these states is coupled to (0, 0) or to (0, 0)* but not to both. The lowering of either (0, 1) or (1, 0) below (0, 0) and (0, 0)* stops the stochastic current and puts the circuit in a blocked state. By modulating U_1 and U_2 with dephased periodic signals, the energy of these intermediate states can be cyclically lowered below that of the (0, 0) and (0, 0)* states while avoiding the cascade configuration of Fig. 5b (Fig. 5c–e). One starts from the situation where both (1, 0) and (0, 1) are above (0, 0) and (0, 0)*. The circuit is in a blocked state with no excess electrons on the islands. At first, an increase of U_1 lowers (1, 0) below (0, 0) and (0, 1). An electron goes through the leftmost junction and the circuit adopts a new blocked state with an extra electron on the first island (Fig. 5c). Then U_2 increases while U_1 decreases: this lowers (0, 1) below (1, 0) and (0, 0)*. A tunnel event consequently takes place through the middle junction, and the circuit now adopts a blocked state with an extra electron on the second island (Fig. 5d). Finally U_2 is decreased to its initial value, making (0, 1) pass above (0, 0)*. An electron goes through the rightmost junction and, apart for a charge e having crossed the entire device, the circuit returns to its initial blocked state (Fig. 5e). If the transport voltage V is reversed, the same modulation cycle will still carry electrons in the same direction, provided the energy difference eV between (0, 0) and (0, 0)* stays small compared with the energy excursions of (0, 1) and (1, 0). The charge now flows in a direction opposite to that imposed by V. Energy conservation is of course not violated. The work done to "charge" the transport voltage source is provided by the control voltage sources. We have therefore nicknamed this three-junction device the single-electron "pump." The pump is reversible: a time-reversed modulation cycle will transfer electrons from right to left.

The actual operation of a physical device is shown in Fig. 6. We first set U_1 and U_2 to the static values $U_1^{dc} = e/C_1$ and $U_2^{dc} = e/C_2$ corresponding to a maximum zero-voltage conductance (center curve marked "no r.f."). Two periodic signals with the same frequency f but dephased by $\Phi \simeq \pi/2$ are then superimposed on the static components U_1^{dc} and U_2^{dc}. This implements the cycle shown in Fig. 5c–e, and a current plateau is observed (see Fig. 6a). One can easily reverse the cycle, leaving all other conditions the same, by changing Φ to $\Phi + \pi$. A current plateau is again observed, with the same absolute value at $V = 0$ but with opposite sign. The height of the plateau is plotted against frequency on Fig. 6b. The relation $I = ef$ is well verified, providing further confirmation that our device does indeed implement the pump principle.

We have seen how two control voltages can transfer electrons one by one in

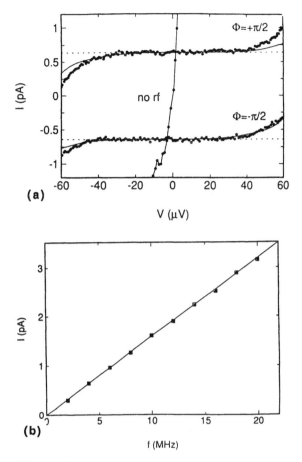

Figure 6   (a) Current–voltage characteristic of the pump with and without a $f = 4$ MHz control voltage modulation. The two modulation signals were phase-shifted by Φ. Dashed lines indicate $I = \pm ef$. Full lines are the result of numerical simulations taking into account quantum fluctuations of the island electron number. (b) Current measured at the inflexion point of the current plateau as a function of the frequency f. Full line is $I = ef$.

a three-junction device. The transfer of a single electron using only one control voltage is possible but needs at least four junctions. In Fig. 7a we show the schematic of a four-junction three-island circuit that we have nicknamed the "turnstile" (41). A gate capacitance, with roughly half the value of the capacitance of the junctions, is connected to the central island. Because the gate voltages of the side islands have only to be set to a constant value of zero (in

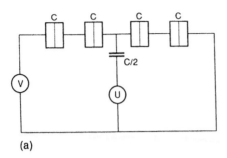

(a)

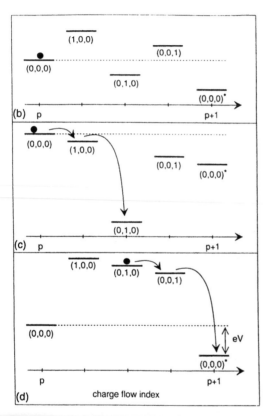

Figure 7 (a) Schematic of single-electron turnstile. (b–d) Turnstile cycle obtained by modulating the control voltage U, which transfers one electron around the circuit of (a).

practice, the external gate voltage must be adjusted to compensate for random offset charges (28)). No gate lines have been represented in the figure. The turnstile can be described as a SET with two junctions in the entrance and exit channels. The intermediate islands create energy barriers whose effect is to suppress the stochastic conduction that takes place in the SET for small V, and for U such that $C_g U = e/2$ (see Fig. 7b–d). For these conditions, the circuit can exist in two states characterized by the presence or absence of an extra electron on the central island. Suppose one starts with no electron in the central island. As U is increased, all the energies of the intermediate states decrease, although the state with an electron on the central island remains the lowest of the intermediate states (see Fig. 7b). Consequently, one electron enters the central island. If now one decreases U, all the energies of the intermediate states increase and at one point the state with an extra electron on the central island is no longer the lowest of the intermediate states. An electron then leaves the central island (Fig. 7d). It is easy to see that after one cycle of modulation of U, a charge of one electron has passed through the whole device. Like the pump, the turnstile produces a current $I = ef$, where f is the modulation frequency. Unlike the pump, however, the turnstile is an irreversible device, the sign of the current being imposed by the sign of the bias voltage V.

VI. METROLOGICAL APPLICATIONS

We have seen that the pump and the turnstile can produce a current determined only by the frequency f and the quantum of charge e. Because frequencies can be accurately determined, these devices would provide in principle a standard of current. The standard is obtained at present by the combination of the Josephson effect (35), which relates a frequency to a voltage through the flux quantum $\Phi_0 = h/2e$, and the quantum Hall effect discovered by von Klitzing (42), which relates current to voltage through the resistance quantum $R_K = h/e^2$. It is important for metrologists to check whether a direct definition of the ampere using the charge quantum e provided by single electron devices is compatible with the "Josephson/Klitzing" definition, which combines Φ_0 and R_K. The value of the fine-structure constant $\alpha = e^2/(2h\epsilon_0 c)$, where c and ϵ_0 denote the speed of light and the electrical permittivity in vacuum, is another important metrological issue that would benefit from the new access to the charge quantum provided by single-electron devices (43). This latter application would not necessitate measuring directly the very low current produced by single-electron devices; one would simply charge a calibrated capacitor with a known number of electrons and compare its voltage with the Josephson volt.

The first experiments carried out to test the precision of the pump and the turnstile were chiefly limited by the precision of current measurements, and it is important to investigate the intrinsic limitations of the devices. One problem is to ensure that the devices are sufficiently cold while passing current. In that respect the pump principle is better than the turnstile, as the pump is reversible and can operate at zero bias voltage. Theoretical analyses show that the fundamental limitation on the accuracy of the devices is due to co-tunnelling events (44) during which several tunnel events take place simultaneously on different junctions. These higher-order processes are a manifestation of the quantum fluctuations of island electron number discussed above. Fortunately, it can be demonstrated that the rate of co-tunnelling events decreases exponentially with the number of junctions in a device. Detailed calculations have shown that an accuracy better than 10^{-8} in the number of transferred electrons is achievable with a pump with five junctions operating at temperatures of 100 mK or less (45,46). An important step towards the practical realization of high-accuracy transfer devices is to show experimentally that the number of electrons on an island is well determined when this island is connected to a charge reservoir through four junctions that block the quantum fluctuations of electron number. We have made a direct measurement of the charge of such an island by using a SET electrometer (47). In Fig. 8 we show single tunnelling events in and out of the island occurring on a time scale of a tenth of a second. Although this time scale is still shorter than expected theoretically, we believe that if more and smaller junctions were used, the spontaneous tunnel rate could be lowered by two orders of magnitude and thus permit metrological experiments. An important step in this direction

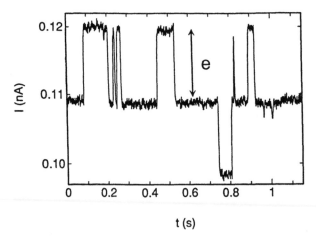

Figure 8 Time variations of the current through a SET electrometer measuring the charge on an island linked to a charge reservoir through a series of four tunnel junctions. Each jump corresponds to an electron tunnelling into or out of the island.

has recently been made by Martinis et al. who have operated a five-junction pump with a 10^{-6} accuracy (48).

VII. SINGLE COOPER PAIR TRANSFER

Up to now we have considered metallic nanojunction circuits in the normal state. For circuits in the superconducting state, one might naively expect that single electron transfer could be transposed into single Cooper pair transfer, e being simply replaced by $2e$; but several features of the superconducting state complicate this direct transposition, and early experiments on Cooper pair transfer in superconducting nanojunction circuits showed unexpected results (49–51,8) that we begin to understand in detail only now. Let us go back to our basic circuit, the electron box of Sec. II, and examine the simplest case, where only the island is in the superconducting state. The energy of the circuit as a function of the number n of electrons in the island is now $E = E_c(n - C_s U/e)^2 + (n \bmod 2)\,\tilde{\Delta}$ + terms independent of n. The first term is the same electrostatic energy as in the normal state, i.e., the electrostatic energy of C and C_s and the work of the voltage source U. The superconducting nature of the island manifests itself in the second term, which is the internal energy of the island, which we suppose for the moment to be at $T = 0$. This internal energy depends on n only through its parity, the parameter $\tilde{\Delta}$ denoting the minimum energy of a quasiparticle excitation. Such an odd–even difference is expected for a superconductor, since with an odd number of electrons, one of them cannot be paired and must remain as a quasiparticle excitation (52). However, it is crucial to realize that the energy cost of this remaining quasiparticle excitation coincides with the superconducting energy gap Δ only for an ideal BCS superconductor in zero magnetic field. From this circuit energy we can predict the ensemble average $< n >$, which we suppose to be equal to the temporal average $\bar{n}$ measured in the experiment.

In Fig. 9a we show as a function of U the energy of the different n states, for the nonsuperconducting case $\tilde{\Delta} = 0$. As we have shown in Sec. II, n will adopt the value of the integer closest to $C_s U/e$, which corresponds to the lowest energy state. We thus get the staircase pattern of Fig. 9b, which is identical to the full line curve of Fig. 2b. In Fig. 9c we show the case of a superconducting island such that $\tilde{\Delta} < E_c$. The effect of the odd–even difference is simply to reduce the span of U over which the system will adopt an odd n ground state and, conversely, to increase the span of U where an even n state will be favored. We thus get an asymmetric staircase that again has e-steps but that is $2e$-periodic (see Fig. 9d). Finally, in Fig. 9e we show the case of a superconducting island such that $\tilde{\Delta} > E_c$. In that case, for every value of U, the ground state of the circuit always corresponds to an even n, which explains the doubling in Fig. 9f of the height and length of the steps with respect to Fig. 9b.

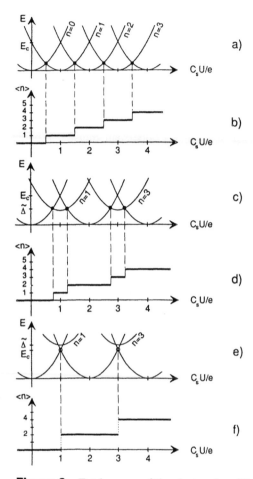

Figure 9 Total energy of the electron box (Fig. 2a) as a function of the polarization $C_s U/e$, for several values of the excess number n of electrons in the island, in the nonsuperconducting state (a) and the superconducting state (c, e). E_c is the electrostatic energy of one excess electron on the island for $U = 0$. The minimum energy for odd n is $\tilde{\Delta}$ above the minimum energy for even n. Panels (c) and (e) differ by the relative magnitude of $\tilde{\Delta}$ and E_c. The black dots correspond to level crossings where a single electron tunnels into and out of the island. The hollow circles correspond to level crossings where the only allowed process is the simultaneous tunnelling of two electrons into the island to form a pair (Andreev process). The equilibrium value $< n >$ versus $C_s U/e$ is shown in the nonsuperconducting (b) and superconducting (d, f) states, at $T = 0$. The Andreev process is shown in (f) by a vertical dashed line to distinguish it from the single-electron tunnelling process shown in (b) and (d) by a vertical continuous line.

These theoretical predictions can be extended at temperatures T such that k_{BT} $\ll E_c$, provided one replaces the odd–even energy difference $\tilde{\Delta}$ by the odd–even *free* energy difference $\tilde{\Delta}(T) = \tilde{\Delta} - k_BT \ln N + O (T^2)$ (53,54), where N $\approx 10^4$, the total number of electron states in the island participating in the superconductivity, is a measure of the degeneracy of the odd ground state with one unpaired electron. In Fig. 10 we show our experimental results for a superconducting aluminum island at $T = 28$ mK. In this experiment we vary $\tilde{\Delta}$ by means of a magnetic field applied to the sample. The evolution of the staircase as the field is varied provides a complete confirmation of the predictions of Fig. 9.

A remarkable point that is not completely understood is why the odd–even symmetry breaking, which manifests itself in traces (b) and (c) of Fig. 10, corresponds to a $\tilde{\Delta} (T \to 0) \simeq \Delta_{BCS}$. The superconducting islands are far from ideal: they contain many defects like impurities, grain boundaries, and surface states. Apparently, none of these defects provides available states near the Fermi energy for an unpaired electron, thus ruining the ideal BCS behavior. However, at the time of this writing, the superconducting box experiment has been unsuccessful when both sides of the junctions are superconducting. A possible explanation is that in the expression for the odd–even free energy difference $\tilde{\Delta} (T)$, the temperature that enters is the temperature of the quasiparticles, not the phonon temperature. In an all-superconducting circuit at low temperature, quasiparticles

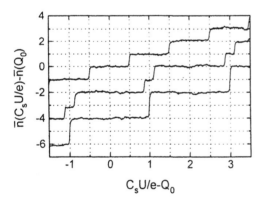

Figure 10   Variations of the average charge of the island, in units of e, with the polarization C_sU/e, at $T = 28$ mK, for three values of the magnetic field H applied to the sample. For the top trace ($H = 0.2$ T) the island is nonsuperconducting. For the middle ($H = 0.05$ T) and bottom ($H = 0$) traces the island is superconducting. For clarity, the middle and bottom traces have been offset vertically by two and four units, respectively. The symbol Q_0 refers to the offset charge on the island, an uncontrolled parameter that drifts on the time scale of hours.

can have a very long lifetime, and thus it is possible that the remaining out-of-equilibrium quasiparticles completely suppress the odd–even asymmetry. Having normal metal on one side of the junction provides an efficient way of relaxing out-of-equilibrium quasiparticles, since they can diffuse into the normal metal but not out.

It is also important to note that when both sides of the junction are superconducting, the Andreev process of Fig. 9e, by which two electrons from the normal side tunnel coherently to form a pair in the island, is replaced by Josephson tunnelling. In contrast with single-electron tunnelling or Andreev two-electron tunnelling, which are irreversible processes characterized by a rate, Josephson tunnelling is a reversible process characterized by a macroscopic coupling energy $E_J \sim \Delta(R_K/R_t)$. When $k_B T \ll E_J$, the superconducting electron box polarized at $C_s U/e = 1$ should be in the macroscopic coherent superposition of charge states $| n = 0 > + | n = 2 >$ (55). This has been indirectly observed in recent experiments in which the critical current of the superconducting SET has been measured as a function of gate voltage (56,57). In particular, the experiment of Joyez et al. (57) demonstrates that the naive picture of single Cooper pair transfer in nanojunction circuits can only hold if the characteristic energies of the superconductors are set properly, i.e., $\tilde{\Delta}(T) > E_c \gg E_J \gg k_B T$

VIII. FUTURE PROSPECTS

It has been suggested (27,28) that single-electron devices might find applications in digital electronics. A single electron would code for one bit, obviously the most economical way to store information. In fact, the electron pump is already very similar to the shift registers found in computers. The SET would be the building block of this "single electronics." A problem, however, is that metallic SETs made using today's technology have no "engineer gain": one transistor can barely feed one other transistor in the chain of signal processing, once the dispersions on parameters are accounted for. And no one understands how to get rid of random offset charges (28), which at present ruin any attempt to have more than a few transistors on one chip. In semiconductor devices, single-electron effects may even appear to be a nuisance because they imply that the electrons go through the dots or channels one at a time, a slowing down of the conventional FET operation. The main benefit of understanding single-electron effects in semiconductor nanotechnology may be just to provide the knowledge to fight them efficiently.

The real virtue of single-electron devices, as far as industrial applications are concerned, is that they teach us how to produce digital functions using only tunnelling and the Coulomb interaction, basic ingredients that are available down to the molecular level. In the as yet undeveloped "molecular electronics" technology (58), basic time constants are very short, there is no dispersion in the parameters of individual components, and there are few electrons to work with anyway. There, the principles underlying the devices that have been discussed in this article may be fruitfully implemented. The ultimate computer imagined by Feynman (59), in which elementary information is carried by a single electron on a single atom, would then cease to be a mere theoretical construction and become a reality.

NOTE ADDED IN PROOF

We would like to emphasize that this chapter is not an exhaustive account of single electron phenomena, nor is the list of references by any means complete and up-to-date. One important development that has taken place since this chapter was written is the observation of single energy levels in a nanometer-size metallic particle by Ralph et al. [*Phys. Rev. Lett.*, *74*,3241 (1995)]. We hope nevertheless that we have listed enough references to enable interested readers to find their way in the literature.

ACKNOWLEDMENTS

We thank P. Joyez, P. Lafarge, P. F. Orfila, and H. Pothier, with whom the experimental results described in this article have been obtained, for helpful discussions and help with the figures. This work has been partly supported by the Bureau National de la Métrologie.

REFERENCES

1. Eigler, D. M., and Schweizer, E. K., *Nature*, *344*, 524 (1990).
2. Wineland, D. J., Itano, W. M., and Van Dyck, R. S., Jr., *Adv. At. Mol. Phys.*, *19*, 135 (1983).
3. Van Dyck, R. S., Jr., Schwinberg, P. B., and Dehmelt, H. G., *Phys. Rev.*, *D34*, 722 (1986).
4. Millikan, R. A., *Phys. Rev.*, *32*, 349 (1911).
5. Solymar, L., *Superconductive Tunneling*, Chapman and Hall, London, 1972, Chap. 2.
6. Büttiker, M., and Landauer, R., *Phys. Rev. Lett.*, *49*, 1739 (1982).
7. Personn, B. N. J., and Baratoff A., *Phys. Rev.*, *B38*, 9616 (1988).
8. Lafarge, P., Pothier, H., Williams, E. R., Esteve, D., Urbina, C., and Devoret, M. H., *Z. Phys.*, *B85*, 327 (1991).
9. Matveev, K. A., *Zh. Eksp. Teor. Fiz.*, *99*, 1598 (1991) [*Sov. Phys. JETP*, *72*, 892 (1991)].
10. Grabert, H., *Phys. Rev.*, *B50*, 17364 (1994).

11. Gorter, C. J., *Physica*, *17*, 777 (1951).
12. Neugebauer, C. A., and Webb, M. B., *J. Appl. Phys.*, *33*, 74 (1962).
13. Giaver I., and Zeller, H. R., *Phys. Rev. Lett.*, *20*, 1504 (1968).
14. Lambe, J., and Jaklevic, R. C., *Phys. Rev. Lett.*, *22*, 1371 (1969).
15. Kulik, I. O., and Shekter, R. I., *Zh. Eksp. Teor. Fiz.*, *68*, 623 (1975) [*Sov. Phys. JETP*, *41*, 308 (1075)].
16. Dolan, G. J., and Dunsmuir, J. H., *Physica*, *B152*, 7 (1988).
17. Fulton, T. A., and Dolan, G. J., *Phys. Rev. Lett.*, *59*, 109 (1987).
18. Likharev, K. K., and Zorin, A. B., *J. Low. Temp. Phys.* 59, 347 (1985).
19. Averin, D. V., and Likharev, K. K., *J. Low Temp. Phys.*, *62*, 345 (1986).
20. Widom, A., Megaloudis, G., Clark, T. D., Prance, H., and Prance, R. J., *J. Phys.*, *A15*, 3877 (1982).
21. Ben-Jacob, E., and Gefen Y., *Phys. Lett.*, *A108*, 289 (1985).
22. Nazarov, Yu. V., *Pis'ma Zh. Eksp. Teor. Fiz.*, *49*, 105 (1989) [*JETP Lett.*, *49*, 126 (1990)].
23. Devoret, M. H., Esteve, D., Grabert, H., Ingold, G.-L., Pothier, H., and Urbina, C., *Phys. Rev. Lett.*, *64*, 1824 (1990).
24. Girvin, S. M., Glazman, L. I., Jonson, M., Penn, D. R., and Stiles, M. D., *Phys. Rev. Lett.*, *64*, 3183 (1990).
25. Cleland, A. N., Schmidt, J. M., and Clarke, J., *Phys. Rev. Lett.*, *64*, 1565 (1990).
26. Kuzmin, L. S., Nazarov, Yu. V., Haviland, D. B., Delsing, P., and Claeson, T., *Phys. Rev. Lett.*, *67*, 1161 (1991).
27. Likharev, K. K., *IBM J. Res. Dev.*, *32*, 144 (1988).
28. Averin, D. V., and Likharev, K. K., in *Quantum Effects in Small Disordered Systems* B. L., Altshuler, P. A., Lee, and R. A., Webb, eds.), Elsevier, Amsterdam, 1991.
29. Schön, G., and Zaikin, A. D., *Phys. Rep.*, *198*, 237 (1990).
30. Grabert, H., and Devoret, M. H., eds., *Single Charge Tunneling*, Plenum Press, New York, 1992.
31. Koch, H., and Lübbig, H., eds., *Single Electron Tunneling and Mesoscopic Devices*, Proc. 4th Int. Conf. SQUID '91, Springer-Verlag, Berlin, 1992.
32. Geerligs, L. J., Harmans, C. J. J. M., and Kouwenhoven, L. P., eds., *The Physics of Few-Electron Nanostructures* , Proc. of the Noordwijk NATO ARW, 1992 North-Holland, Amsterdam, 1993.
33. Cleland, A. N., Esteve, D., Urbina, C., and Devoret, M. H., *Appl. Phys. Lett.*, *61*, 2820 (1992).
34. Fraser, D. A., *The Physics of Semiconductor Devices*, Clarendon, Oxford, 1986.
35. Barone, A., and Paterno, G., *Physics and Applications of the Josephson Effect*, John Wiley, New York, 1982.
36. See Beenakker, C. W. J., *Single Charge Tunneling*, (Grabert H. and M. H. Devoret, eds.), Plenum Press, New York, 1992, Chap. 5.
37. Wilkins, R., Ben-Jacob, E., and Jaklevic, *R. C. Phys. Rev.Lett.*, *63*, 801 (1989).
38. Schönenberger, C., van Houten, H., and Beenakker, C. W. J., *Physica*, *B189*, 218 (1993).
39. Nejoh, H., *Nature*, *353*, 640 (1991).

40. Pothier, H., Lafarge, P., Urbina, C., Esteve, D., and Devoret, M. H., *Physica*, *B 169*, 573 (1991); *Europhys. Lett.*, *17*, 259 (1992).
41. Geerligs, L. J., Anderegg, V. F., Holweg, P. A. .M., Mooij, J. E., Pothier, H., Esteve, D., Urbina, C., and Devoret, M. H., *Phys. Rev. Lett.*, *64*, 2691 (1990).
42. von Klitzing, K., *Rev. Mod. Phys.*, *58*, 519 (1986).
43. Williams, E. R., Gosh, R. N., and Martinis J. M., *J.Res. Natl.Ins.Stand.and Technol.*, *97* (1992).
44. Averin, D. V., and Odintsov, A. A., *Phys. Lett.*, *A149*, 251 (1989); Averin, D. V., Odintsov, A. A., and Vyshenskii, S. V., *J. Appl. Phys.*, *73*, 1297 (1993).
45. Jensen, H. D., and Martinis, J. M., *Phys. Rev.*, *B46*, 13407 (1992).
46. Pothier, H., Lafarge, P., Esteve, D., Urbina, C., and Devoret, M. H., *IEEE Trans. Instr. Meas.*, *42*, 324 (1993).
47. Lafarge, P., Joyez, P., Pothier, H., Cleland, A., Holst, T., Esteve, D., Urbina, C., and Devoret, M. H., *C. R. Acad. Sci. Paris*, *314*, 883 (1992).
48. Martinis, J. M., Nahum, M., and Dalsgaard Jensen, H., *Phys. Rev. Lett.*, *72*, 904 (1994).
49. Fulton, T. A., Gammel, P. L., Bishop, D. J., and Dunkleberger, L. N., *Phys. Rev. Lett.*, *63*, 1307 (1989).
50. Geerligs, L. J., Anderegg, V. F., Rommijn, J., and Mooij, J. E., *Phys. Rev. Lett.*, *65*, 377 (1990).
51. Geerligs, L. J., Verbrugh, S. M., Hadley, P., Mooij, J. E., Pothier, H., Lafarge, P., Urbina, C., Esteve, D., and Devoret, M. H., *Z. Phys.*, *B85*, 349 (1991).
52. Averin, D. V., and Nazarov, Yu. V., *Phys. Rev. Lett.*, *69*, 1993 (1992).
53. Tuominen, M. T., Hergenrother, J. M., Tighe, T. S., and Tinkham, M., *Phys. Rev. Lett.*, *69*, 1997 (1992).
54. Lafarge, P., Joyez, P., Esteve, D., Urbina, C., and Devoret, M. H., *Phys. Rev. Lett.*, *70*, 994 (1993).
55. Lafarge P., Ph.D. thesis, Paris 6 (1993).
56. Eiles, T. M., and Martinis, J. M., *Phys. Rev.*, *B50*, 627 (1994).
57. Joyez, P., Lafarge, P., Filipe, A., Esteve, D., and Devoret, M. H., *Phys. Rev. Lett.*, *72*, 2458 (1994).
58. Aviram, A., and Ratner, M., *Chem. Phys. Lett.*, *29*, 27 (1974); *Molecular Electronic Devices* (F. L. Carter, ed.), North Holland, Amsterdam, 1991.
59. Feynman, R. P., *Optics News*, *11* (1985).

7

Dynamical Repertoire of Interacting Networks

Günter Mahler

University of Stuttgart
Stuttgart, Germany

Volkhard May

Humboldt University of Berlin
Berlin, Germany

Michael Schreiber

Technical University
Chemnitz, Germany

In this section we turn our attention to networks. One would expect that the complexity of a network would be significantly enlarged by interactions between its subunits. On the other hand, such interactions would tend to destroy the local nature of states and make this composite system more and more "bulklike"; and bulk material, despite its obviously complicated many-particle features, does usually not support any dynamics suitable for device operation. Can these opposing trends be reconciled to give rise to a qualitatively new dynamical repertoire? This section is concerned with (collective) dynamical modes as they are constrained by structure and controled by the environment.

Ensslin et al. investigate the transition region between bulk and network behavior in semiconductor nanostructures. Aspects of this transition have been intensively studied in the context of "low-dimensional physics," going from three dimensional bulk objects via two-dimensional layers and one-dimensional wires and finally to zero-dimensional dots. For our present discussion a different approach is preferable, namely to consider a fixed (mesoscopic, say) system with varying "compartmentalization" features. A concrete example of this type is the transition from a two-dimensional electron sea separating a matrix of nonconducting islands (a so-called antidot array) to a matrix of isolated electron pools (a so-called dot array, reminiscent of local defects). In the former structure the electronic system monitored by its transport properties behaves classically, with the electron trajectories repeatedly reflected at the forbidden islands ("electron billiard"). In the latter case the electron transfer between the dots dies out and so does the dc conductivity. The respective high-frequency conductivity is then dominated by an ensemble of dots each approximated as a parabolic potential confining a few electrons. Intermediate scenarios (i.e., with electron transfer between the dots) can also be realized.

With sufficiently small interdot distance (but still too large for direct charge transfer) the dots begin to interact via Coulomb forces. This kind of interaction can be enhanced by proper design, e.g., in the form of dot superstructures.

The following two papers are concerned with theoretical models "only." In the system envisaged by Knoll and Mehring the structure is further scaled down so that the network nodes become individual donor–acceptor segments (separated by a spacer) on molecular chains or films. Consequently the relevant electronic energies form a finite discrete spectrum (containing at least three states), if phonon side bands, i.e., coupling to quasicontinua resulting from other degrees of freedom, can be neglected. The respective excitation energies from the ground state are in the optical regime if the participating molecules are appropriately chosen. The number of electrons per node is not fixed: there are two different charge states that can be addressed by optically induced electron relaxation between the nodes combined with an applied voltage at external electrodes.

With respect to those charge states the system can be tailored to approximate a molecular shift register, but only if an appropriate superstructure (i.e., comprising alternating subunits) is used and if the transfer is restricted to nearest neighbors; another perturbing source is thermally activated hopping.

These problems are indicative of the delicate balance between a structure with well-defined local states (used to represent information) and a structure allowing for interactions (required for molecular information processing), which, as stated before, at the same time tend to destroy locality. A *hierarchical* structure of more levels of internal length scales than found in small molecules or bulk semiconductor lattices appears to be mandatory.

Addressing the individual nodes is another severe problem. For the networks discussed by Ensslin et al. the pertinent states emerge from one single band and could be controlled by contacts. In the Mehring model the states are in a way immobilized by large excitation energies as barriers, which can then be overcome by external light pulses. Even though the light field as part of the controlling environment is spatially homogeneous, its effect is selective in frequency space and therefore also in the local-state space. Such a property is fundamental also to spectral hole burning; in the present case, however, it gives rise to conditional dynamics, i.e., to local rules. Unfortunately, this selectivity tends to get lost at higher temperatures.

Similar ideas are employed also in the last example by Körner and Mahler. Here the internode charge transfer is assumed to be suppressed completely: each node has a fixed electron number and there are no longer external contacts. Instead, nodes with optically induced charge transfer states are considered (e.g., semiconductor dots with internal structure), which, via Coulomb interaction, modify the transition energies of the neighboring nodes. The basic principles have already been discussed in Chap. 3. Again, a transition between bulk and localized-subsystem behavior can be observed, but this time in terms of optical properties, which may give evidence for intermolecular coupling.

Three different scenarios are considered: firstly, the system can be made to simulate an Ising model of given temperature and antiferro- or ferromagnetic interaction. The second example, optical image processing, exploits the fact that the optical properties depend on neighborhoods, i.e., on local correlations. Finally, a chemical sensor model makes use of the transient dynamics that occur at the first-order phase transition connected with the Ising behavior. These examples indicate that the dynamical repertoire of such a network can indeed be considerably larger than that of a conventional bulk material.

In all three papers a macroscopic environment (heat bath, magnetic field, electric contacts, laser light field) is used to control microstates, which in turn determine the response of the system. Under favorable conditions the relationship between control (input) and response (output) can be taken to implement a nontrivial function.

8

Electron Motion in Lateral Superlattices on Semiconductors

K. Ensslin, W. Hansen, and J. P. Kotthaus

Ludwig-Maximilians-University
Munich, Germany

I. INTRODUCTION

The understanding of the electronic properties of a crystalline solid starts from the periodic potential that arises from the ordered arrangement of the atoms in a lattice. The fundamental characteristics of an insulator, semiconductor, or metal can be understood by a careful analysis of its periodic potentials that are essentially determined by the chemical elements involved and therefore by nature itself. Modern semiconductor technology opens the path to tailor artificial super-lattices in tune with the benefit to design synthetic patterns according to the needs of fundamental physics or applied science. Sample fabrication starts in most cases from a high-mobility two-dimensional electron gas generally imbedded in a GaAs/AlGaAs heterostructure. In such systems the electrons are confined in one spatial dimension constituting the growth direction of the heterolayer system. (1) The electron motion along the remaining two dimensions parallel to the heterojunction interface is almost free. In high-quality samples the average distance between two scattering events is of the order of several microns. This results from the advantage of modulation doping that positions the doping atoms in a plane away from the electron gas and therefore reduces the Coulomb scattering rate dramatically, especially at low temperatures.

The third dimension perpendicular to the plane of the two-dimensional electron gas (2DEG) lends itself to impose a lateral potential modulation that can be realized by various means. Here two fabrication methods will be discussed, i.e., electrostatically induced potential modulation and the creation of locally insulating regions via irradiation with low-energy ions through a suitable mask. In order to dominate electron motion by the artificial potential landscape, characteristic lateral feature sizes in the submicron regime are required. A basic tool to shape suitable fabrication masks is holographic lithography. The sample surface is coated with photoresist and exposed to the interference grating created by two laser beams. Periods down to 200 nm can be realized this way. The patterns are inherently periodic with high accuracy. Features with sizes below 100 nm and arbitrary shape can be accomplished by electron beam lithography and other modern nanolithography tools. (For an introduction see special issues of *Physics Today*, Feb. 1990 and June 1993.)

Important intrinsic length scales of semiconductor nanostructures are the Fermi wavelength λ_F, typically of the order of 40 nm, the electron mean path l being typically many microns; the phase coherence length λ_P strongly depends on temperature. Since superlattice periods p are much smaller than the mean free path, $p \ll l$, electron transport can safely be regarded as ballistic. Single quantum dots or wires can be fabricated where quantum confinement effects due to the lateral potential are clearly observable. (2) One challenge, however, is the fabrication of artificial lateral superlattices with periods small enough that a lateral band structure becomes observable and electron motion is dominated by quantum effects. (3) Presently, the scientific community in the field of nanostructure physics is close to achieving this goal, but no convincing experiment has been presented so far that unambiguously proves the existence of a lateral band structure. Nevertheless, new physical phenomena have been discovered in lateral superlattices that have attracted widespread attention. Many unpredicted effects have been experimentally observed, and most can be explained by classical ballistic electron transport. The topic of this article is the presentation and explanation of the main results in this field.

This chapter is organized as follows. The first section is devoted to the description of fabrication procedures and is followed by sections on electron transport in square and rectangular antidot lattices. The next sections summarize high-frequency studies on lateral superlattices containing geometries such as dot, coupled dot, and antidot structures as well as arrays of large electron disks. The focus will be on interaction phenomena in array-type devices.

II. FABRICATION

Some basic ingredients for the fabrication of lateral superlattices on semiconductors will be discussed in this section. Details will be omitted for the sake of clarity. Fabrication often starts from modulation doped heterojunctions in which mobile electrons populate the subbands created by the triangular shaped potential close to the heterointerface and form a two-dimensional electron gas. Photosensitive resists are spun onto the wafer material that contains a heterointerface as close as possible to the surface of the sample. Using light and electron beams as lithographic tools, well-defined areas of the sample can be controllably exposed. Via suitable development conditions the exposed photoresist is removed, whereas the unexposed areas serve as a dielectric that is laterally modulated in thickness on the semiconductor surface. This pattern can be electrostatically transferred onto the electron gas by applying a gate voltage to a front gate electrode that is evaporated directly onto the patterned photoresist. (4) Alternatively, the photoresist pattern can be used as a mask for ion irradiation or etching processes. (5) Figure 1 shows an electron micrograph of a typical photoresist pattern as is used for antidot lattices. The detailed shape of the laterally confining potential for the electrons will depend on the fabrication process. (6) For heterostructures that are not intentionally doped, the low-dimensional electron system can be induced via a laterally periodic gate electrode.

Lateral confinement of the electrons can be achieved in one or two lateral dimensions, i.e., quantum wires or quantum dots can be realized. (7) Quantum wires are accessible for transport experiments in the direction along the wires (8), or, if the wires are weakly coupled, also in the perpendicular direction. (9–11) Quantum dots have to be accessed by capacitance (12,13) or tunnelling techniques (14), because standard transport experiments are no longer possible. Alternatively, the concept of antidots lends itself for passing current through the system (15–19). Here a locally repulsive periodic potential is superimposed onto the electron gas to simulate the effect of ionized atoms on lattice sites in a bulk crystal.

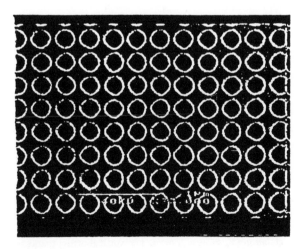

Figure 1 Electron micrograph of a photoresist pattern for antidot fabrication with lateral period $p = 300$ nm.

III. SQUARE ANTIDOT LATTICES

Potential pillars are superimposed onto an electron gas to produce a locally repulsive potential with the same periodicity in the two lateral dimensions. Antidot lattices are realized once the potential maxima exceed the Fermi energy. Figure 2 presents typical experimental results for the magnetoresistance as observed on a lateral superlattice in the shape of a Hall bar geometry. In this particular case the laterally periodic potential modulation is imposed via a gate voltage to study the transition from a homogeneous two-dimensional electron gas to a system with weak and with strong potential modulation. The lowest curve (Vg = 0) shows the usual Shubnikov–de Haas oscillations of an unmodulated two-dimensional system. For intermediate gate biases (here $V g = -200$ mV) the weak potential modulation leads to the occurrence of maxima in the low-field magnetoresistance that are related to the commensurability of the classical cyclotron diameter at the Fermi energy and the lattice period (20,21). The phase of these oscillations is well-defined for one-dimensional superlattices but depends on details of the potential landscape in the case of two-dimensional superlattices

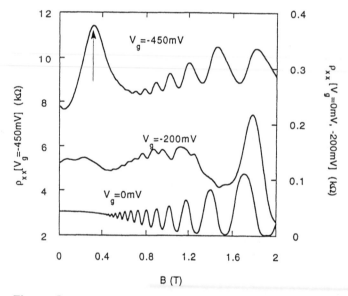

Figure 2   Magnetoresistance of an electrostatically induced antidot lattice with period $p = 460$ nm. The curves show experimental traces (from bottom to top) for the homogeneous 2D case ($V_g = 0$), the case of weak potential modulation ($V_g = -200$ mV), and the case of strong potential modulation ($V_g = -450$ mV). The vertical arrow marks the position of the resistance maximum that corresponds to an electron orbit around a single antidot. (From Ref. 6.)

(21). Quantum-mechanically this can be understood by considering that in a laterally periodic potential the Landau levels split into bands whose line width at the Fermi energy oscillates in a magnetic field (10,11). This leads in turn to the oscillations in the magnetoresistance. Classically, the cyclotron diameter is proportional to $1/B$,

$$2R_c = \frac{2mv_F}{eB} = \frac{2\hbar\sqrt{2\pi N_S}}{eB} \tag{1}$$

N_s being the areal carrier density and v_F the Fermi velocity. The commensurability oscillations are thus also periodic in $1/B$, similar to the Shubnikov–de Haas oscillations, but quite different in period. In a classical picture the average drift velocity is reduced each time the classical cyclotron diameter matches an integer number of lattice periods (22). This in turn leads to the commensurability oscillation of the magnetoresistance periodic in $1/B$.

For very negative gate biases, $V_g = -450$ mV (see Fig. 2), the existence of the antidot lattice is manifested through a strong maximum in the magnetoresistance that is related to a pinned electron orbit around a single antidot (6). Before the details of this phenomenon will be discussed we present the experimental magnetoresistance of an antidot lattice prepared by low-energy ion irradiation (see Fig. 3). Here two pronounced maxima occur at low magnetic fields, while simultaneously plateau like structures in the Hall effect arise (18,23). In addition, the Hall effect is quenched for very low magnetic fields, $B<0.1$ T. As long as the mean free path of the electrons is longer than multiple periods of the antidot lattice, the electron motion can safely be regarded as ballistic. In contrast to a one-dimensional superlattice, the electrons follow chaotic trajectories in a two-dimensional lattice (21,24). In a magnetic field, pinned electron orbits around groups of antidots arise that remain stable under finite electric fields. The more antidots an orbit encloses, the more antidot potentials the electron encounters on its way. Those trajectories might therefore be deformed and deviate from circular symmetry (24). As the electrons that travel on chaotic trajectories approach a pinned orbit, their trajectories follow the pinned orbit for a while until they continue on their chaotic path. This changes the conductivity of the system and leads to the maxima in the magnetoresistance. A prerequisite for the occurrence of a given maximum is that the electron mean free path be longer than the perimeter of the pinned orbit, otherwise the electron will scatter before the orbit is completed and lose the information along which orbit it was travelling before.

The size of a possible pinned orbit is also limited by the aspect ratio of the lattice potential and the steepness of an individual potential pillar. The smaller the size of an antidot with respect to the lattice period, and the steeper the potential, the larger are the pinned orbits that can occur in an antidot lattice (24). In linear response, the magnetoresistance can be calculated based on ballis-

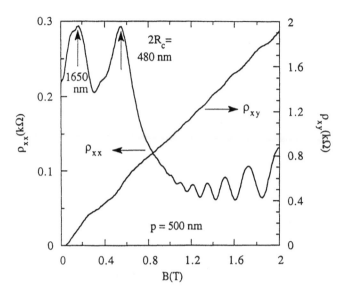

Figure 3 Magnetoresistance and Hall resistance of an antidot lattice fabricated by low-energy ion irradiation. Two pronounced peaks occur in the magnetoresistance while simultaneously weak plateaulike structures arise in the Hall resistance. In addition, the Hall resistance is quenched for low magnetic fields ($B < 0.1$ T). (From Ref. 23.)

tic electron trajectories, and a close agreement with the experimental data was obtained. Different classes of electron trajectories are responsible for the quenching of the Hall effect and the plateaulike structures, but again all essential features of the experimental trace can be explained within the framework of ballistic electron transport (25). It is important to note that electron–electron interactions apparently are not important to explain the experimental results. The conceptual idea of an electron billiard can be applied, where the electrons move without friction and without electronic collisions in the potential landscape and are reflected specularly at the potential pillars.

In a narrow regime of coupling strengths, interesting interaction features may even be expected in the static transport properties of weakly coupled electron dots. Electron transport through such a superlattice takes place by tunnelling processes, and the tunnel resistances between adjacent cells should be high enough to localize the electron wave functions in the respective lattice cells between the tunnel processes, rather than to spread them over several superlattice periods. In this regime, interaction is also very interesting because the charging of the cells is hampered by Coulomb gaps if the capacitance between adjacent cells is very small. (26,27) Although such Coulomb blockade has been observed on metallic systems and on single electron dots generated in semiconducting systems (28), corresponding transport experiments on superlattices at semiconductor interfaces in this regime are still to be done.

IV. RECTANGULAR ANTIDOT LATTICES

A square lattice is fundamentally isotropic in its resistance properties. The diagonal components of the resistivity tensor are identical. If the periods in the two lateral dimensions are different, this isotropy can be broken. Furthermore, a whole class of system configurations can be realized by various parameter sets of the rectangular lattice. For example, if the antidots are closely spaced in one direction so that their potential tails almost overlap, while the period in the perpendicular direction is much larger, then the transition to a wirelike system can be investigated (29).

Figure 4 presents an electron micrograph of a photoresist pattern that is used as a mask for a rectangular antidot lattice with periods $p_x = 1.0$ μm and $p_y = 0.5$ μm. The holographic lithography used for patterning in this case leads to ellipsoidally shaped antidots on a rectangular lattice. The mean free path of the electrons is of the order of 5 μm, i.e., even much larger than the larger period of the lattice. Two Hall geometries are superimposed onto the lattice and oriented along the two main axes of the lattice (see inset of Fig. 5). This facilitates the experimental determination of all four components of the resistivity tensor. Figure 5 presents the experimental results of the magnetoresistance for the two possible

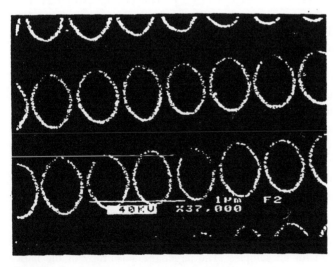

Figure 4   Electron micrograph of a rectangular antidot lattice with periods $p_x = 1.5$ μm and $p_y = 0.5$ μm.

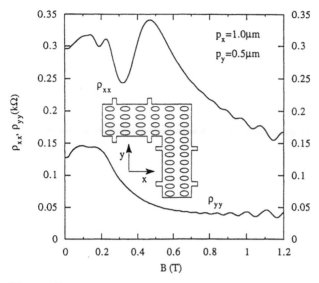

Figure 5 Magnetoresistance of a rectangular antidot lattice with $p_x = 1.5$ μm and $p_y = 0.5$ μm. The inset shows the direction of current flow with respect to the lattice orientation. Several maxima arise for ρ_{xx} (current flow through the closely spaced antidots), while only one pronounced structure is observed in ρ_{xx}. (From Ref. 23.)

directions of current flow. It is obvious that a pronounced anisotropy exists in many respects. Firstly, the resistance at $B=0$ is larger for current flow along the long period of the lattice, $\rho_{xx}(B=0) > \rho_{yy}(B=0)$. This is a direct consequence of the wide open channels along which the electrons travel for the $B=0$ resistance ρ_{yy} and the narrow constrictions between two closely spaced antidots that dominate $\rho_{xx}(B=0)$. In the low-field magnetoresistance a series of maxima arises similarly as in a square lattice but with positions that strongly depend on the direction of current flow.

For a clearer understanding of the observation a series of samples was investigated where the anisotropy of the two in-plane periods of the lattice was systematically changed from $p_x:p_y = 1:1$ to $3:1$. The short period was kept constant at $p_y = 0.5$ μm, and the long period p_x was varied accordingly. Figure 6 contains a summary of magnetoresistance measurements for both directions of current flow. For current flow between the closely spaced antidots (ρ_{xx}) a strong maximum persists around $B = 0.5$ T for all periodicity ratios. The positions of these maxima correspond to a cyclotron diameter of the size of the short period of the lattice $p_y = 0.5$ μm for all samples. The larger the anisotropy of the lattice becomes, the more maxima arise at lower magnetic fields. In a square lattice, electron orbits around one or four antidots are favored by the symmetry of the

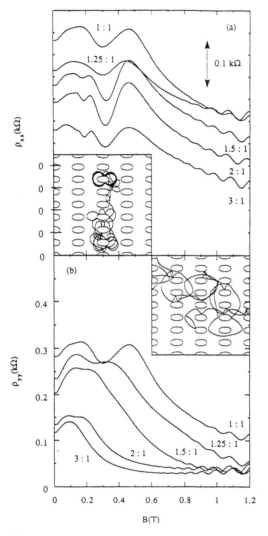

Figure 6 (a) Magnetoresistance ρ_{xx} for current flow along the long period p_x of the lattice for a series of samples with different values of p_x. The carrier density is $N_s = 5 \times 10^{11}$ cm^{-2}, and the short period is $p_y = 0.5$ μm for all samples. The curves are vertically offset for clarity. The inset shows a typical calculated electron trajectory for a magnetic field where the classical cyclotron diameter equals the short lattice period $2R_c = p_y$. The antidots are indicated by the contour lines where the potential energy is identical to the Fermi energy. (b) Magnetoresistance ρ_{yy} for current flow along the short period p_y of the lattice for the same series of samples as in (a). The inset presents a typical electron trajectory for a magnetic field where the classical cyclotron diameter equals $2R_c = 3.3 \times p_y$. (From Ref. 23.)

lattice. In a rectangular lattice, the occurrence of orbits around two antidots is also possible. A corresponding peak in the magnetoresistance arises for large anisotropies (see Fig. 6a). In the limit of extremely large anisotropies, orbits around any number of antidots become possible until the orbit touches the next row of antidots. Here the situation is conceptually similar to electron focusing. (30) In that case the maxima in the resistance arise from backscattering of the electrons each time an electron trajectory fits into the next possible opening between two antidots. In fact the resistance trace as displayed in Fig. 6a for $p_x{:}p_y = 3{:}1$ is already very similar to the results of electron focusing experiments.

For current flow along the wide open channels of a rectangular lattice the situation is completely different. The experimental data in Fig. 6b clearly show that the magnetoresistance maximum corresponding to the orbit around a single antidot vanishes for large anisotropies. Simultaneously the maxima at lower magnetic fields increase in strength, and finally only one maximum survives for $p_x{:}p_y = 3{:}1$. The existence of pinned electron orbits obviously does not depend on the direction of current flow. However, the electrons travelling on chaotic trajectories in the wide channels between the rows of antidots will probably feel little of the small pinned orbits around single antidots. With decreasing magnetic field as soon as the orbit size becomes comparable to the large period of the lattice ρ_{yy} will also be influenced by the pinned trajectories. In the limit of large anisotropies of the periodicities (here $p_x{:}p_y = 3{:}1$), one pronounced maximum in the magnetoresistance persists (see lowest curve in Fig. 6b). In the geometry considered here, the potential is very much like that of a quantum wire where the correlation length of the edge roughness is determined by the distance between two closely spaced antidots. Indeed, the magnetoresistance as indicated in Fig. 6b behaves similarly to other observations on single quantum wires. In that case the occurrence of the maximum is explained by diffuse boundary roughness scattering (31).

In contrast to the magnetoresistance, the Hall resistance does not depend on the direction of current flow. Figure 7 presents experimental traces for three gate voltages. Within the accuracy of the experiment, the curves for one gate voltage are identical for both directions of current flow. This is a consequence of the symmetry relations as derived by Büttiker (32). According to Büttiker a four-terminal resistance $R_{ij,kl} = V_{kl}/I_{ij}$ obeys a generalized symmetry relation $R_{ij,kl}(B) = R_{kl,ij}(-B)$. In the case of a square geometry with voltage probes in the middle of each side, an exchange of current and voltage probes (each set lying on opposing sides of the square) symbolizes two Hall measurements with current directions perpendicular to each other. If the rectangular lattice is oriented with its main axes parallel to the sides of the square, there is no voltage drop perpendicular to the current direction for $B = 0$. The additional symmetry relation $R_{ij,kl}(B) = -R_{ij,kl}(-B)$ explains the experimentally observed behavior. These generalized

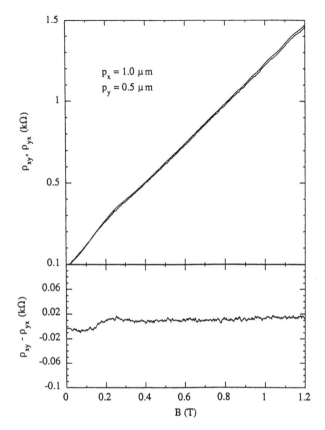

Figure 7 Hall resistance ρ_{xy} and ρ_{yx} for both directions of current flow for three gate voltages $V_g = 0$, -400 mV, and -800 mV. Within the accuracy of the experiment, ρ_{xy} and ρ_{yx} are indistinguishable. The periods are $p_x = 1.0$ μm and $p_y = 0.5$ μm. (From Ref. 23.) The lower part of the figure presents the calculated difference $\rho_{xy} - \rho_{yx}$.

Onsager relations (33) do not hold any more if the lattice axes are not oriented parallel to the directions of current flow. It is therefore a nontrivial consequence of the symmetry of a rectangular antidot lattice that the Hall effect is isotropic for current flow along the main axes of the system.

It is important to note that for large anisotropies in rectangular antidot lattices the conceptual picture of electron motion changes dramatically. For large values of $p_x{:}p_y$ electron transport can be conceived as backscattering of electrons via magnetic field induced electron focusing ρ_{xx} and boundary roughness scattering in a quantum wire (ρ_{yy}). Conversely, for values of $p_x{:}p_y$ close to one, the magnetoresistance is explained by the existence of pinned electron trajectories around groups of antidots and their influence on the conductivity of the electrons on chaotic trajectories.

All concepts mentioned above rely on classical physics. However, symmetry relations play a distinct role, and they will persist into the quantum-mechanical regime. In rectangular lattices, experiments were done with the direction of current flow tilted with respect to the main axes of the lattice. Although the cyclotron diameter in the quantum Hall regime was more than one order of magnitude smaller than any lithographic features, pronounced overshoots before and after quantum Hall plateaus were observed. (34) This leads to the important conclusion that even today, where the smallest features that can be realized technologically are usually larger than the Fermi wavelength, quantum-mechanical behavior (e.g., the quantum Hall effect) can be modified by a lateral superlattice. Future experiments will rely on processes that aim at periods below 100 nm. It is expected that the existence of a quantum-mechanical miniband structure will be probed and investigated. (3) This will open the door to new physical concepts and probably enable the nanostructure community to propose new ideas for possible applications.

V.   DYNAMIC CONDUCTIVITY

Whereas in the preceding sections we have focused on static transport experiments, we shall now proceed to the dynamic properties of lateral superlattices. We shall demonstrate that the interaction between the neighboring lattice cells is an essential issue for the understanding of high-frequency conductivity in lateral superlattices. Generally, we may distinguish between two types of interaction mechanisms: interaction by charge transfer and that by polarization fields. As depicted in Fig. 8, an antidot superlattice transforms into a matrix of isolated

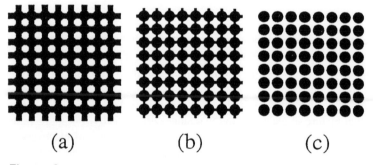

(a)   (b)   (c)

Figure 8   Sketch of the gradual transition from an antidot superlattice (a) to a matrix of isolated dots (c). Intermediate is a system of weakly coupled dots as illustrated in (b).

dots if the cells of the superlattices become gradually decoupled. Whereas charge transfer is possible in antidot superlattices, it is hampered by large tunnel barriers in the case of a superlattice consisting of an electron dot matrix. The static conductivity decreases dramatically as soon as transport between adjacent cells of the superlattice takes place via tunnelling processes and finally vanishes completely.

In the limit of weakly coupled dot arrays, the conductivity should exhibit resonances at finite frequencies that reflect the way the isolated dots are formed by the lateral potential modulation. It is quite suggestive to discuss this situation in analogy to the excitations of isolated impurity states in semiconductors or even spectroscopic properties of electron states in atoms. We shall see, however, that too close a comparison to those cases is dangerous, since here the electron system generally consists of a very high number of interacting electrons coherently reacting to the radiation field that is spatially coherent over distances much larger than the superlattice period.

VI. MATRIX OF ISOLATED DOTS

We first discuss the dynamic properties of a matrix of isolated dots. Presently, a lively discussion exists on whether the Coulomb interaction between polarized cells in the superlattice can be exploited for neural networks and new computing device concepts (35). Special focus will therefore be on the signatures of interaction effects that have so far been found in experiments on electron dot matrices. Subsequently, we discuss experiments performed on antidot superlattices. Those can be understood at least qualitatively on the basis of the same model that has previously been introduced for the explanation of excitations in isolated electron dots.

In our discussion of dynamic conductivity, we shall focus on a frequency regime in which only electronic excitations within the conduction band are important. There is another very interesting regime at higher energies where carrier excitations across the bandgap become possible. So far, only a few experiments on antidot or electron dot arrays (36) or single electron dots (37) are performed in this regime, however. We thus refer the interested reader to the pertinent publications.

VII. QUANTUM DOT ARRAYS

The experiments discussed in the following are performed on electron systems generated either in GaAs heterojunctions or in MOS devices prepared on (100) Si surfaces. These are favored not only because of the existence of very well developed preparation technologies but also because of the relatively simple conduction band structure for electrons moving in the plane of two-dimensional confinement. Thus for the following we assume an isotropic and constant effective mass of the conduction band electrons. The simplest model to describe the high frequency conductivity of an electron system in a homogeneous excitation field is the well-known Drude model. It is remarkable that measurements of the high frequency conductivity in two-dimensional electron systems very often can be explained to a very good approximation by this model even if a magnetic field is applied perpendicular to the plane of the 2DEG. This may be understood in that a spatially homogeneous excitation field accelerates all the electrons simultaneously without changing the interelectron distances. As shown by Kohn (38), the magnetic field applied perpendicular to the electron system does not change this situation. Now all electrons move phase-locked with the cyclotron frequency on circular orbits. Thus according to the famous Kohn theorem, electron–electron interaction does not affect the cyclotron resonance frequency. (38) Indeed, Kohn's theorem can be generalized to all cases where the electrons move in a parabolic confinement potential. (39) Then a homogeneous excitation field again couples only to the center of mass motion of the electrons, whereas effects of electron–electron interaction resulting from relative electron motion are not visible. Thus in the limits of the generalized Kohn theorem—a parabolic confinement potential and a homogeneous excitation field—the response of electrons in a quantum dot array is easily described as the response of their center of mass motion.

We should like to note that the potential that according to the generalized Kohn theorem has to be parabolic is generated by the external charges only, and that the interaction potential of the electron system itself is excluded. This potential is thus often called the "bare potential," while the potential that includes the electron–electron interaction is the so-called "effective potential." In a Hartree-type description the effective potential would be used to calculate the single-electron states. Although a parabolic bare potential might appear to be rather special, it turns out that it is a quite reasonable approximation for many experimental systems investigated so far. The reason is that the external charges defining the bare lateral confinement are very often located relatively far away from the electron system and thus cause a smooth confinement.

To illustrate this point, in Fig. 9 the cross-section of a sample is depicted, where the potential modulation is generated by a split gate evaporated on top of the crystal surface. Since the heterojunction is modulation doped in the AlGaAs barrier region, there exists a homogeneous two-dimensional electron system at the semiconductor heterojunction at zero gate voltage. At negative gate biases, the electrons will be pushed out of the regions beneath the gate metal. At sufficiently low gate bias, the areas below the gate are totally depleted. The biased gate metal now induces a potential landscape in the plane of the heterojunction interface that confines the electrons beneath the gap of the gate. With proper

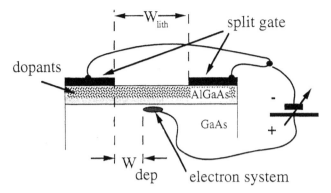

Figure 9 Sketch of a modulation doped AlGaAs/GaAs heterojunction with a split front gate used to induce a lateral confinement potential via field effect. The lithographic width W_{lith} of the gap in the gate and the depletion width W_{dep} are indicated.

assumption about the potential in the gap regions of the device surface, it is straightforward to approximate analytically the potential minimum where the electrons gather. (40–43) Self-consistent calculations show (44) that the effective width of the confined electron system is considerably smaller than the lithographically defined width W_{lith} of the gap in the gate. Thus a rather large depletion width W_{dep} separates the electron systems from the electrodes, and consequently the potential can be described by a parabolic term in good approximation. This situation does not change fundamentally if the confinement potential is defined by etching methods rather than by field-effect electrodes. It seems that in AlGaAs/GaAs heterojunctions the main effect of etching into the crystal surface is the generation of uncompensated charges on the surface. This means that rather than charges on the gate electrodes now charges on the etched surfaces generate the lateral confinement potential with no essential difference with respect to the depletion width.

The present experimental investigations of the dynamic conductivity in quantum dot matrices are thus to first order best understood with a high frequency conductivity derived for an electron dot consisting of N electrons confined in a parabolic potential. A particle of charge eN and mass $m*N$ confined in a symmetric parabolic potential

$$V(x, y) = \frac{1}{2} N{\cdot}m^*\Omega_o (x^2 + y^2) \tag{2}$$

has a high frequency conductance $\Gamma \pm (\omega)$ in a radiation field of positive or negative circular polarization (45):

$$\Gamma_\pm(\omega) = \frac{e^2 N \tau_e}{m^*} \cdot \frac{1}{1 + \left(\dfrac{\Omega_o^2}{\omega} - \omega \pm \omega_c\right)^2 \tau_e^2} \tag{3}$$

In this equation we assume that the radiation field is polarized in the x–y plane and a magnetic field $B = m^*\omega_c/e$ has been applied in the z direction perpendicular to the 2DEG. Furthermore, a phenomenological relaxation time τ_e is introduced. For linearly polarized light one obtains from the above equation a single resonance at zero magnetic field that splits into two modes at finite fields:

$$\omega_\pm = \sqrt{\Omega_o^2 + \left(\frac{\omega_c}{2}\right)^2} \pm \frac{\omega_c}{2} \tag{4}$$

We note that these results may be derived from either classical or quantum mechanics. Although the quantum levels of a harmonic oscillator show a complicated behavior as a function of a magnetic field (46), the dipole selection rules for absorption in a homogeneous radiation field lead to the above simple absorption spectrum. If the confinement is asymmetric,

$$V(x,y) = \frac{1}{2} N \cdot m^* \cdot (\Omega_x x^2 + \Omega_y y^2) \tag{5}$$

the spectrum is slightly modified so that two resonance frequencies are found already at zero magnetic field (47,48):

$$\omega_\pm^2 = \frac{1}{2} \{\Omega_x^2 + \Omega_y^2 + \Omega_c^2 \pm [(\Omega_x^2 + \Omega_y^2 + \omega_c^2)^2 - 4\,\Omega_x^2\,\Omega_y^2]^{1/2}\} \tag{6}$$

Deviations from these predictions found in the experiments are discussed as effects of interdot interaction or nonparabolic terms in the bare confining potential.

State-of-the-art values for the lateral extension of artificial electron dots fabricated in semiconducting systems are in the 100 nm range. The electrons are thus localized by the electrostatic potentials to areas quite comparable to the areas defined by the cyclotron orbit in a magnetic field of roughly a Tesla (see Eq. (1)). Thus, like the cyclotron resonances of homogeneous two-dimensional systems, the resonances of todays's electron dot matrices are found in the far-infrared (FIR) spectral regime. Unless the optical constants of the host device have a strong spatial modulation, the assumption of a homogeneous radiation field over several periods of the superlattice is quite realistic.

As an example, in Fig. 10a FIR transmission spectra of a Si-MOS device with a dual stacked gate are shown. (49). The cross-section in Fig. 10b depicts the dual stacked gate configuration prepared on the Si (100) surface; it consists of a mesh type bottom gate and a homogeneous top gate separated from the

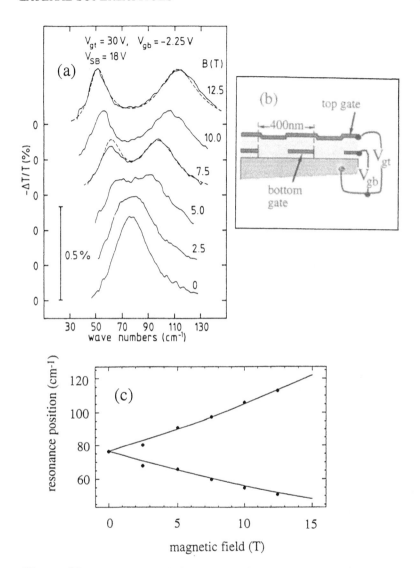

Figure 10 (a) FIR transmission spectra of a matrix of 0D electron systems on a stacked gate Si-MOS device recorded at different magnetic fields applied perpendicular to the sample surface. The dotted lines are calculated according to Eq. (4). (b) Cross-section of the gate configuration on the surface of the MOS device. The bottom gate consists of a mesh with period 400 nm and symmetric openings of 150 nm diameter. (c) Resonance positions plotted vs. magnetic field. The straight lines are calculated according to Eq. (5). (From Ref. 49.)

bottom gate by an insulating spacer. With positive voltages applied at the top gate, electrons are induced beneath the holes of the bottom gate mesh. To generate a dot matrix, the bottom gate is biased sufficiently negative so that no electrons are induced beneath the mesh. The dual gate technique is very versatile because it allows us to control the number of electrons that occupy the dots and the strength of the confinement potential almost independently. From the oscillator strength at zero magnetic field it was determined that each dot of the matrix contains about 140 electrons. Quantum dot matrices containing only a few electrons have been realized on InSb MOS devices and GaAs heterojunctions. (50,51) The overall behavior of the resonance positions is nicely explained by the harmonic oscillator model. Small deviations of the resonance positions from Eq. (4) are generally explained with nonparabolic terms in the confinement potential. (52–55) Less well described by the harmonic oscillator model are often the line widths of the modes. Obviously, the assumption of a phenomenological relaxation time independent of magnetic field and resonance frequency seems to be oversimplifying. (49,50,56)

VIII. COULOMB INTERACTION EFFECTS

Coulomb interaction between neighboring dots may affect the absorption spectrum, as has been addressed in several publications. (57–63) With decreasing interdot separation, the external charges that constitute the bare potential of a dot start to influence also the neighboring dots. The following simple model demonstrates that this interaction is the main contribution of the interdot interaction if the dot separation is sufficiently large and the excitation field homogeneous. If the separation of the dots is not too small compared to their diameter, the potential of the external charges again can be expanded up to quadratic terms. Under this condition the Kohn theorem still applies. The situation is illustrated in Fig. 11. In each dot N electrons are bound by a parabolic potential, and thus the electron systems can be represented by charges of massNm^* and charge Ne.

Figure 11 Visualization of the effect of particle interaction in an electron dot superlattice if the excitation is phase coherent. The superlattice consists of particles bound in parabolic potential minima. Interdot electron–electron interaction is represented by springs between adjacent dots.

The electron distance both within and between dots does not change if their motion is excited phase coherently throughout the superlattice. In this case the only difference vis-à-vis noninteracting dots can be understood as a renormalization of the bare potential arising from the external charges of the neighboring dots.

Let us assume that the external charges defining the confinement potential in the electron dot superlattice consist of positive point charges Ne forming a rectangular lattice with periods a and b in x and y direction, respectively. Taylor expansion up to quadratic terms of the contribution of the ions at lattice sites n, $m \neq 0$ to the bare potential of the electron dot at site $n = m = 0$ is then

$$V_i(x, y) = \frac{Ne^2}{8\pi\epsilon\epsilon_0 a^3} \sum_{n,m}^{\infty} \frac{(m^2\beta^2 - 2n^2)}{(n^2 + m^2\beta^2)^{5/2}} x^2 + \frac{(n^2 - 2m^2\beta^2)}{(n^2 + m^2\beta^2)^{5/2}} y^2 \qquad (7)$$

$$= -\frac{Ne^2}{2\epsilon\epsilon_0 a^3} (\xi_x x^2 + \xi_y y^2)$$

Here $\beta = b/a$ is the ratio of the lattice constants, and ϵ is a dielectric constant of the surrounding medium. The potential $V_i(x, y)$ adds to the bare potential $m^* \Omega_0 (x^2 + y^2)/2$ of a single dot and thus renormalizes the curvature. Correspondingly, a modified resonance frequency of the superlattice

$$\Omega_{x,y}^2 = \Omega_0^2 - \frac{Ne^2}{\epsilon\epsilon_0 m^* a^3} \xi_{x,y} \qquad (8)$$

results. In a quadratic lattice ($\beta = 1$) the "lattice tensor" components are equal: $\xi x, y = 0.36$. Thus in a quadratic superlattice the interaction effect just leads to a reduction of the resonance frequency without additional change of the magnetic field dispersion. This is difficult to quantify from experimental data because of the uncertainty with which other parameters are known.

However, in a rectangular array with different lattice constants, the situation changes. Even if the electron disks are perfectly symmetric, the asymmetric arrangement in the rectangular lattice leads to an absorption spectrum that equals the spectrum of noninteracting elliptic electron dots with dispersion according to Eq. (6). Here the mode splitting directly reflects the interdot Coulomb interaction. Such an experimental verification of the interdot coupling has recently been achieved with experiments on rather large electron disks in a rectangular array with strongly different periods. (64) The disks were circular, 37 μm in diameter, and arranged in a matrix with periods of 40 μm and 80 μ in two perpendicular directions. Positions of the resonances measured on this dot array are presented in Fig. 12. In (64) the interaction effect is modelled by the polarization fields of point dipoles on the lattice sites, which leads to the same results as described here. (64) The splitting of the resonance positions found in the experiment can

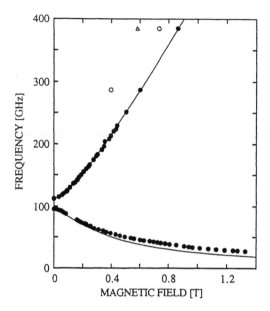

Figure 12 Resonance frequencies measured on circular electron disks arranged in a rectangular matrix of 40 and 80 μm period, respectively. At zero magnetic field, two resonances occur with a separation that is a measure of the Coulomb interaction between disks. The solid line reflects the magnetic field dispersion according to Eq. (7). Open symbols denote higher-order modes. (From Ref. 64.)

be nicely explained within this model. The magnetic field dispersion is roughly described by Eq. (6) with small deviations of the low-frequency mode at high magnetic fields, which are related to anharmonicity of the confinement in such large dots. The occurrence of weak higher-order modes (open symbols in Fig. 12) can also be associated to anharmonicity of the confinement in the relatively large dots used in this experiment.

For suitably chosen dot shape and interdot separation, the reduction of the resonance frequency might become so large that the resonance frequency of the dot array becomes zero. This corresponds to a polarization mode frozen into a ferroelectric state, as pointed out by Kempa et al. (65) They showed that, allowing for a phase shift between the polarizations at different lattice sites, the most stable configuration would correspond to a transversal mode in an antiferroelectric arrangement. It is easy to see that the energetically most favorable arrangement of dipoles in a two-dimensional square lattice will be rows with alternating polarization of the dipoles parallel or antiparallel to the row as shown in Fig. 13a. (65) The realizability of a dot matrix that exhibits a phase transition into an antiferroelectric state has been discussed in many subsequent

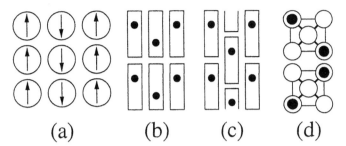

Figure 13   Ordered ground state of coupled dot arrays. (a) Antiferroelectric arrangement of dipoles in a square matrix of symmetric dots. (b) Rectangular lattice of electron
dashes as proposed by Bakshi et al. (61). Here point charges reside in a positively charged
jelly. (c) Arrangement of electron dashes in a face-centered lattice as proposed by Dahl
et al. (64). (d) Two unit cells consisting of five quantum dots connected by transparent
tunnel barriers as proposed by Lent et al. (35). Two electrons are filled into each unit
cell. The depleted cells are in different polarization states that are bistable as long as
there is no interaction between neighboring cells.

publications. (35,61,62,64) Bakshi et al. (61) point out that spontaneous polarization is enhanced in a lattice of elongated dots, so-called dashes (see Fig. 13b),
and that nonparabolic terms in the confining potential determine the polarization
of the dashes in the antiferroelectric state. Also the arrangement in a rectangular
lattice might not be optimal, as pointed out by Dahl et al. (64), who propose a
face-centered lattice of dashes as sketched in Fig. 13c.

It was noticed soon (35,61) that such a superlattice of polarized dots could
be of great interest for information processing. For this purpose an array of more
complex unit cells may be advantageous. Figure 13d depicts a polarizable unit
cell suggested by Lent et al. for the construction of an array of cellular automata.
(35) Each cell consists of five weakly coupled dots, i.e., tunnelling is assumed
to be possible between the five dots inside a unit cell but forbidden between
neighboring cells. Lent et al. show that with two electrons filled into a unit cell
the polarizability becomes highly nonlinear and saturates into one of two bistable
states. (35) Thus depending on the polarization of neighboring cells a cell snaps
into a desired complete polarization. This behavior can be used for fast information transport as well as data processing. No electrical interconnections are
needed between the active components of the array, which makes possible a
vast improvement of integration density.

IX. STRONGLY COUPLED DOTS AND ANTIDOTS

Thus far only the interdot interaction by Coulomb forces has been discussed.
However, already the unit cell proposed by Lent et al. demonstrates that a

matrix of dots coupled so strongly that charge transfer is possible might be very interesting. As already sketched in Fig. 9, the dot matrix then gradually transforms into an antidot superlattice. A square lattice of electron dots, where such a transition could be investigated just by tuning a gate voltage, has been realized by Lorke et al. (66) Here a metal electrode set back from the electron system by a corrugated insulator is used to control the dot separation.

Figure 14a presents FIR transmission spectra of the dot matrix recorded at different gate voltages. The dot separation increases with decreasing gate voltage, and the dots are electrically isolated at gate voltage $V_g = -3.1\ V$. Accordingly, the resonances of the high frequency conductivity shows modes characteristic for an array of isolated dots. The magnetic field dispersion of the modes depicted in Fig. 14b is similar to the behavior of the modes of electron dots in silicon in Fig. 10. As the gate voltage is increased above $V_g = -2.9$ V, charge transfer between neighboring dots becomes possible. This is reflected in two characteristic

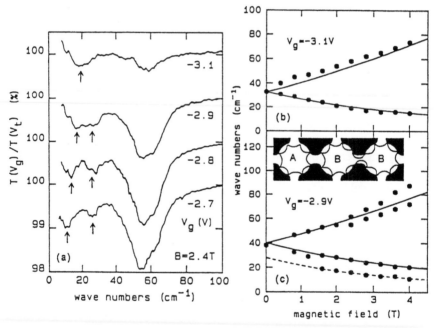

Figure 14 (a) FIR transmission spectra of an electron dot array induced in a AlGaAs/ GaAs heterojunction by a nanostructured gate. A magnetic field of $B = 2.4$ T is applied perpendicular to the sample surface. (b) Resonance positions vs. magnetic field at gate voltage $V_g = -3.1\ V$, where the electron dots are isolated from each other. (c) Resonance positions of coupled electron dots. The inset depicts trajectories with (B) and without (A) charge exchange between neighboring dots. (From Ref. 66.)

features of the FIR spectrum. The prominent alteration is observed at the low-frequency branch of the spectrum (arrows in Fig. 14). In a magnetic field a second low-field mode appears. The resonance positions are depicted in Fig. 14c. As shown by the straight lines in Fig. 14c, the upper modes still fit reasonably well to the dispersion of an isolated dot mode (Eq. (5)).

The evolution of the low-frequency mode can be understood with the trajectories depicted in the inset of Fig. 14c. The classical single-particle trajectories (a) on the left represent the motion that corresponds to the mode with negative magnetic field dispersion in an isolated dot at high magnetic fields. For obvious reasons this mode is often called a perimeter mode, and it can be shown in a hydrodynamic model that at zero magnetic field the resonance frequency of this mode is proportional to $1/R$ where R is the dot radius. (67) In addition to the trajectories (a), also trajectories (b) indicated on the left become possible once tunnelling between adjacent dots is permitted. The dog-bone-shaped trajectories propagating in two neighboring dots now sense roughly twice the circumference. The corresponding magnetic field dispersion is entered as the dashed line in Fig. 14c and fits well the observed resonance position.

The second feature arising with increasing coupling between adjacent dots is a high-frequency shoulder at the mode with positive magnetic field dispersion, as can be seen from the resonance positions above $B = 2$ T in Fig. 14c. This shoulder is identified as a magnetoplasmon resonance that propagates through the dot matrix as soon as charge transfer is possible. Again, such a mode can exist only if charge transfer between the dots is possible.

With increasing coupling, the dot matrix transforms into an electron mesh or antidot superlattice. Again the FIR transmission properties change dramatically once the mesh is formed, i.e., electron motion around antidots is not hampered any more. (17,68,69) In Fig. 15a and b we plot positions of FIR resonances measured on a GaAs heterojunction device in which either an electron dot matrix or an antidot superlattice can be generated, depending on the voltages applied on a gate configuration consisting of two stacked electrodes. (70) The bottom electrode consists of a metal mesh that—if positively biased—induces the antidot superlattice. A homogeneous top electrode biased negatively then controls the potential hills forming the antidots. If top and bottom electrodes are biased with opposing polarities, electron dots are induced beneath the gaps of the bottom gate, which then controls the barriers between the dots.

The resonance positions plotted in Fig. 15a are determined for gate voltages where electron dots are formed. The data of Fig. 15b represent the behavior of the resonances in an antidot superlattice. The magnetic field dispersion of the modes in Fig. 15a again is nicely described by Eq. (4) with $\Omega_o = 33$ cm^{-1}. The prominent difference of the modes in an antidot superlattice is evident in the magnetic field dispersion at low magnetic fields ($B < 3$ T in the case of Fig. 15b). At the cyclotron resonance positions (squares) a strong resonance occurs with

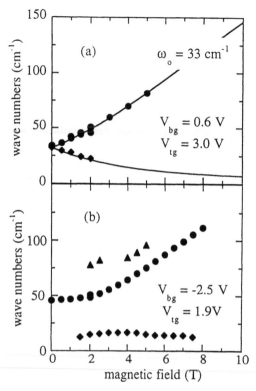

Figure 15 Resonance positions measured on a MISFET-type heterojunction with dual stacked gate biased in subgap mode (a) to create an electron dot matrix and in the subgrating mode (b) to create an antidot superlattice. (From Ref. 70.)

positive magnetic field dispersion. The positive dispersion saturates at higher fields, exhibits a maximum at about 3 T, and then decreases with increasing fields beyond 3 T. At the same time, the oscillator strength is gradually transferred to the high-frequency mode.

Encouraged by the success of the single-particle center of mass motion picture applied to discuss the high-frequency modes in an electron dot superlattice, equivalent models have also been proposed to describe qualitatively the character of the excitations in the antidot superlattice. (24,53,56,68) At low magnetic fields, this resonance corresponds to classical trajectories enclosing several periods of the superlattice. The motion deviates from the cyclotron orbit due to interaction with the superlattice potential, and a resonance frequency smaller than the cyclotron frequency arises. At the maximum position of the dispersion ($B = 3$ T), the cyclotron orbit is comparable to the antidot size. At higher magnetic fields, the cyclotron orbit is smaller, and thus the low-frequency mode corresponds to trajectories orbiting around the antidots. In this regime, the trajectories enclosing an antidot become edge modes, and in fact the magnetic field dispersion at high fields is similar to the low-frequency mode of electron dots. The high-frequency mode with resonance position 50 cm^{-1} at $B = 0$ T is identified as a plasmon with wavelength given by the antidot superlattice period. At high magnetic field the corresponding trajectories propagate between the antidots without encircling them. (53,68)

Although a discussion of the antidot modes with the aid of single-particle trajectories helps to understand their character, such arguments cannot be used for a quantitative description. The assumptions of the Kohn theorem are certainly not valid when the electron system forms an antidot superlattice. A description of the antidot modes that includes collective interaction effects has been published recently by Zhao et al. (69) They point out that the behavior of the modes can be understood as an anticrossing of the cyclotron resonance mode with an edge magnetoplasmon at a magnetic field where both modes would be degenerate if there were no interaction.

X. CONCLUSION

In this survey we have presented experimental results on static and dynamic electron transport in state-of-the-art artificial lateral superlattices that are imposed onto two-dimensional electron systems confined in semiconductor heterostructures. We have tried to demonstrate that ballistic electron motion both within and across several unit cells of the periodic potential landscape governs electron transport in these systems. In addition to ballistic charge transport, one has to consider the effects of intracell and intercell Coulomb interactions to understand the essential ingredients of electron interactions in such artificial network-type devices. Since todays' artificial superlattices usually have lateral periods well above the Fermi wavelength of the mobile electrons, band structure effects are hardly discernible. Quantum phenomena enter clearly only via density-of-states effects as visible in high field magnetoresistance oscillations as well as infrared excitations beyond the dominating center of mass modes.

Within the limits discussed above, recent experimental and theoretical studies of electronic transport and interactions have given us the basic understanding that is required to start considering the implementation of lateral superlattices into artificial computing networks. One promising example of how one may incorporate quantum dot arrays into an edge-fed cellular automaton has recently been proposed by Lent et al. (35). However, both more refined fabrication technologies and a more detailed understanding of quantum transport phenomena are required before one may realistically consider fabricating even the most rudimentary computing structures based on quantum dot arrays interconnected by tunnelling and Coulomb interactions. Conceptually, such devices appear attractive since they combine the possibilities of highest integration with low power consumption and few external interconnects and are also based on a well-established semiconductor technology. At present, the various physical phenomena that are observed can be explained within the framework of classical ballistic electron motion in combination with Coulomb interaction between neighboring dots, dashes, or wires. Even without band structure effects, the variety of new physical effects in lateral superlattices proves the mature and well-developed stage of this field.

ACKNOWLEDGMENTS

This review is based on collaborative work with many colleagues, as identified in the references. We wish to thank those colleagues as well as Pradip Bakshi, David Broido, Alik Chaplik, Detlef Heitmann, Kris Kempa, Rolf Landauer, Craig Lent, Günter Mahler, and Ulrich Merkt for stimulating discussions. We gratefully acknowledge financial support from the Deutsche Forschungsgemeinschaft, the ESPRIT Basic Research Action, and the Volkswagen-Stiftung.

REFERENCES

1. Ando, T., Fowler, A. B., and Stern, F., *Rev. Mod. Phys.*, *54*, 437 (1982).
2. Kastner, M., *Physics Today*, January 1993.
3. Bryant, G. W., *Phys. Rev.*, *B40*, 1620 (1989).
4. Hansen, W., Horst, M., Kotthaus, J. P., Merkt, U., Sikorski, Ch., and Ploog, K., *Phys. Rev. Lett.*, *58*, 2586 (1987).
5. Scherrer, A., Roukes, M. L., Craighead, H. G., Ruthen, R. M., Beebe, E. D., and Harbison, J. P., *Appl. Phys. Lett.*, *51*, 2133 (1987).
6. Ensslin, K., Häusler, K. T., Lettau, C., Lorke, A., Kotthaus, J. P., Schmeller, A., Schuster, R., Petroff, P. M., Holland, M., and Ploog, K., *New Concepts in Low Dimensional Physics* (G. Bauer, F. Kuchar, and H. Heinrich, eds.), Springer-Verlag, Berlin, 1992, p. 45.
7. Hansen, W., Merkt, U., and Kotthaus, J. P., in *Semiconductors and Semimetals* (R. K. Williardson, A. C. Beer, and E. R. Weber, eds.), Vol. 35 (M. Reed, ed.), Academic Press, San Diego, 1992, p. 279.
8. Berggren, K.-F., Thornton, T. J., Newson, D. J., and Pepper, M., *Phys. Rev. Lett.*, *57*, 1769 (1986).
9. Weiss, D., v. Klitzing, K., Ploog, K., and Weimann, G., in *High Magnetic Fields in Semiconductor Physics II*, (G. Landwehr, ed.), Springer, Berlin, 1988, p. 357.
10. Gerhardts, R. R., Weiss, D., and v. Klitzing, K., *Phys. Rev. Lett.*, *62*, 1173 (1989).

11. Winkler, R. W., Kotthaus, J. P., and Ploog, K., *Phys. Rev. Lett.*, *62*, 1177 (1989).
12. Hansen, W., Smith, T. P., III, Brum, J. A., Lee, K. Y., Knoedler, C. M., Hong, J. M., and Kern, D. P., *Phys. Rev. Lett.*, *62*, 2168 (1989).
13. Ashoori, R. C., Störmer, H. L., Weiner, J. S., Pfeiffer, L. N., Pearton, S. J., Baldwin, K. W., and West, K. W., *Phys. Rev. Lett.*, *68*, 3088 (1992).
14. Reed, M. A., Randall, J. N., Aggarwal, R. J., Matyi, R. J. Moore, T. M., and Wetsel, A. E., *Phys. Rev. Lett.*, *60*, 535 (1988).
15. Fang, H., Zeller, R., and Stiles, P. J., *Appl. Phys. Lett.*, *55*, 1433 (1989).
16. Ensslin, K., and Petroff, P. M., *Phys. Rev.*, *B41*, 12307 (1990).
17. Lorke, A., Kotthaus, J. P., and Ploog, K., *Superlattices and Microstructures*, *9*, 103 (1991).
18. Weiss, D., Roukes, M. L., Menschig, A., Grambow, P., v. Klitzing, K., and Weimann, G., *Phys. Rev. Lett.*, *66*, 2790 (1991).
19. Berthold, G., Smoliner, J., Rosskopf, V., Gornik, E., Böhm, G., and Weimann, G., *Phys. Rev.*, *B45*, 11350 (1992).
20. Gerhardts, R. R., Weiss, D., and Wulf, U., *Phys. Rev.*, *B43*, 5192 (1991).
21. Lorke, A., Kotthaus, J. P., and Ploog, K., *Phys. Rev.*, *B44*, 3447 (1991).
22. Beenakker, C. W. J., *Phys. Rev. Lett.*, *62*, 2020 (1989).
23. Schuster, R., Ensslin, K., Kotthaus, J. P., Holland, H., and Beaumont, S. P., *Superlattices and Microstructures*, *12*, 93 (1992).
24. Fleischmann, R., Geisel, T., and Ketzmerick, R., *Phys. Rev. Lett.*, *68*, 1367 (1992).
25. Fleischmann, R., Geisel, T., and Ketzmerick, R. *Europhys. Lett.*, *25*, 219 (1994).
26. van der Zant, H. S. J., Geerligs, L. J., and Mooij, J. E., in Proc. NATO ASI, *Quantum Coherence in Mesoscopic Systems* (B. Kramer, ed.), Series B: Vol. *254*, Plenum Press, New York, 1991, p. 511.
27. Averin, D. A., and Schön, G., in Proc. NATO ASI, *Quantum Coherence in Mesoscopic Systems* (B. Kramer, ed.), Series B: Vol. *254*, Plenum Press, New York 1991, p. 531.
28. van Houten, H., Beenakker, C. W. J., and Staaring, A. A. M., in *Single Charge Tunneling*, NATO Advanced Study Institute, Series B: Physics, (H. Grabert and M. M. Devoret, eds.), Plenum Press, New York, 1992.
29. Schuster, R., Ensslin, K., Kotthaus, J. P., Holland, M., and Stanley, C., *Phys. Rev.*, *B47*, 6843 (1993).
30. van Houten, H., Beenakker, C. W. J., Williamson, J. G., Brockaart, M. E. I., van Loodsrecht, P. H. M., van Wees, B. J., Mooji, J. E., Foxon, C. T., and Harris, J. J., *Phys. Rev.*, *B39*, 8556 (1989).
31. Thornton, T. J., Roukes, M. L., Scherer, A., and Van de Gaag, B. P., *Phys. Rev. Lett.*, *63*, 2128 (1989).
32. Büttiker, M., *Phys. Rev. Lett.*, *57*, 1761 (1986).
33. Onsager, L., *Phys. Rev.*, *38*, 2265 (1931).
34. Salzberger, F., Schuster, R., Ensslin, K., and Kotthaus, J. P., to be published.
35. Lent, C. S., et al., *Appl. Phys. Lett.*, Feb. 15 (1993).
36. Hirler, F., Strenz, R., Küchler, R., Abstreiter, G., Böhm, G., Smoliner, J., Tränkle, G., and Weimann, G., *Semicond. Sci. Technol.*, *8*, 617 (1993).

37. Brunner, K., Bockelmann, U., Abstreiter, G., Walther, M., Böhm, G., Tränkle, G., and Weimann, G., *Phys. Rev. Lett.*, *69*, 3216 (1992).
38. Kohn, W., *Phys. Rev.*, *123*, 1242 (1961).
39. Brey, L., Johnson, N. F., and Halperin, B. I., *Phys. Rev.*, *B40*, 10647 (1989).
40. Kotthaus, J. P., and Heitmann, D., *Surf. Sci.*, *113*, 481 (1982).
41. Wulf, U., *Phys. Rev.*, *B35*, 9754 (1987).
42. Davies, J. H., *Semicond. Sci. Technol.*, *3*, 995 (1988).
43. Hansen, W., in *Physics of Nanostructures* (J. H. Davies and D. A. Long, eds.), SUSSP and IOP, Bristol, 1992, p. 257.
44. Laux, S. E., Frank, D. J., and Stern, F., *Surf. Sci.*, *196*, 101 (1988).
45. Wilson, B. A., Allen, S. J., Jr., and Tsui, D. S., *Phys. Rev.*, *B24*, 5887 (1981).
46. Fock, V., *Z. Phys.*, *47*, 446 (1928).
47. Peeters, F. M., *Phys. Rev.*, *B42*, 1486 (1990).
48. Yip, S. K., *Phys. Rev.*, *B43*, 1707 (1991).
49. Alsmeier, J., Batke, E., and Kotthaus, J. P., *Phys. Rev.*, *B41*, 1699 (1990).
50. Sikorski, Ch., and Merkt, U., *Phys. Rev. Lett.*, *62*, 2164 (1989).
51. Meurer, B., Heitmann, D., and Ploog, K., *Phys. Rev. Lett.*, *68*. 1371 (1992).
52. Demel, T., Heitmann, D., Grambow, P., and Ploog, K., *Phys. Rev. Lett.*, *64*, 788 (1990).
53. Lorke, A., *Surf. Sci.*, *263*, 307 (1992).
54. Gudmundsson, V., and Gerhardts, R. R., *Phys. Rev.*, *B43*, 12098 (1991).
55. Pfannkuche, D., and Gerhardts, R. R., *Phys. Rev.*, *B44*, 13132 (1991).
56. Lorke, A., Jejina, I., and Kotthaus, J. P., *Phys. Rev.*, *B46*, 12845 (1992).
57. Que, W., and Kirczenow, G., *Phys. Rev.*, *B38*, 3614 (1988).
58. Que, W. M., *Phys. Rev. Lett.*, *64*, 3100 (1990).
59. Dempsey, J., Johnson, N. F., Brey, L., and Halperin, B. I., *Phys. Rev.*, *B42*, 11708 (1990).
60. Bakshi, P., Broido, D. A., and Kempa, K., *Phys. Rev.*, *B42*, 7416 (1990).
61. Bakshi, P., Broido, D. A., and Kempa, K., *J. Appl. Phys.*, *70*, 5150 (1991).
62. Chaplik, A. V., and Ioriatti, L., *Surf. Sci.*, *263*, 354 (1992).
63. Que, W. M., *Phys. Rev.*, *B45*, 11036 (1992).
64. Dahl, C., Kotthaus, J. P., Nickel, H., and Schlapp, W., *Phys. Rev.*, *B46*, 15590 (1992).
65. Kempa., K., Broido, D. A., and Bakshi, P., *Phys. Rev.*, *B43*, 9343 (1991).
66. Lorke, A., Kotthaus, J. P., and Ploog, K., *Phys. Rev. Lett.*, *64*, 2559 (1990).
67. Fetter, A. L., *Phys. Rev.*, *B33*, 5221 (1986).
68. Kern, K., Heitmann, D., Grambow, P., Zhang, Y. H., and Ploog, K., *Phys. Rev. Lett.*, *66*, 1618 (1991).
69. Zhao, Y., Tsui, D. C., Santos, M., Shayegan, M., Ghanbari, R. A., Antoniadis, D. A., and Smith, H. I., *Appl. Phys. Lett.*, *60*, 1510 (1992).
70. Hertel, G., Drexler, H., Hansen, W., Schmeller, A., Kotthaus, J. P., Holland, M., and Beaumont, S. P., in Proc. of the 6th International Conference on Modulated Semiconductor Structures, Garmisch Partenkirchen, 1993, to be published in *Solid State Electronics*.

9

Controlled Electronic Transfer in Molecular Chains and Segments

Hanspeter Knoll

ComTech GmbH
Waiblingen, Germany

Michael Mehring

University of Stuttgart
Stuttgart, Germany

I. INTRODUCTION

Under the notion of ''molecular electronics'' a number of proposals have appeared on how to use molecules to perform information processing functions (1–6). The suggested molecular functions range from mimicking known semiconductor devices like diodes and transistors or charge storage devices up to molecular shift registers (4,7). Less ambitious proposals considered molecules as simple storage elements, e.g., in a holographic memory, or devised a more passive role to molecules by using them in nonlinear optic devices (3,6).

No realization of these proposals has yet reached the marketplace. Besides some demonstrations of molecular memories, not even demonstration devices exist. Nevertheless, the notion of molecular electronics is exciting and has influenced the way of thinking of scientists working with molecular structures. In this chapter we want to introduce some simple concepts, starting from well-known molecular functions that, combined in the proper way, may lead to information processing units.

We shall briefly review in Sec. II the structure of aromatic(unsaturated) and aliphatic (saturated) molecules. Their electronic structure, including their charge excitations, are summarized in Sec. III. Aromatic molecules can be considered as basic building blocks of more functionalized molecular segments. Since we are dealing here predominantly with charge transfer processes, the essential properties of these building blocks will be their redox states. Accordingly they will be grouped into donor (**D**), acceptor (**A**), and bridge (**B**) molecules, where the bridge molecule is positioned intermediate between the donor and the acceptor. Every aromatic molecule will be separated by an aliphatic (saturated) segment from the next aromatic molecule. **DA** and **DBA** molecular segments constructed in this way will be discussed in Sec. IV. Information processing capabilities may be achieved by grouping **DBA** cells into molecular chains leading to a form of molecular cellular automata (**MCA**). The controlled charge transfer function in those chains will be treated in Sec. V.

II.   MOLECULAR CHAINS AND SEGMENTS

In this section we introduce the basic forms of molecular chains and segments that might be applicable to molecular electronic devices to be discussed in Secs. IV and V. The basic distinction is based on the electronic structure and will be made between "saturated" and "unsaturated" molecular chains. By this we mean hydrocarbons where all bonds are either saturated in terms of sp^3 orbitals or unsaturated in the sense of residual π orbitals that allow low-energy excitations and can lead to charged states.

A.   Saturated and Unsaturated Molecular Chains

Figure 1 displays two representatives of saturated (a, b) and unsaturated (c, d) molecular chains. As an example of a saturated molecular chain we have chosen an alkyl chain $(CH_2)_n$ that consists of n molecular segments composed of a carbon atom (C) and two hydrogen atoms (H) in an sp^3 bonding configuration. Two forms, namely the transform and the cis form, are shown. Due to the rotational freedom around the C–C bond, a large flexibility arises and a number of different conformations are possible. The two configurations shown are just two out of many. In a polymer like polyethylene the chains are coiled randomly to a large extent. However, it is well known that ordered phases exist in crystalline form and in Langmuir–Blodgett (LB) films, where alkyl chains may play a role as spacers between active (unsaturated) molecules. The electronic state of aklyl chains will be discussed in the next section. We note here that saturated hydrocarbons are electronically inactive in the sense that they merely play the role of a spacer and provide a tunneling barrier for electron transfer. There are other saturated (sp^3 orbitals) molecular chains that can be derived from the alkyl unit and other bridging bonds in order to obtain a stiff (nonflexible) network of sp^3 bonds. These are chemically more difficult to construct but have been sucsessfully applied as spacers in electron transfer molecular units. In the following discussion we shall often refer to the alkyl chain for simplicity but imply that stiff saturated chains like steroids or ada-

(a)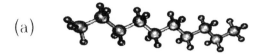

(b)

(c)

(d)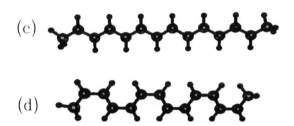

Figure 1   Saturated (a, b) and unsaturated (c, d) molecular chains. Alkyl chains (a, b) will be used in the following as synonyms for any type of saturated ("insulating") molecular chain, whereas polyenes (c, d) symbolize conjugated ("semiconducting") molecular chains.

mantane molecular units would be used instead. Their electronic states are essentially the same as those of the alkyl chains.

In Fig. 1 (c, d) we also display the trans and cis form of polyene $(CH)_n$ chains as representatives of unsaturated (conjugated) molecular chains. There is a number of variants of conjugated molecules, to be discussed later, that have similar electronic structures to the polyenes (8–10). The difference with respect to the alkyl chains is tremendous, and it is due not only to the missing of one proton but also to the completely different electronic structure including delocalized π bonds. The C–C bond is built from sp^2 orbitals (σ bonds) spanning the xy plane and π orbitals made from p_z orbitals (not shown in Fig.1). This type of bonding gives a certain rigidity and planarity to these types of molecular chains and segments and at the same time allows low-energy excitations of the delocalized π orbitals. Moreover these conjugated molecules can easily be oxidized (electron hole doping) and reduced (electron doping) by chemical or electrochemical means. The generated charges can be localized on a molecular segment or can become delocalized in a highly doped molecular chain, leading to extremely large conductivities as in *trans*-polyacetylene-$(CH)_n$. In this sense conjugated chains can be viewed as "molecular wires" allowing one-dimensional charge transport over large distances on the molecular scale. See (11–16) for further reference.

B. Conjugated (Aromatic) Molecular Segments

There is a number of conjugated aromatic molecules that can be used to construct molecular chains or segments (called oligomers). Some of them are displayed in Fig. 2. They are composed of C–C bonds made from sp^2 orbitals resulting in a π bonding network of planar geometry. In addition there is π bonding (not shown in Fig. 2) as in the polyenes leading to similar electronic states as for the polyenes. In the following section we discuss the electronic states of the polyenes. It should be kept in mind, however, that the electronic states of conjugated molecules like the ones shown in Fig. 2 are in fact quite similar to the ones of the polyenes. Polyenes are treated here mainly for simplicity in order to provide a simple language for the discussion of the electronic states of conjugated molecular chains in general (14–16). The molecular segments shown in Fig. 2 are stiff and nearly planar. The low lying π–π^* excitations of these molecules are in the visible range of the spectrum and decrease in energy with increasing chain length. Simply by varying the chain length, tailoring of the optical transition and the redox potentials is possible. Different charge excitations are possible in these molecular segments, which make them extremely useful candidates as active molecules in controlled electron transfer molecular assemblies, as will be discussed in the next sections.

III. ELECTRONIC STATES

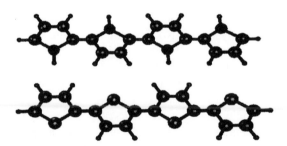

Figure 2 Typical examples of conjugated molecular segments (oligomers), namely oligopyrrole (top) and oligothiophene (bottom) with "semiconducting" properties.

A. Tight Binding Description of Unsaturated (Conjugated) Molecular Chains

To provide some insight into the electronic states of molecular chains and segments, we summarize some simple concepts based on one-electron theories (10,15). We consider a chain of sp^2 bonded C–C bonds where each carbon atom contains an additional p_z electron as is sketched in Fig. 3a as synonyms for conjugated molecular segments. If electron–phonon interaction is ignored for a moment, the p_z (π) electrons are mobile and can delocalize over the whole segment and can be viewed within a "free-electron" model of these molecules as particles in a one-dimensional box. The electronic states can be readily calculated and allow some insight into the excitations of these molecules. If electron–phonon interaction is explicitly taken into account, the energy can be lowered by forming alternating single and double bonds as shown in Fig. 3b. The π electrons are no longer mobile in this case, but they still give rise to low-energy excitations and lead in particular to different charged states. In order to set up a simple tight-binding one-electron theory, Su, Schrieffer, and Heeger (SSH) (11,12) have formulated the Hamiltonian

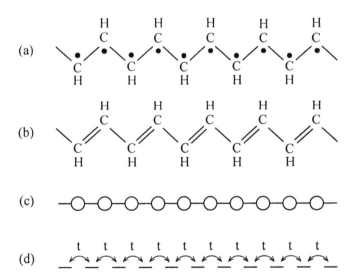

Figure 3 Different schematic views of a polyene chain. (a) τ-bonding with π-electrons represented by dots; (b) alternating double bond representation leading to a HOMO-LUMO gap; (c) symbolic representation as an atomic chain; (d) simplified electronic representation in a tight-binding model where only the singly occupied quantum state of the π-electrons is shown with nearest neighbor hopping integral t.

$$H = H_0 + H_e + H_n \tag{1a}$$

$$H_O = \sum_{l,\sigma} \varepsilon_0 \, c_{l,\sigma}^+ c_{l,\sigma} \tag{1b}$$

$$H_e = -\sum_{l,\sigma} t[c_{l,\sigma}^+ c_{l+1,\sigma} + c_{l+1,\sigma}^+ c_{l,\sigma}] \quad \text{with} \quad t = t_0 + \alpha(u_l - u_{l+1}) \tag{1c}$$

$$H_n = \sum_l \frac{P_l^2}{2M} + \frac{1}{2} K(u_l - u_{l+1})^2 \tag{1d}$$

and discussed its implications for the electronic states of polyacetylene. The corresponding orbital situation is sketched in Fig. 3c, d. Every C atom contributes one orbital with energy ϵ_0, where ϵ_0 can be considered to be the ionization energy of the p_z orbital of the C atom. The transfer integral t describes the electron tranfer between the neighboring orbitals and depends on the exchange parameter t_0 and the electron–phonon coupling constant α. Second quantization language has been used where the operators $c_{l,\sigma}^+$ create an electron in the local orbital at site l with spin σ, whereas $c_{l,\sigma}$ destroys an electron in this orbital. Lattice (bond) deformation is taken into account by the linear deformation parallel to the chain axis $u_l - u_{l+1}$ between neighboring atoms. Because of the corresponding change in overlap of the wave functions, the transfer integral changes too. Kinetic and elastic energy of the nuclei in the σ bond frame are represented by H_n. The SSH Hamiltonian is derived from the usual Hückel description of aromatic molecules (8,10,9,15). SSH have analyzed the electronic energy and soliton excitation in *trans*-polyacetylene within this approximation.

Here we are interested in a simple one-electron description of the electronic states of molecular segments. If we ignore bond alternation, the Schrödinger equation for the SSH Hamiltonian results in the eigenvalues (15)

$$\epsilon_j = \epsilon_0 - 2t \cos(j\frac{\pi}{n+1}) \qquad 1 < j < n \text{ linear chain} \tag{2a}$$

$$\epsilon_j = \epsilon_0 - 2t \cos(j\frac{2\pi}{n}) \qquad 1 < j < n \text{ cyclic chain} \tag{2b}$$

The continuous energy spectrum of an infinite chain (ring structure in order to avoid end effects) can readily be expressed as

$$\epsilon_j = \epsilon_0 - 2t \cos k \qquad \text{with} \qquad -\pi < k < \pi \tag{2c}$$

The ground state is obtained by filling each of the lowest levels with two electrons, according to the Pauli principle, with opposite spin. The corresponding energy levels for different chain lengths including those of infinite chain length are plotted in Fig. 4. Note that odd-numbered segments have a singly occupied energy level at ϵ_0, whereas an energy gap occurs always between the highest occupied molecular orbital (HOMO) and the lowest unoccupied molecular orbital (LUMO) for finite length segments. This gap, however, decreases towards zero with increasing chain length, leading to a half-filled band case (metal) for the infinite chain. The HOMO-LUMO gap can be expressed as (15)

$$E_g = 4t \sin\left(\frac{\pi}{2(n+1)}\right) \qquad n \text{ even; linear chain} \tag{3a}$$

$$E_g = 4t \sin\left(\frac{\pi}{n}\right) \qquad \text{cyclic chain} \tag{3b}$$

The dispersion curve shown in Fig. 4. for the infinite chain length is typical for a one-dimensional metal. We note that already for molecular segments of less than ten sites the energy spectrum spans essentially the same range as for an infinite chain. We therefore apply sometimes in the following the "band picture" (see Fig. 4) when referring to the occupied and unoccupied molecular levels in a molecular segment. It has been shown in the literature that bond alternation cannot be neglected and gives rise to a HOMO-LUMO gap opening even for the infinite chain. Polyacetylene therefore is not a metal. If one introduces the bond alternation order parameter Δ_0 and varies the transfer integral at site l

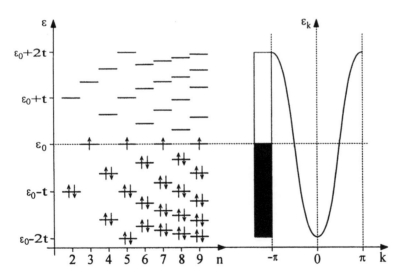

Figure 4 Energy level diagram for the π-electrons of polyenes versus chain length n according to Eq. (2). Note the closing of the HOMO-LUMO gap in the infinite chain (right). A gap opens at $k = \pi/2$ if bond alternation is included (see text).

according to $t(l, l+1) = t_0 - (-1)^l \Delta_0/2$ one arrives at energy levels that are somewhat modified with respect to Eq. (2) as (15)

$$\epsilon_j = \epsilon_0 \pm \sqrt{t_1^2 + t_2^2 - 2t_1 t_2 \cos(j\frac{4\pi}{n})} \quad \text{with} \quad t_{1,2} = t_0 \pm \frac{\Delta_0}{2} \qquad (4)$$

which holds only for an even-numbered cyclic chain of n atoms or odd-numbered linear chain with $n\text{-}l$ atoms. For the linear chain there is in addition one nonbonding singly occupied orbital at energy ϵ_0. The HOMO-LUMO gap results for an infinite chain in

$$E_g = 2(t_1 - t_2) = 2\Delta_0$$

In the continuum limit the energy dispersion curve is readily obtained as

$$\epsilon_k = \sqrt{4t_0^2 \cos^2 k + \Delta_0^2 \sin^2 k} \quad \text{with} \quad -\pi/2 < k < \pi/2 \qquad (5)$$

Note that the energy gap is again $2\Delta_0$ and the total bandwidth is $W = 4t_0$. In Fig. 5 we have sketched the dispersion curve for the reduced Brillouin zone together with a simplified sketch of the filled valance band and the unfilled conduction band. This particular description is useful even for conjugated(aromatic) molecular segments that contain more than ten carbon–carbon bonds. For the shorter segments it still presents a gross description of the possible energy spectrum of the occupied and unoccupied orbitals. We shall use this picture to discuss possible charge and other excitations of conjugated molecules. In order to emphasize the

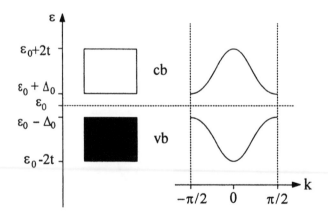

Figure 5 Band structure description for long chain conjugated molecules including bond alternation. The band structure model appears to be applicable for chain lengths in excess of $n = 15$.

importance of the simple C–C bond tight binding approach to conjugated molecular segments, we have sketched in Fig. 6 two representative conjugated molecules, namely oligomers of pyrole and thiophene, in such a way that the C–C bonds are enhanced. This should support our proposal that the whole class of conjugated molecules can be treated on the same footing. It should be noted that in the discussion presented here electron–electron interactions have been ignored. They do in fact play a role in these molecules, but the general outline of the energy levels is not changed drastically besides a few exceptions. These and more advanced theoretical treatments are discussed in (17).

B. Spin and Charge Localization in Molecular Segments

In the previous section we have seen that *even*-numbered molecular chains can be described by filled valence orbitals (valence band) and unoccupied virtual orbitals (conduction band). In this section we discuss the different spin and charge excitations that can be obtained by either oxidizing (hole doping) or reducing (electron doping) conjugated molecular segments. In the case of *odd*-numbered polyene chains we saw already in the previous section that always a single occupied orbital at energy ϵ_0 exists, which carries a spin ½. This spin ½ carries no charge, since the whole chain is neutral. It can therefore be viewed as a quasiparticle with charge $Q = 0$ and spin ½. This was termed a topological soliton in (11) and corresponds to the Pople-Wamsley defect in polyenes (9). It

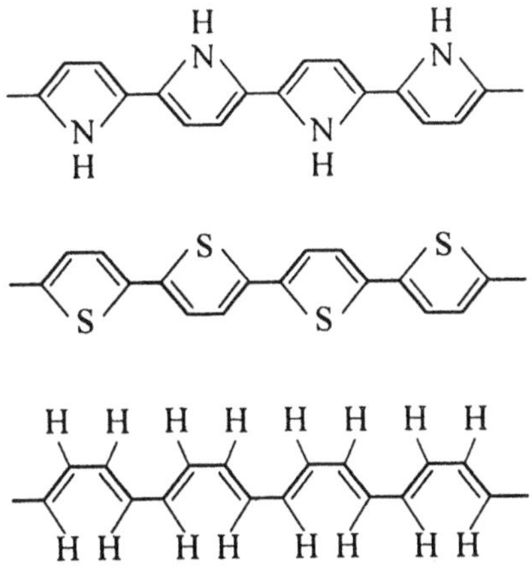

Figure 6 Comparison of different conjugated molecular segments (top: oligopyrrole; middle: oligothiophene; bottom: *cis*-oligoacetylene) emphasizing the alternating carbon bond structure. The corresponding electronic states are very similar for these different oligomers.

is schematically drawn as a dot in Fig. 7a. Physically speaking it is a topological defect in the polyene chain that represents a domain wall between two degenerate segments with opposite bond order parameter. In an infinite chain this "soliton" is mobile and can be used to transport bond order along the chain. In Fig. 7b the positively charged soliton is drawn that is obtained from the neutral soliton by oxidation (electron removal). This quasiparticle contains charge $Q = +e$ and spin $S = 0$. Its negative counterpart also exists. Strictly speaking, the topological solitons are not localized to one carbon atom, as is schematically drawn in Fig. 7. ESR and ENDOR experiments have shown that its half-width extends over 14–22 carbon atoms in trans polyacetylene (18). In a finite polyene segment it therefore extends over the whole molecular chain.

In addition, we have drawn the polaron (c) and bipolaron (d) structure for an oligothiophene in Fig. 7 (15,17). Again the drawing is purely schematic. The extension of polarons and bipolarons is expected to occur over four thiophene rings. For compactness we have reduced this to two thiophene rings in Fig. 7. Their wave function is of course a smooth function and will level off monotonically. Spin $S = \frac{1}{2}$ and charge $Q = +e$ are not separated and are distributed equally over about four thiophene rings in the case of the polaron. The polaron can be oxidized once more to a bipolaron with spin $S = 0$ and charge $Q = +2e$. Both polarons and bipolarons are mobile along the chain and can be transported over large distances, as has been demonstrated by the electrical conductivity of conjugated polymers (19).

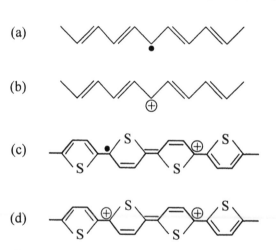

Figure 7 Schematic representation of the neutral (a) and positively charged (b) soliton in polyenes and the polaron (c) and bipolaron (d) in thiophenes. The extension of the corresponding wave function is larger than drawn here (see text).

For a controlled electron transfer in molecular segments it is of importance to know the electronic energy of the spin and charge excitations. In order to discuss this on a general level we have drawn schematic energy diagrams for all possible spin charge excitations in Fig. 8. The filled valence orbitals and the empty virtual orbitals are always characterized by energy bands with a band gap between them. Due to the characteristic spin charge excitations, localized levels appear within the gap. Let us begin with Fig. 8a, which represents the soliton in polyenes. The neutral and the positive soliton levels in the middle of the gap are shown as discussed previously. The corresponding negative soliton can be readily obtained by adding an electron (and a reversed spin) to the neutral soliton. These solitons are quite exceptional, since they only occur in degenerate ground state systems like *trans*-polyene. Already in *cis*-polyene they cannot exist. Instead polarons are formed as in the other conjugated molecular segments. Their electronic states are sketched in Fig. 8b. Positive and negative polarons exist and are characterized by their split off levels from the valence and virtual orbital bands due to electron–phonon coupling. The bipolaron (Fig. 8c) can be obtained from the polaron state by either extracting (oxidation) or adding (reduction) an electron to the molecule. Both polaron and bipolaron are charged quasiparticles that will be the essential objects in the following when we discuss controlled charge transfer in molecular chains and segments. Extensive optical spectroscopy on these "nonlinear excitations" of conjugated polymers have been performed and are reviewed in (20). Moreover, magnetic resonance experiments have demonstrated the theoretically proposed spin and charge relations in these materials (21).

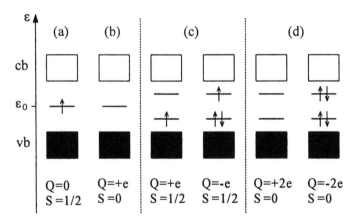

Figure 8   Energy level scheme of solitons (a), polarons (b,c), and bipolarons (d) in conjugated molecular chains.

C. Molecular Quantum Wells

Because of the large difference in redox states of aromatic molecules and unsaturated chain segments, molecular quantum wells may be constructed from segmenting different aromatic oligomers with unsaturated molecular groups. An example is shown in Fig. 9, where we have combined two equivalent oligothiophenes separated by an alkyl chain. This should serve as just one out of many possible examples for molecular quantum wells. The general structure will be $M_1-S-M_2-S-M_3-S-$, where M_j stands for any aromatic oligomer and $-S-$ is a spacer group with a large HOMO-LUMO separation and with ionization potential and electron affinity well separated from the ones of the aromatic group. The spacer group should serve as a tunnelling barrier that allows electron transfer mediated by superexchange coupling. As mentioned before, the alkyl chain is not the ideal spacer because of its molecular flexibility but is drawn here as a synonym for related unsaturated molecular segments. The different redox states of aromatic oligomers as discussed in the previous sections could be used to carry localized charged states, which could be controlled by light excitation and/or by electrode induced charges. In the example shown in Fig. 9, a positive polaron on the thiophene oligomer on the left-hand side could be shifted to the right-hand side by optical excitation. This electron transfer could be detected either by hole burning experiments or by the change in dipole moment.

It is of course obvious to extend such a molecular quantum well to a chain containing a number of similar segments. In the example shown here, no directed electron transfer occurs, however. To achieve this, the molecular units must be properly functionalized, as will be discussed in the next section.

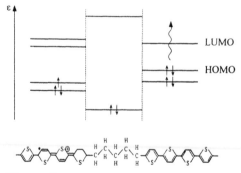

Figure 9 Molecular quantum well structure and corresponding energy level diagrams of two quaterthiophenes connected by a saturated spacer. The alkyl chain is a symbolic representation for a more rigid saturated spacer molecule.

IV. BISTABLE ELECTRON TRANSFER

In this section we discuss the possibilities for a bistable electron transfer in the sense that an electron can reside either on molecule A or on molecule B, which are connected to each other to allow controlled electron transfer between them. Moreover, we want to achieve vectorial electron transfer in a given direction (7). In the next sections we discuss some of the principles involved. Electron transfer processes are an intensively studied subject, and an enormous amount of literature is available. Here we summarize only some of the experimental (22–30) and theoretical publications (31–33, 4, 5, 34–36) that have contributed to the understanding of electron transfer processes in molecular systems.

A. Donor-Acceptor (*DA*) Molecular Segments

In the following we consider two unsaturated (aromatic) molecules coupled via a saturated spacer. We label one of the molecules donor (*D*) and the other acceptor (*A*) depending on their respective redox states. A simplified level diagram is shown in Fig. 10. Only HOMO and LUMO are sketched together with the charged (polaron) states. Characteristic energies often referred to are the ionization potential (*IP*) and the electron affinity (*EA*). They may be identified with the HOMO or LUMO energies according to the Koopmans theorem, as is indicated in

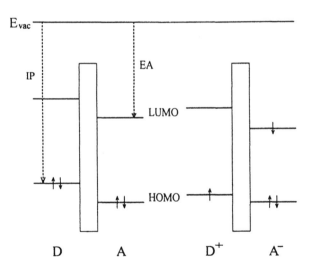

Figure 10 Simplified energy level diagram for a donor (D)–spacer–acceptor (A) molecular segment in the ground state (left) and the charge separated state (right). Electron affinity (*EA*) of the acceptor and ionization potential (*IP*) of the donor molecule are indicated. E_{vac} corresponds to the vacuum level.

Fig. 10. The spacer is drawn as a tunnelling barrier. Note that the ground state DA is represented by a different level diagram than the electron transfer state D^+A^-. The rearrangement of the HOMO-LUMO levels takes electron–phonon and electron–electron interaction into account in an average manner. Donor and acceptor molecules must be chosen so that no electron transfer occurs in the ground state. Only after excitation of either the donor or the acceptor molecule is electron transfer accomplished. Under donor excitation, electron transfer proceeds from the donor to the acceptor, whereas under acceptor excitation, hole transfer proceeds from the acceptor to the donor. The result is the same in both cases, although the pathways and the corresponding transfer rates are different. Since the D^+A^- state is a metastable state, electron–hole recombination eventually occurs, closing the cycle. In nature this cycle is realized in the photosynthetic process. The complete cycle may be expressed in a grossly simplified form as

$$DA \quad \overset{\overset{P}{\rightarrow}}{\underset{k_l}{\leftarrow}} \quad D^*A \quad \overset{k_s}{\rightarrow} \quad D^+A^- \quad \overset{k_r}{\rightarrow} \quad DA$$

The ground state DA is excited by light with energy $E_{ex} = h\nu$, e.g., at the donor singlet–singlet transition leading to the excited state D^*A with pumping rate P. Neglecting the fluorescence decay, electron–hole separation occurs with rate k_s, resulting in the metastable state D^+A^-, which after electron–hole recombination with rate k_r relaxes to the ground state. The energetics of the process is sketched in Fig. 11, where the energy hypersurfaces and the relevant energy scales are drawn schematically (31,32). The following sum rule holds

$$E_{ex} + \Delta G_s^0 + \Delta G_r^0 = 0 \tag{6}$$

where ΔG_r^0 is the free energy for electron–hole recombination, and the free energy for electron–hole separation ΔG_s^0 can be expressed as

$$\Delta G_s^0 = IP - EA - E_{ex} + \Delta G(D^+) + \Delta G(A^-) - \frac{e^2}{4\pi\epsilon\epsilon_0 a} \tag{7}$$

where $\Delta G(D^+)$ and $\Delta G(A^-)$ are the solvation enthalpy of the donor and acceptor ions and where the last term represents the Coulomb energy of the separated charges in a polarizable environment with dielectric constant ϵ and average charge separation a. This equation has been successfully applied to electron transfer reaction in solution. In the rigid lattice situation, as discussed in this article, the solvation enthalpy must be replaced by a conformational relaxation energy that is much smaller than the corresponding energies in solution. Experi-

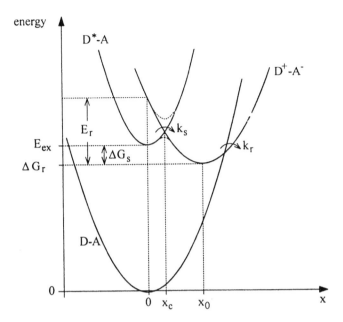

Figure 11 Configuration energy diagram of the D-A molecular segment. The vibronic energies of the ground state (D-A), the excited state (D*-A), and the charge separated state (D$^+$-A$^-$) are represented by parabolas with respect to the configuration coordinate x. Excitation energy E_{ex}, lattice reorganization energy E_r, charge separation enthalpy ΔG_s, and recombination enthalpy ΔG_r are indicated together with the rate constants k_s for charge separation and k_r for charge recombination, respectively.

mentally the redox potentials $E(D/D^+)$ and $E(A/A^-)$ are more readily accessible than the ionization potential and the electron affinity. By applying the relation

$$IP = E(D/D^+) - \Delta G(D^+)_{sol}$$

$$EA = E(A/A^-) + \Delta G(A^-)_{sol}$$

Eq. (7) can therefore be replaced for practical purposes by (22,23)

$$\Delta G^0{}_s = E(D/D^+) - E(A/A^-) - E_{ex} - \frac{e^2}{4\pi\epsilon\epsilon_0 a} \tag{8}$$

Note that $\Delta G < 0$ if the final state of the process has lower energy than the state from which charge separation starts.

Research in this area has been performed extensively by Weller and coworkers (22,23) and others (24–30). Marcus (31) and Jortner (32) have paved the way for a theoretical understanding of the basic mechanisms involved. There have been numerous investigations of electron transfer processes in biology that are treated along these lines (33).

B. Electron Transfer Rates in *DA* Segments

After excitation of one of the molecules, either the donor or the acceptor, the electron transfer process proceeds from the excited state into the charge separated state with rate k_s, as is outlined in Fig. 11. In order to be specific, we assume in the following that the donor molecule was excited. Equivalent relations hold if the acceptor molecule was excited. The relevant tunnelling matrix element is

$$V(x) = <D*A|H(x)|D^+A^->$$ (9)

where $H(x)$ is the coupling Hamiltonian between excited state and the electron transfer state. Together with Fermi's golden rule this results in the following electron transfer rate k_s for electron–hole separation (31,32):

$$k_s = \frac{2\pi}{h}|V(x)|^2 FC$$ (10a)

where

$$FC = \frac{1}{\sqrt{4\pi E_r kT}} \exp\left[-\frac{(E_r + \Delta G_s^0)^2}{4E_r kT}\right]$$ (10b)

is the Frank–Condon factor, which takes the reorganization energy E_r of the charge separated state into account. This energy is usually referred to as the Marcus parameter λ. By taking molecular vibrational modes more explicitly into account, the Frank–Condon factor can be expressed as (32)

$$FC = \frac{1}{\sqrt{4\pi E_s kT}} \sum_{m=0}^{\infty} \frac{S^m}{m!} e^{-s} \exp\left[-\frac{(E_s + \Delta G_s^0 + mh\omega)^2}{4E_s kT}\right]$$ (10c)

$$\text{with} \quad S = \frac{E_i}{h\omega}$$

where ω is the appropriate phonon frequency and also a distinction is made between the intramolecular reorganization energy E_i and the solvent reorganization energy E_s. The electron–hole recombination rate may be calculated in a similar manner by replacing the corresponding parameters. Experimentally vastly different charge separation and recombination rates have been observed in model compounds and in biological systems ranging from $10^9 < k_s < 10^{12}$ and $10^6 < k_r < 10^{11}$ depending on the molecules, their coupling strength, and the polarizable surrounding (reorganization energy).

C. Electron Transfer Kinetics in DA Segments

Let us consider the kinetics in DA segments in a simplified manner in order to discuss the key elements of electron transfer in these molecular units. We label the three states involved in the photoinduced electron transfer process as $DA - > 1$, $D^*A - > 2$, and $D^+A^- - > 3$, as shown in Fig. 12. The most general description of the dynamics of such a multilevel system would proceed via a density matrix treatment (37) to a stochastic Liouville equation. Since coherent processes (off diagonal elements of the density matrix) die out on the timescale of femtoseconds in the molecular systems considered here, we restrict ourselves to the relaxation of the diagonal part. This leads to a set of rate equations

$$\frac{d\vec{p}}{dt} = \Gamma\vec{p} \quad \text{with} \quad \Gamma = \begin{pmatrix} -P \, k_l & & k_r \\ P & -(k_l + k_s) & 0 \\ 0 & k_s & -k_r \end{pmatrix} \quad (11)$$

where $\vec{p} = (p_1, p_2, p_3)$ is the probability vector for finding the DA segment in one of the three states with normalization condition $p_1 + p_2 + p_3 = 1$. The pumping rate of the donor molecule is expressed by P and its luminescence rate by k_l. The electron–hole separation rate k_s and the electron–hole recombination rate k_r have been discussed in the last section. An analytic solution of Eq. (11) is readily obtained with eigenvalues

$$\lambda_0 = 0 \quad (12)$$
$$\lambda_{1,2} = \Gamma_1 \pm \Gamma_2$$

where

$$\Gamma_1 = -\frac{k_r + k}{2} \quad \text{and} \quad \Gamma_2 = \sqrt{\frac{(k - k_r)^2}{4} - Pk_s}$$

and with $k = k_1 + k_s + P$. By using the initial conditions $p_1 = 1$, $p_2 = p_3 = 0$, the following solution results:

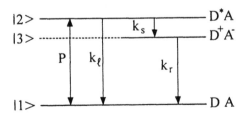

Figure 12 Simplified level diagram for the three active states in the **DA** segment, namely the ground state DA→|1>, the donor excited state D*A→|2>, and the charge separated state D$^+$-A$^-$→|3>.

$$p_1(t) = -\frac{k_r(2\Gamma_1 + P + k_r)}{\Gamma_1^2 - \Gamma_2^2} + \left[1 + \frac{k_r(2\Gamma_1 + P + k_r)}{\Gamma_1^2 - \Gamma_2^2}\right] \cosh(\Gamma_2 t) e^{\Gamma_1 t}$$

$$-\left[\frac{\Gamma_1 + P}{\Gamma_2} + \frac{k_r(2\Gamma_1 + P + k_r)\Gamma_1}{(\Gamma_1^2 - \Gamma_2^2)\Gamma_2}\right] \sinh(\Gamma_2 t) e^{\Gamma_1 t} \tag{13a}$$

$$p_2(t) = \frac{k_r P}{\Gamma_1^2 - \Gamma_2^2} - \frac{k_r P}{\Gamma_1^2 - \Gamma_2^2} \cosh(\Gamma_2 t) e^{\Gamma_1 t} + \left[\frac{P}{\Gamma_2} + \frac{k_r \Gamma_1 P}{(\Gamma_1^2 - \Gamma_2^2)\Gamma_2}\right] \sinh(\Gamma_2 t) e^{\Gamma_1 t} \tag{13b}$$

$$p_3(t) = 1 + \frac{k_r(2\Gamma_1 + k_r)}{\Gamma_1^2 - \Gamma_2^2} - \left[1 + \frac{k_r(2\Gamma_1 + k_r)}{\Gamma_1^2 - \Gamma_2^2}\right] \cosh(\Gamma_2 t) e^{\Gamma_1 t} \tag{13c}$$

$$+\left[\frac{\Gamma_1}{\Gamma_2} + \frac{k_r(2\Gamma_1 + k_r)\Gamma_1}{(\Gamma_1^2 - \Gamma_2^2)\Gamma_2}\right] \sinh(\Gamma_2 t) e^{\Gamma_1 t}$$

Of particular importance is the equilibrium population of the charge separated state D^+A^-, which is obtained as

$$p_3(\infty) = \frac{1}{1 + k_r/k_s + k_r/P + k_i k_r/P k_s} \tag{14}$$

If illumination of the DA segment is terminated, the D^+A^- population $p_3(t)$ decays nearly with rate k_r.

D.   Donor–Bridge–Acceptor (*DBA*) Segments

Due to the close distance of D^+ and A^- in the DA segment, electron–hole recombination is rather likely to occur with high recombination rates k_r. It therefore seems advisable to include an aromatic bridge (*B*) molecule between the donor and the acceptor molecule. A typical example of a *DBA* segment is shown in Fig. 13 (7b). The alkyl spacers should be replaced by stiff saturated chains as discussed before. To achieve a long-lived electron transfer, the saturated spacer molecules must be made much longer than is sketched in Fig. 13. Light excitation can be used to bring the donor, the bridge, or the acceptor molecule into the excited state. In general this will occur at different wavelengths. The result, namely the separation of electron and hole, will be the same. The transfer rates may be different, and one has to choose which pathway is the most advantageous. The different scenarios are sketched in Fig. 14.

In fact, nature follows this scheme in photosynthetic reactions, where intermediate molecular units transfer the electron energetically downhill, which results in a stable charge separation. The *DBA* architecture is therefore bound to provide more efficient and more stable charge separation. In the *DA* segment, the initial step of charge separation is accompanied by nonequilibrium vibronic excitations.

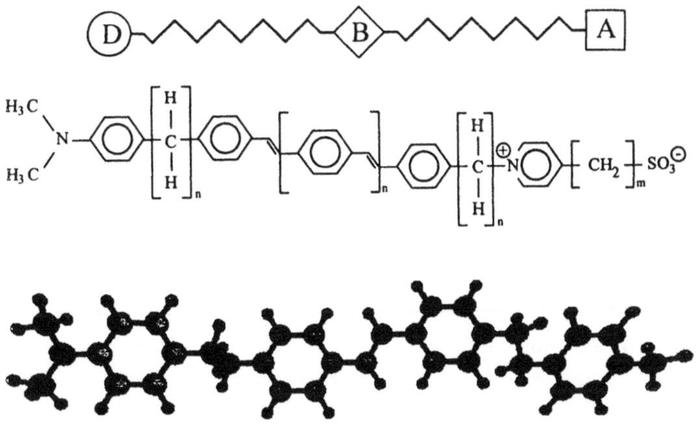

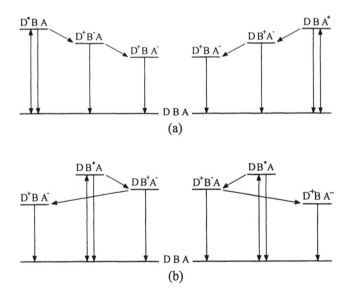

Figure 13   A typical example of a **DBA** segment, consisting of dimethylanilin donor (D), a phenylene vinylene bridge (B), and a pyridinium acceptor (A). The alkyl chain spacers shown are synonyms for any saturated spacer molecule.

It is therefore not legitimate to neglect the back reaction to the excited state completely, as was done in the previous section. If the donor molecule (D) in a DBA segment is excited (Fig. 14a) with electron transfer to the B molecule,

Figure 14   Different scenarios for excitation and electron transfer in a **DBA** molecular segment. (a) Donor (left) and acceptor excitation (right). (b) Bridge excitation.

the back transfer from the B molecule to the D molecules must compete with the rapid (ps timescale) electron transfer to the A molecule. This makes the "forward" electron transfer to the A molecule much more likely. The same argument holds if the acceptor molecule (A) is excited (Fig. 14a). In this case a hole is transferred via the B molecule to the donor (D), which is complementary to the previous situation where an electron was transferred.

In case the bridge molecule (B) is excited (Fig. 14b), two states are available for electron transfer, namely the electron and the hole. Their rate might differ, but both favor charge separation and will increase the efficiency considerably. In all these cases the stabilization of the charge separated state is mainly due to the exponential decrease of the tunnelling matrix element for electron–hole recombination with distance. A further stabilization can be achieved by embedding the DBA segment in a highly polarizable medium and/or by working at low temperatures.

The electron transfer kinetics can be described basically in the same way as is done for the DA segments. Again we restrict ourselves to the relaxation of the diagonal part of the density matrix. A master equation for the population p_i of state i can be set up as

$$\frac{dp_i(t)}{dt} = - \sum_j W_{ji}\, p_i\,(t) + \sum_j W_{ji}\, p_j(t) \qquad (15)$$

where W_{ji} is the transition rate from state i to state j. The pumping rate P, the luminescence decay rates, and different charge separation and recombination rates must be included. The resulting set of equations can be solved numerically.

The efficiency of electron transfer in a DBA segment can be expressed by the probability $P(D^+B^0A^-, t=\infty)$ to find the charge separated state after long enough light excitation. This probability depends on the overall charge separation rate k_s, where we made no distinction between electron and hole transfer rates. Fig. 15 shows the calculated rates required to obtain certain values for $P(D+B^0A^-, t=\infty)$. In the calculation shown in Fig. 15 the pumping rate was kept fixed to $P = 10^9\ \mathrm{s}^{-1}$ and the luminescence rate to $k_l = 10^9\ \mathrm{s}^{-1}$. Note that the charge separation efficiency is mainly dictated by the recombination rate k_r, once the separation rate k_s is larger than about $10^9\ \mathrm{s}^{-1}$. The advantage of the DBA segment over the DA segment is its reduced recombination rate k_r, which results in a much higher charge separation efficiency. The parameter set represented by the dotted lines in Fig. 15 was used in the simulations to be discussed in the next section.

In summary, the DBA segment undergoes the cycle of excitation, charge separation, and recombination under light excitation with certain rates that are dictated by the molecular properties and that can be tailored by choosing the appropriate molecular building blocks. In the next section we shall investigate how DBA segments could possibly be combined in order to fulfill certain information processing functions.

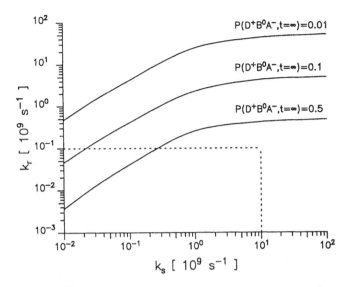

Figure 15 Parameter space for the charge separation rate k_s, the recombination rate k_r, and the probability of finding the charge separated state $D^+B^0A^-$.

V. MOLECULAR COMPUTING

The idea behind ''molecular electronics'' is to use molecular units to perform information processing functions similar to those implemented in current computers. Therefore basic electronic elements like ''molecular diodes'' and ''molecular transistors'' were proposed as essential elements for such a technology. We deviate here from this concept of trying to mimic standard semiconductor technology on the molecular level. First of all, single-molecular elements as proposed by Aviram and Ratner (1) are not fail-safe under environmental hazards like cosmic radiation; secondly, their interconnections, in order to achieve some processing function, are rather complex if done on a molecular level or else are macroscopic, which would make the proposal of single-molecular functions obsolete. We therefore propose to use an ensemble of similar molecular elements performing similar functions (7). These functions are performed in a probabilistic way and must be described in terms of a density matrix formalism (38). The ensemble properties of the density matrix will result in a stochastic logic, and the information processing capabilities of molecular segments should be considered more like those of ''fuzzy logic'' than of the Boolean logic used in current computers. In the following we shall touch on some of these aspects.

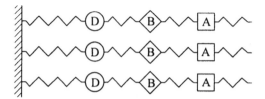

Figure 16 Langmuir–Blodgett (LB) **DBA** segments connected to an electrode.

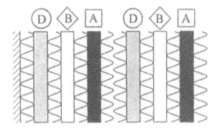

Figure 17 Thin film layers of **DBA** segments connected to an electrode. The wavy lines represent saturated spacer layers.

A. Molecular Cellular Arrays

In order to use *DBA* segments in a computational environment one has to build ordered structures. We distinguish the following types of structures, namely

1. Chain ordered structures
2. Layer ordered structures

Chain ordered structures, as outlined in Fig. 16, can be constructed e.g. with Langmuir–Blodgett (LB) techniques, whereas layer ordered structures as shown in Fig. 17 are more easily built with standard thin film techniques. Note that the active molecules labelled *D*, *B*, and *A* are separated by insulating chains or layers in both cases. This is important in order to prevent electron–hole recombination and to localize the charges on the active molecules. Both chain ends are attached to electrodes made of either metals or semiconductors. The first molecular layer could be either directly connected with the electrode, if the potential change of this molecular layer is important, or disconnected from the electrode by saturated spacer layers, if only electron transfer is requested. The insulating sheets between the active molecules within a *DBA* segment must be tailored properly in order to obtain the desired charge separation and recombination rates, whereas each *DBA* segment should be separated enough from the other *DBA* segments to allow electron transfer, but to prevent back transfer. Investigations are currently under way to find the proper spacings.

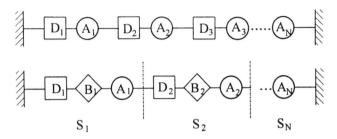

Figure 18   **DA** (top) and **DBA** (bottom) molecular cellular automata (**MCA**) consisting of an array of N cells each.

Chain ordered structures could in principle allow controlled electron transfer along a single molecular chain if the neighboring chains are well separated. It might be interesting, however, to allow also a lateral electron transfer in order to improve redundancy and efficiency due to electron delocalization. This is naturally provided in the layer ordered structures. Here even lateral structuring must be applied in order to confine the molecular information processing function to restricted segments.

In Figs. 16 and 17 only one electrode is drawn (left hand side) on which the molecular structures are grown. It is understood, however, that the other end is also connected to an electrode, both of which could be either a metal or a semiconductor as discussed before. These electrodes should provide charges or potentials depending on the applied voltage and the electrode surface layer. This way an electrical signal is finally transmitted to a semiconductor structure, which allows us to embed the molecular unit in a classical semiconductor structure. This semiconductor or metallic structure can also provide connections with other molecular segments in order to perform specific molecular computing functions.

B. Molecular Cellular Automata

We consider in the following molecular chains consisting of *DBA* or simply *DA* segments, as discussed in Sec. IV, which are connected to electrodes as sketched in Fig. 18. Such a device will be called a molecular cellular automaton (MCA)(7) because of its information processing function similar to ordinary cellular automata (39). Electrons or holes can be introduced into the MCA from the electrodes and are shifted by a clock pulse provided by incoherent light (7b).

1. Occupation Probabilities and Transfer Rates

A molecular chain can be considered as a quantum-mechanical system with n relevant global states Φ_i with energies E_i. Each chain consists of N cells. The molecular chain is embedded in an environment that consists of two electrodes, an incoherent light field, and a heat bath. Every cell j is represented by L_j states $\phi_{l_j}^j$. The states ϕ_i of the whole chain can be expressed by the product of cell states as

$$\Phi_i = \prod_{j=1}^{N} \varphi_{l_j}^j \tag{16}$$

Every global state Φ_i can be represented by a unique set of cells states $(\varphi_{l_1}^1, \dots, \varphi_{l_j}^j, \dots, \varphi_{l_N}^N)$ that leads to a total number

$$n = \prod_{j=1}^{N} L_j \tag{17}$$

of chain states in case all states of an individual cell are accessible independent of the states of the residual cells. Rate equations can be set up for the diagonal elements p_i of the density matrix analogous to the molecular segments discussed in the previous section. This type of description implies a large number of molecular chains that can be considered as an ensemble. For a small number of chains the global occupation numbers p_i take on discrete values and vary stochastically. We introduce the local occupation numbers $p_{l_j}^j$ of cell j that are connected with the global occupation number p_i by

$$p_i = \bigcap_{j=1}^{N} p_{l_j}^j \tag{18}$$

Due to the cellular structure of the molecular chain we must distinguish two different dynamical processes:

Intracellular Dynamics. The electron transfer dynamic within the *DBA* and *DA* cells has been discussed already in the previous section. The local description used there can be applied to the *MCA* under the assumption that all transition rates between intracellular states are unaffected by the neighboring cells. This requires long spacers ($d > 1,0$ nm) or a highly polarizable medium with a large dielectric constant. The transition of $\phi_{l_j}^j \rightarrow \phi_{k_j}^j$, where k labels all possible states within cell j, is represented by the cumulate decay rate

$$W_{l_j}^{\text{intra}}(j) = \sum_{k=1}^{L_j} W_{k_j l_j}^{\text{intra}}(j) \tag{19}$$

Intercellular Dynamics. The electronic states and rate constants of the cell elements j are designed in such a way that the light pulse does not lead to a change of the cell state unless the neighboring cell $j+1$ is in an oxidized, reduced, or charge separated state (7b). Fig.19 demonstrates the case of a directed electron transfer after light pulse excitation. For simplicity the process is sketched only for neighboring *DA* cells. It should be obvious that similar reasoning holds for *DBA* cells. The donor excitation before the electron transfer (Fig. 19 (left)) leads to an electron

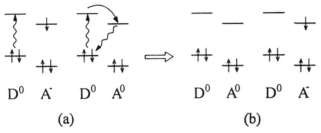

Figure 19 (a) Simultaneous excitation of $D^0A^- - D^0A^0$ neighboring cells. Electron transfer occurs when the right-hand donor D^0 is excited. (b) Final state with electron shifted by one cell.

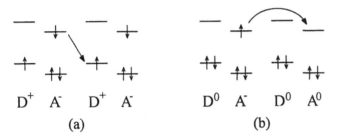

Figure 20 Different scenarios for unwanted electron transfer. (a) Electron–hole recombination in a D^+A^- pair. (b) Electron hopping from D^0A^- to D^0A^0.

transfer cycle in which the left-hand acceptor (A^-) is blocked until a hole is created at the right-hand donor (D^0). At this instance an electron transfer occurs from the left-hand acceptor to the right-hand donor, and the coupled pair reaches the final transfer state shown in Fig. 19 (right). This corresponds to the shift of a logical I to the neighboring cell in the sense of information processing. Simultaneous excitation in Fig. 19a is not required but only considered as the worst case.

There are, however, competing processes like those shown in Fig. 20. In Fig. 20a a recombination between two charge separated states is shown. Such a process leads to a separation of electrons and holes, which travel to the opposite electrodes. This process adds to the noise and eventually leads to charge accumulation at the electrodes, thus preventing further charge transfer. Another possible (unwanted) intercellular electron transfer process is visualized in Fig. 20b. If one of the cells is in the ground state, a hole or an electron can be transferred to it from the neighboring cell by electron tunnelling. This electron transfer is undirected, because both cells can be considered nearly equivalent if they consist of the same molecules. However, the (A^-) and (A^0) states are not degenerate because of the polaron distortion discussed in the last section. In addition, electric fields due to Coulomb interaction detune the energy levels differently. The process shown in Fig. 20 is therefore not very likely but may still lead to some "random hopping" rate as will be discussed later.

In the following we consider only near neighbor electron transfer with rates $W^{inter}_{l'_j l'_{j+1} l_j l_{j+1}}$. If the final states l'_j and l'_{j+1} are uniquely determined by the inital states, as will be assumed in the following, these labels can be dropped. This also leads to a simplification in the Monte Carlo algorithm to be discussed below. A set of differential equations for the temporal evolution of the occupation probabilities p_i of the global *MCA* states i can be derived as

$$\frac{d}{dt}p_i = \frac{d}{dt}p(l_1, \ldots, l_j, \ldots, l_N) = - \sum_{j=1}^{N} W^{intra}_{l_j} p(l_1, \ldots, l_N)$$

$$- \sum_{j=1}^{N-1} W^{inter}_{l_j l_{j+1}} p(l_1, \ldots, l_j, \ldots, l_N) + \sum_{j=1}^{N} \sum_{k_j}^{L_j} W^{intra}_{l_j k_j} p(\ldots, k_j, \ldots) \tag{20}$$

$$+ \sum_{j=1}^{N-1} \sum_{k_j}^{L_j} \sum_{k_{j+1}}^{L_{j+1}} W^{inter}_{l_j l_{j+1} k_j k_{j+1}} p(\ldots, k_j, k_{j+1}, \ldots)$$

This set of equations can be solved, e.g., by using the Runge–Kutta algorithm, for a small number of cells and for typical rate constants W.

The relevant parameters for the information processing function of the *MCA* are the local occupation probabilities p^i_{lj} rather than the global occupation probabilities. They must be determined according to the relation

$$p^i_{lj} = \sum_{i \neq j} \sum_{k_j \neq ij} p(\ldots, k_i, \ldots, l_j, \ldots) \tag{21}$$

It should be evident that the straightforward solution of the rate equations by the Runge–Kutta algorithm is a formidable task. The memory requirement alone places stringent limits on the maximum size of the *MCA* that can be handled with such a procedure. The occupation probabilities alone for a chain consisting of 16 cells with 6 different states each requires about 11 terabyte of memory if single precision is used. In order to store the transfer matrix in the current example, in addition $4n^2$ byte, which amounts to 3×10^{25} byte, are needed. As a way out of this dilemma we proceed in the following by using a Monte Carlo type algorithm (40).

2. Monte Carlo Simulation of the MCA

Let us consider m ($m = 1, \ldots, M$) equivalent chains that will be treated in a stochastic manner. In every chain, stochastic jumps (Markoff process) occur between the different cell states $(\varphi^j_1, \ldots, \varphi^j_j, \ldots, \varphi^j_N)$, dictated by the occupation and transition probabilities. The current state at time t of cell j in the mth chain is characterized by the random variable $\hat{f}_m$ (t), which equals the current cell state l_j. These functions describe the random walk through all accessible local cell states. The Monte Carlo algorithm provides those functions for the ensemble of m equivalent cells. Projecting the random function onto the requested cell state l_j by

$$p^j l_j (t) = \lim_{M \to \infty} \frac{1}{M} \sum_{m=1}^{M} \delta(l_j, \hat{f}_m (t)) \qquad (22)$$

where $\delta(l_j, \hat{f}_m, (t)$ is the Kronecker symbol, and summing over all M chains, leads to the time evolution of the occupation probability of cell j in state l_j. The sum projects out only those cells that are in state l_j at time t. For large M this corresponds to the cellular occupation probabilities defined in Eq. (21).

The Monte Carlo procedure used here can be summarized as follows:

1. Determine the transition time T when a change in the chain state occurs.
2. Determination of the new state out of all possible states:
   a. Chain ensemble: select chain m^* in which the change should occur.
   b. Selected chain m^*: determine cell j in which an intracellular transition occurs or the pair of cells $j, j + 1$ that take part in an intercellular transfer.
   c. Selected cell j: determine the relevant intracellular state l_j. *This step is not required for intercellular transitions.*
3. Steps 1 and 2 determine the functions $f_m(t)$ at time $t + T$. The whole procedure is repeated all over again.

By repeating steps 1 to 3, the time evolution of $f_m(t)$ for all the cells j of all chains m is determined. The memory requirements are drastically reduced. Details of the Monte Carlo algorithm applied here are discussed in (41).

C. Molecular Computing Functions

The most obvious application of a molecular chain is that of a shift register. Its function is to shift a logical 1 from one cell to the next at each clock pulse. The logical 1 can be represented by electrons or holes, respectively. We shall discuss here only the electron shift register, although the hole version can be implemented in a similar way.

The molecular chain is composed of molecular cells that represent one bit of information. Input and output of the molecular chain are connected to electrodes whose potential can be varied. An incoherent light pulse that affects the whole chain serves as a clock pulse. In the ideal realization of a molecular shift register, a single electron would be shifted at each clock pulse along the chain. Even in an ensemble of chains a useful function might be obtained by shifting many electrons at each clock pulse. In the following we want to discuss how one can achieve this with molecular chains.

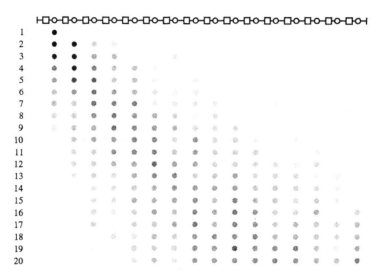

Figure 21 Monte Carlo simulation of the time evolution of a molecular shift register based on **DBA** segments. Squares represent donor and circles acceptor molecules. The charged state is represented by filling the corresponding symbol with a gray scale, where black corresponds to maximum charge. All donor molecules stay in the ground state all the time, and the first acceptor molecule is fully charged during the initialization process. Time evolution proceeds downwards in steps after every clock pulse (see text). The charge distribution is shown before each clock pulse $n = 1, 2, \ldots$

Since we are dealing with an ensemble of molecular chains, the logical state of a cell can be represented by a number Z of electrons. The logical 0 may be represented by a maximum of Z_0 electrons, whereas the logical 1 corresponds to at least Z_1 electrons within an ensemble of cells. If the number of electrons within a cell ensemble is Z with $Z_0 < Z < Z_1$, its state is undetermined. The different rates in combination with the duration of the light pulse must be chosen appropriately in order to secure a unique logical state (0, 1) of every cell. This would imply that electron transfer must only occur from a $D^0 B^0 A^-$ cell state to the neighboring cell, whereas the electron transfer from a $D^+ B_0 A^-$ and thermally activated hopping processes from one cell to the next cell are prohibited. In reality this situation is difficult to realize, and we shall discuss a few implications of those processes. Moreover we present a modification of the chain structure that remedies part of these problems.

In the following we assume a *DBA* type *MCA* as represented in Figs. 16 and 17. The molecular chain is symbolically represented by a chain of squares (D) and circles (A) as shown in Fig. 21 top. The bridge molecule (B) is omitted for simplicity. An empty symbol represents the ground state, whereas the number of charges Z in a cell is represented by shading with a gray scale in which "black" corresponds to the maximum number of charges, i.e., logical I. In the situation shown in Fig. 21. All D^+ states are empty. The temporal development of the population of the A^- states (circles) is shown in Fig. 21, where time increases downwards in steps of $n\tau$, where $n = 1, 2$, as given by the light (clock) pulses. Each line of circles represents the population of the A^- states prior to the light pulse. At $t = 0$ (top line) a logical 1 is set in the first cell by a concurrent light and electrode pulse. The intention is to shift this bit to the right after every light pulse. Applying the internal rate equations to this situation leads, however, to a diffusion of charges as is shown in Fig. 21. This is caused by the intercellular transfer rates, which cannot be made arbitrarily small.

A reduced set of differential equations for the population of the A^- states can be derived under the assumption of a highly efficient charge separation within each cell, leading to

$$\dot{p}^1_{A-}(t) = -Rp^1_{A-}(t) \tag{23a}$$

$$\dot{p}^j_{A-}(t) = -Rp^j_{A-}(t) + Rp^{j-1}_{A-}(t) \qquad \text{with } j = 2, \ldots, N-1 \tag{23b}$$

$$\dot{p}^N_{A-}(t) = +Rp^{N-1}_{A-}(t) \tag{23c}$$

Here we assume that no charge can leave the output electrode. The rates R represent only the effective rates, since not all neighboring cell states required for charge transfer are equally populated. Eqs. (23) are valid only during the light pulse, i.e., $0 < t < \tau$ where τ is the length of the light pulse. After n light pulses, the solution of the rate equations leads to

$$p^j_{A-}(n\tau) = \frac{1}{(j-1)!} (n\tau R)^{j-1} \exp\{-nR\,\tau\} \tag{24a}$$

$$p^N_{A-}(n\tau) = 1 - \sum_{j=1}^{N} \frac{1}{(j-1)!} (n\tau R)^{j-1} \exp\{-nR\tau\} \tag{24b}$$

This is a Poisson distribution of charges that corresponds to a shift of the center of gravity and leads to an increase in dispersion with an increasing number of pulses (see Fig. 21). This behavior is not appropriate for a discrete molecular shift register but may be applicable to more continuous logical states as in fuzzy Logic.

1. Alternating Bridge (B-B')$_n$ Molecular Chains

The dispersion of charges in the ordinary *DBA-MCA* may be circumvented if a slight modification is introduced, namely the alternating variation of the bridge molecule (B). In the following we assume that all *odd*-numbered cells contain a certain bridge molecule (B), whereas all even-numbered molecules contain a different bridge molecule (B') that has distinctly different absorption bands com-

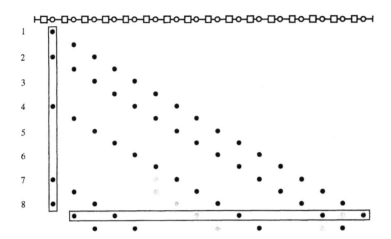

Figure 22 Monte Carlo simulation of the time evolution of an 8 bit molecular shift register based on double cell **DBA-DB′A** segments (see text). Symbols and representations are similar as in Fig. 21. Note that two different clock pulses at different transitions (B, B′) are required to shift the charge by one double cell. The numbering corresponds to clock pulse pairs. The stored bit sequence is 11010011. The vertical rectangle represents the input, whereas the horizontal rectangle corresponds to the stored bit pattern.

pared with molecule (B). Two different light sources are therefore required that emit light pulses at different frequencies. This allows us to excite the *even*-numbered cells separately from the *odd*-numbered cells. Even and odd numbered cells must be excited alternately in order to fulfill the requirement

$$
p^j_{D^+D^0_{A^-}} = \begin{cases} \simeq 0 \text{ if } p^{j+1}_{D^+_B{}^0_{A^-}} > 0 \\ > 0 \text{ if } p^{j+1}_{D^+_B{}^0_{A^-}} \simeq 0 \end{cases} \tag{25}
$$

necessary for stabilizing the charge within a double cell (DBA-$DB'A$). Furthermore, the pulse timing must be chosen in such a way that every even- or odd-numbered cell completes a full cycle consisting of excitation, separation, and recombination before the next light pulse occurs. Fig. 22 displays the electron transfer processes that can be achieved in such an alternating (B-B′) molecular chain (called $BB'MCA$).

Every odd-numbered pair of D (square)-A (circles) segments in Fig.22 represents a DBA molecular segment, whereas in every even-numbered pair B is replaced by B. First a logical 1 is written into the first cell (from the left) by applying a potential pulse to the left electrode during the B light pulse. The next B′ light pulse then stores the electron in the second cell of the first DBA-$DB'A$

molecular segment. During this process the *DBA* cell is inactive. The next pair of B, B' light pulses shifts the electron to the next *DBA-DB'*A molecular segment. The odd numbered *DBA* cells work as an intermediate storage for electrons, which are then finally stored in the *DB'A* cell during the B' light pulse while the *DBA* cell is inactive. A clock pulse cycle consists of two alternating B and B' light pulses that shift the charge by one double cell.

Fig. 22 shows a scenario of such a shift register consisting of 16 double cells, i.e., 8 bits of information storage; n corresponds to the number of clock cycles (two light pulses). In Fig. 22 the bit sequence (11010011) is read into the shift register from the left in every clock cycle. In the last step the first bit (first in first out) is read out at the right electrode by applying an electrode potential pulse concurrently with the last B' light pulse.

In order to achieve such a reliable shift of charges with a high transfer rate, certain requirements must be fulfilled. Let us define the delay time between two consecutive light pulses of the same sort (B or B' respectively) as τ_3. All relaxation and transfer processes within a cell must die out during this time, i.e., every $D^0B^0A^0$ cell goes through the excitation cycle and returns to the ground state before the next light pulse occurs. The duration of the light pulse τ_l must be shorter than that of τ_3. In the examples discussed here we set $\tau_l = 20$ ns and $\tau_3 = 200$ ns. The sequence for the B' light pulse is shifted by $\tau_3/2 = 100$ ns with respect to the B light pulse cycle. In order to achieve stable charge transfer we demand in addition that every $D^+B^0A^-$ cell returns to the ground state during a time τ_2 defined by $\tau_l + \tau_2 = \tau_3/2$, which assures that the *DBA* cell is in a definite state when the B' light pulse appears. If no charge is introduced into the chain, all cells complete their internal cycle before the next light pulse appears. The charge separated state decays approximately as

$$p^j_{D^+B^0A^-}(t) = \exp(-k_R^{\text{eff}} t) \tag{26}$$

where k_R^{eff} is the effective recombination rate, which must fulfill the condition $k^{\text{eff}}_R \tau_2 \gg 1$. The pulse width τ_l must be long enough in order to achieve a well-defined charge separated state. This depends on the pumping rate P as well as on the charge separation rate k_s. Also the charge transfer from cell j to cell $j + 1$ depends on the internal parameters of the *DBA* cells as well as on the pumping rate. The Monte Carlo simulations show that it can be approximated by

$$p^{j+1}_{D^0B^0A^-}(t) = 1 - \exp(-W_{eff} t) \tag{27}$$

when $l_j = D^0B^0A^-$ and $l_j + 1 = D^0B^0A^0$ at time $t = 0$. In order to achieve a definite shift of the charges from one cell to the next the light pulse should fulfill the condition $W_{eff} \tau_l \gg 1$. This effective transfer rate must be distinguished from the intercell transfer rate W^{inter}, which is the bare electron transfer rate between neighboring cells. The connection between these different parameters

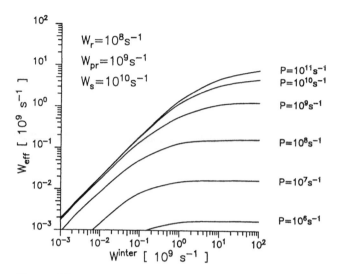

Figure 23 Parameter space of the effective intercellular transition rate W_{eff}, the molecular intercellular transition rate W^{inter}, and the pumping rate P. Fixed values for the molecular recombination rate W_r, the cellular charge separation rate W_s, and the primary recombination rate W_{pr} are given in the insert.

as a result of the Monte Carlo simulation is shown in Fig. 23. If W^{inter} exceeds 10^9 s^{-1}, the effective transfer rate W_{eff} is basically dictated by the pumping rate.

In order to achieve an efficient and rapid shift operation the following conditions should be obeyed:

$$k_s \gg P > k_r, k_l$$
$$W^{inter} \approx k_s$$

2. Thermally Activated Hopping Processes

Up to now we have neglected random hopping processes, which can occur at elevated temperatures. The question arises whether such a molecular shift register can still operate in a useful manner if random hopping processes are allowed. It is of course to be expected that random hopping processes always lead to a loss of information. A dispersion of charges over the neighboring cells will occur. The degree of dispersion and the corresponding loss of information depends on the different rate constants of the cells, the duration of the light pulse, and the repetition rate of the sequence. Table 1 summarizes the different rates used in the following simulations.

The intercellular rate was set to $W^{inter} = 10^9$ s^{-1}, which is slightly smaller than the intracellular transfer rates. This is accomplished by using slightly longer

Table 1   Intracellular Rates

Initial state	Final state	Rate in 10^9 s^{-1}	Type of process
$D^0B^0A^0$	D^0B*A^0	1.0	absorption
D^0B*A^0	$D^0B^0A^0$	1.0	fluorescence
D^0B*A^0	$D^0B^+A^-$	10.0	charge separation
D^0B*A^0	$D^+D^-A^0$	10.0	charge separation
$D^0B^+A^-$	$D^+B^0A^-$	10.0	charge separation
$D^+B^-A^0$	$D^+B^0A^-$	10.0	charge separation
$D^0B^+A^-$	$D^0B^0A^0$	1.0	charge recombination
$D^+B^-A^0$	$D^0B^0A^0$	1.0	charge recombination
$D^+B^0A^-$	$D^0B^0A^0$	0.1	charge recombination

spacers between the cells than within the cells. The light pulse duration was chosen to be 20 ns, whereas the repetition time was 200 ns. All simulations discussed here were performed by using four double cells and 1000 chains for a single shift register. Results of such a simulation are shown in Fig. 24. In order to demonstrate the effect of hopping processes we have chosen large hopping rates. Already the initial write process leads after several light pulses to a delocalization of the charge (hatched area in Fig. 24). Immediately after the inital light pulse the charge is initially fairly localized in the first cell. However, already during the first cycle a diffusion of charge to the other cells in the chain occurs.

In Fig. 24 we compare the population of the first and the last cell for every clock pulse, beginning with the insertion of an electron in cell 1 during the first clock pulse(B excitation). The hatched area corresponds to the initial charge, which is stored during the initialization process in cell 1. The other populations (solid line) correspond to the local electron transfer cycle of the *DBA* and *DB'* A cells that does not lead to intercellular electron transfer. The initial decay of population in cell 1 is caused by hopping processes and is nearly exponential with an assumed hopping rate of $W_h = 2 \times 10^7$ s^{-1}. The second clock pulse (*B'* excitation), which appears at 100 ns, transfers almost all electrons to the *B'* cell within cell 1. This process is very efficient (see minimum at 100 ns in cell 1 in Fig. 24) because of the 50 times larger intercellular transfer rate compared with the hopping rate. Back transfer due to hopping then leads to an increase of charge population in cell 1 immediately after the second pulse. After the third pulse there is still some charge left in cell 1, which is shifted to cell 2 during the fourth pulse at 300 ns. The process continues and finally leads to a rise of charge population in cell 8 as seen in Fig. 24 (bottom). This charge can be extracted at the output electrode. As long as the hopping rate can be kept low, the dispersion of charge may be kept at a tolerable level. The logical 1 could still

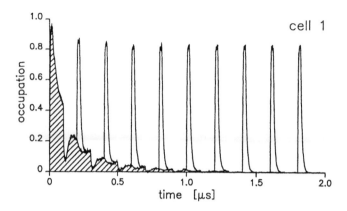

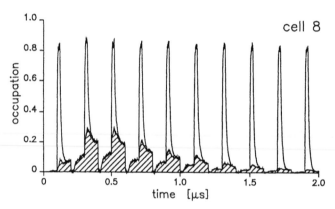

Figure 24 Monte Carlo simulation of the occupation probability of the charged cell state A^- for a **DBA-DB′A** double cell (8 cells total) shift register. Only the charged states of cell 1 (top) and cell 8 (bottom) are shown. Alternating B- and B′-type light pulses are applied. The whole cycle repeats every 200 ns. The solid curves represent the cell internal clock cycle, whereas the hatched area corresponds to the additional charge stored into the first cell at time $t = 0$. The dispersion of charge due to random hopping processes is visible in the evolution of the hatched area (see text).

be recovered in this case by imposing a certain threshold at the output electrode. Further discussion of these and associated problems will be given elsewhere.

VI. SUMMARY

We have discussed the electronic and information processing properties of molecular segments and chains consisting of mixed aromatic (unsaturated) and aliphatic (saturated) molecular units. The basic unit consists of a donor (D), a bridge (B), and an acceptor (A), labelled the DBA unit for simplicity. They all consist of aromatic molecules and are separated by aliphatic spacer molecules, which should have a stiff skeleton in order to prevent drastic conformational changes. A DBA unit could be constructed as a Langmuir–Blodgett film or as a thin film with heteromolecular layers. Essential prerequisites for the functionality of these systems are (1) appropriate selection of redox states for the different donor and acceptor units, (2) the correct selection of the excitation transition of the bridge molecule (B), and (3) the proper design (length and type) of the spacer molecules. By making the appropriate choice, the intra- and intercellular transfer rates can be tailored to the needs of achieving molecular information processing functions. These have been discussed in terms of molecular celluar automata (MCA) and molecular shift registers.

ACKNOWLEDGMENT

We acknowledge financial support by the Deutsche Forschungsgemeinschaft (SFB 329) and the Fonds der Chemischen Industrie.

REFERENCES

1. Aviram, A., and Ratner, M. A., *Chem. Phys. Lett.*, *29*, 277 (1974).
2. Carter, F. L., (a) NRL Memorandum Report, 3960, 121 (1974); (b) in *Proceedings of the Molecular Electronic Device Workshop*, NRL Memorandum Report, 4662 (1981).
3. Haarer, D., *Angewandte Makromolekulare Chemie*, *100/109*, 267 (1982).
4. Hopfield, J. J., *Natl. Acad. Sci. USA*, *71*, 3640 (1974).
5. Hopfield, J. J., Onuchic, J. N., and Beratan, D. N., *Science*, *241*, 817 (1988).
6. Hong, F. T., *Molecular Electronics*, Plenum Press, New York, 1989.
7. Mehring M., (a) in Electronic properties of conjugated polymers III (H. Kuzmany, M. Mehring, and S. Roth, eds.), Springer Series in Solid State Sciences, 91, 242 (1989); (b) *Int. J. Electronics*, *73*, 1073 (1992).
8. Hückel, E., *Z. Phys.*, *76*, 628 (1932).
9. Pople, J. A., and Walmsley, S. H., *Mol. Phys.*, *5*, 15 (1962).
10. Salem, L., *Molecular Orbital Theory of Conjugated Systems*, Benjamin, London, 1966.
11. Su, W. P., Schrieffer, J. R., and Heeger, A. J., *Phys. Rev. Lett.*, *42*, 1698 (1979).
12. Heeger, A. J., Kivelsan S., Schrieffer, J. R., and Su, W. P., *Rev. Mod. Phys.*, *60*, 781 (1988).
13. Bredas, J. L., and Street, G. B., *Acc. Chem. Res.*, *18*, 304 (1985).
14. Skotheim, T., *Handbook on Conducting Polymers*, Marcel Dekker, New York, 1986.
15. Andre, J. M., Delhalle, J., and Bredas, J. L., *Quantum Chemistry Aided Design of Organic Polymers*, World Scientific, Singapore, 1991.

16. Kiess, H., *Conjugated Conducting Polymers*, Springer Series in Solid State Sciences, 102 (1992).

17. Baeriswil, D., Campbell, D. K., and Mazundar, S., *Conjugated Conducting Polymers*, Springer Series in Solid State Sciences, 102, 7 (1992).

18. Kahol, P. K., Clark, W. G., and Mehring, M., in *Conjugated Conducting Polymers*, (H. Kiess, ed.), Springer Series in Solid State Sciences, 102, 217 (1992).

19. Rehwald, W., and Kiess, H. G., *Conjugated Conducting Polymers*, Springer Series in Solid State Sciences, 102: 135 (1992).

20. Kiess, H. G., and Harbeke, G., *Conjugated Conducting Polymers*, *Springer Series in Solid State Sciences*, 102, 175 (1992).

21. Nechtstein, M., Devreux, F., Genoud, F., Veil, E., Pernant, J. M., and Genies, E., *Synth. Metals*, *15*, 59 (1986).

22. Weller, A., *Z. physik. Chemie, N. F.*, *18*, 163 (1958) and in *Progress in Reaction Kinetics* (G. Porter, ed.), Pergamon Press, London, vol. I, 187 (1961).

23. (a) Leonhardt, H., and Weller, A., *Ber. Bunsenges. physik. Chemie*, *67*, 791 (1963); (b) Rehm, D., and Weller, A., *Ber. Bunsenges. physik. Chemie*, *72*, 257 (1968); *Ber. Bunsenges. physik. Chemie*, *73*, 834 (1969).

24. Moore, A. L., Dirks, G., Gust, D., and Moore, T. A., *Photochem. Photobiolo.*, *32*, 691 (1980).

25. Gust, D., and Moore, T. A., *J. Photochemistry*, *29*, 173 (1985).

26. Gust, D., Moore, T. A., Liddell, P. A., Nemeth, G. A., Makings, L. R., Moore, A. L., Barrett, D., Pessiki, P. J., Bensasson, R. V., Rougee, M., Chathaty, F. C., De Schryver, F. C., van der Anweraer, M., Holzwarth, A. R., and Conolly, J. S., *J. Am. Chem. Soc.*, *109*, 846 (1987).

27. (a) Wasielewski, M. R., and Niemczyk, M. P., *J. Am. Chem. Soc.*, *106*, 5043 (1984), (b) *M* Wasielewski, M. R., Niemczyk, M. P., Svec, W. A., and Pewitt, E. B., *J. Am. Chem. Soc.*, *107*, 1080 and 5562 (1985).

28. Closs, G. L., Calcatara, L., Green, N., Penfold, K., and Miller, J., *J. Phys. Chem.*, *90*, 3673 (1986).

29. (a) Heitele, H., and Michel-Beyerle, M. E., *J. Am. Chem. Soc.*, *107*, 8068 (1985); (b) Heitele, H., Michel-Beyerle, M. E., and Finckh, P., *Chem. Phys. Lett.*, *134*, 273 (1987); (c) Finckh, P., Heitele, H., and Michel-Beyerle, M. E., *Chem. Phys.*, *138*, 1 (1989); (d) Heitele, H., Pöllinger, F., Weeven, S., and Michel-Beyerle, M. E., *Chem. Phys.*, *143*, 325 (1990).

30. Harrison, R. J., Pearce, B., Beddard, G. S., Cowan, J. A., and Sanders, J. K. M., *Chem. Phys. Lett.*, *116*, 429 (1987).

31. Marcus, R. A., (a) *J. Chem. Phys.*, *24*, 966 (1956); (b) *J. Chem. Phys.*, *43*, 679 (1965); (c) in *Chemische Elementarprozesse* (H. Hartmann, ed.)., Springer-Verlag, Berlin, 1968, p. 348.

32. Jortner, J., *J. Chem. Phys.*, *64*, 4860 (1976).

33. DeVault, D., *Quantum-Mechanical Tunneling in Biological Systems*, Cambridge Univ. Press. Cambridge, 1984.

34. (a) Beratan, D. N., Onuchic, J. N., and Hopfield, J. J., *J. Chem. Phys.*, *86*, 4488 (1987); (b) Onuchic, J. N., and Beratan, D. N., *J. Chem. Phys.*, *92*, 722 (1990).

35. Cukier, R. I., and Nocera, D. G., *J. Chem. Phys.*, *97*, 7371 (1992).

36. Wang, Z., Tang, J., and Norris, J. R., *J. Chem. Phys.*, 97, 7251 (1992).
38. Blum, K., *Density Matrix Theory and Applications*, Plenum Press, New York, 1981.
39. Wolfram, S., *Rev. Mod. Phys.*, 55, 601 (1983).
40. Binder, K., *Monte Carlo Methods in Stochastical Physics*, Springer-Verlag, Berlin, 1979.
41. Knoll, H. P., Diploma thesis, Stuttgart, 1993.

10

Cooperative Optical Properties of Interacting Charge Transfer Subunits

H. Körner and Günter Mahler

University of Stuttgart
Stuttgart, Germany

I. INTRODUCTION

The direct and detailed observation of quantum dynamics became possible with the preparation of single quantum objects. Examples are single ions in an electromagnetic Paul trap (1), single dopant molecules in a host crystal (2,3), or single defects in small semiconductor devices (4,5). As a result, it was demonstrated that the dynamics of single *open* (i.e., damped) quantum systems is inherently stochastic. The environment causes the system to jump between its discrete quantum levels, which becomes apparent in random telegraph signals of observable quantities, e.g., fluorescence intensity or electric current (1,4).

In further experiments, different aspects of this stochastic dynamics were investigated. Transition rates can be controlled by external fields (6). Interactions between individual quantum objects have been demonstrated in doped crystals (7) or, even more pertinent, between individual defects in small electronic devices, where complex random telegraph signals emerge (8–10).

In this connection, synthetic quantum systems are of particular interest. Here, parameters as transition energies or relaxation rates could be adjusted to optimize the low-level quantum dynamics in certain respects, and different aspects that have been observed in different systems may be combined to give novel phenomena. In a network of interacting quantum objects, complex cooperative dynamics between the simple dynamics of an isolated object and the usual ensemble behavior might emerge. Such networks are appealing systems for an application in molecular electronics. The challenge is to make the complex dynamics represent computer functions.

The network structure is widely discussed for computer architectures beyond the conventional von Neumann type. However, conceptions like neural nets (11), parallel algorithms (12), or deterministic cellular automata (13) only define abstract rules: the implementation on the molecular level is an unsolved problem (14). On the other hand, nanoscopic networks that have been realized so far, e.g., with quantum point contacts (15) or in doped zeolith crystals (16), do not seem to support any computer function. We propose an optically driven quantum network as a prototype system that strongly connects computer function and physical implementation. It seems that a central paradigm of computational science, the independence of computer conceptions from their realization, will be violated in molecular electronics.

In Sec. II, we propose the realization of a molecular network by an array of charge transfer subunits. The dipole–dipole interaction between the subunits should lead to a complex dynamical behavior. In Secs. III.A and III.B, we discuss the quantum dynamics of a single and two coupled subunits and clarify the mechanism that is responsible for the complex dynamics in a large network. In section III.C, we describe in so-called ''pump and probe'' scenarios the optical properties of a two-dimensional network that emerge as a consequence of its underlying dynamics. Qualitative results for the cooperative network behavior are supplied by the mean field treatment in Sec. III.D. In Sec. IV, possible applications in molecular electronics are demonstrated. The network can act as a highly parallel Monte Carlo simulator, as demonstrated for the Ising model in Sec. IV.A. In a second application (Sec. IV.B), we show how simple image processing tasks can be programmed and directly performed by pump and probe experiments. Finally, in Sec. IV.C, the network works as a programmable chemical sensor that recognizes and classifies molecules.

II. CHARGE TRANSFER NETWORKS

We can think of two promising possibilities to realize an optically controllable network of interacting charge transfer subunits. In Sec. II.A we propose an array of charge transfer quantum dots as a prototype system within the semiconductor nanotechnology. An alternative realization with modern methods in organic chemistry is discussed in Sec. II.B. We show that the subunits in both systems can be described by minimal three-level models with the same generic structure, i.e., with the same localization behavior of their states and corresponding hierarchy of transition rates. The model calculations in the later sections thus are independent of the specific realization.

A. Charge Transfer Quantum Dot Array

Semiconductor nanostructuring allows us to built "synthetic molecules," i.e., to design the spectrum and, in particular, the transition rates of a few-electron system. In the $Ga_{1-x}Al_xAs$ technology this can be reached by spatially varying the ratio of Ga and Al in this mixed semiconductor, which leads to a spatial variation of the energy gap for the electrons: the electrostatic potential "seen" by the electrons, in principle, could be shaped in three dimensions (17).

A charge transfer quantum dot, as shown in (18), is built up by three different "disks" of $Ga_{1-x}Al_xAs$ imbedded in a spacer region and confines electrons in all three dimensions. Quantum dots with such complex internal structure indeed have not yet been realized. However, two-dimensional quantum well structures of that type (i.e., no lateral structuring) (19) or simpler zero-dimensional quantum dots (only lateral structuring, simple structure in the axial direction) (20,17) are already under experimental control.

In the axial direction of our charge transfer quantum dot the valence and conducting band both form asymmetric double well potentials (18). The charge transfer dynamics will be described by three relevant states $s = 0, \pm 1$. The highest valence band states -1 and 1, with energies E_{-1} and E_1 ($E_{-1} < E_1$), are localized within the different potential wells, while the lowest state 0 (energy E_0) in the conducting band is delocalized over the whole structure. The p-doped region serves as a source for holes and pins the Fermi level between the states -1 and 1 so that in the ground state all levels in the valence band are occupied up to level -1, and level 1 is free. Charge transfer from state -1 to state 1 is easily induced via optical excitation into the transient state 0 and subsequent spontaneous decay (21). The specific localization behavior of the states stabilizes the charge transfer. This is expressed by the following hierarchy in the spontaneous transition rates $W_{s,s'}$ for the transitions $s' \rightarrow s$ that arise from coupling the system to a (e.g., photon or phonon) heat bath (21):

$$W_{1,-1}, W_{-1,1} \ll W_{-1,0}, W_{1,0} \tag{1}$$

"Switching" back from state 1 to state -1 nevertheless can be induced, again, by optically driving the other transition $1\leftrightarrow 0$. For simplicity we assume a heat bath temperature $k_B T \ll E_0 - E_{\pm 1}$, so that for the remaining transition rates we have also

$$W_{0,-1}, W_{0,1} \ll W_{-1,0}, W_{1,0} \tag{2}$$

The different spatial localization of the three states is associated with different static dipole moments $d_{s,s} = <s \mid \hat{d} \mid s>$ in the axial direction of the dot (22):

$$d_{1,1} = -d_{-1,-1} = d, \qquad d_{0,0} = 0 \tag{3}$$

or, in short,

$$d_{s,s} = sd \tag{4}$$

The state index s thus simply denotes the dipole moment in units of d.

We now consider a square array of N identical charge transfer quantum dots, as depicted in Fig. 1. The state $\{s_n\} = \{s_1, s_2, \ldots, s_N\}$ of this network structure is determined by the states s_n of all subunits $n = 1, 2, \ldots, N$. The energy of a state $\{s_n\}$ is not simply the sum of the energies E_{s_n} of the isolated subunits: the dipole–dipole interaction between two subunits n and m leads to a contribution to the total energy of amount $(d_{sn,sn} \, d_{sm,sm})/(4 \pi \epsilon \epsilon_0 |R_n - R_m|^3)$ (for dipole moments perpendicular to the connecting line between the subunits (cf. Fig. 1)) (22). R_n is the center of subunit n. The energy of a state $\{s_n\}$ thus is (23).

$$E_{\{s_n\}} = \sum_{n=1}^{N} E_{s_n} + \frac{h}{2} \sum_{\substack{n,m=1 \\ n \neq m}}^{N} C_{nm} \, s_n s_m \tag{5}$$

where we have used Eq. (4) and the abbreviation

$$C_{nm} = \frac{d^2}{4 \pi h \, \epsilon \epsilon_0 |R_n - R_m|^3} \tag{6}$$

B. Molecular Array

Besides quasi-molecular semiconductor structuring, one might consider the chemical synthesis and (self-)assembly of functional molecules as an approach starting from the opposite direction.

Donor–acceptor modified molecular chains (*D–C–A*) are promising candi-

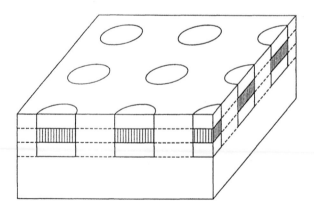

Figure 1   Charge transfer quantum dot array: prototype network of interacting quantum subunits. The single three-layered dot represents an asymmetric double well potential in the valence and conducting band so that a spectrum as in Fig. 2 results (18).

dates for an optically controllable charge transfer subunit (24). Model calculations show that the charge transfer dynamics also can be described within a three-level model that consists of the ground state $(D–C–A)$, the metastable charge transfer state $(D^+–C–A^-)$, and a transient state that mediates the charge transfer (25). These three states qualitatively have the same localization behavior as the three relevant states in the quantum dot, though now in the high-dimensional electron-conformation state space. The spontaneous transition rates thus also meet the hierarchy Eq. (1). The LB technique is a promising approach for the assembly of the molecules into an ordered network (26).

III. NETWORK DYNAMICS

A. Single Subunit: Effective Charge Transfer Dynamics

Both in molecular and semiconductor realizations the charge transfer subunit is described by a three-level system with specific localization behavior as depicted in Fig. 2. We now consider the coupling to a pump light field that controls the charge transfer dynamics and a heat bath that allows for dissipation. For incoherent light field, the dynamics is described by a rate equation for the occupation probabilities $\rho(s; t)$ of the three states $s = 0, \pm 1$ (23):

$$\frac{d}{dt} \rho(0; t) = B_{-1} U_p(\omega_{-1}) \rho(-1; t) + B_1 U_p(\omega_1) \rho(1; t) \tag{7}$$
$$- [W_{-1} + B_{-1}U_p(\omega_{-1}) + W_1 + B_1 U_p(\omega_1)] \rho(0; t)$$

$$\frac{d}{dt} \rho(-1; t) = [W_{-1} + B_{-1} U_p(\omega_{-1})] \rho(0; t) - B_{-1} U_p(\omega_{-1}) \rho(-1; t) \tag{8}$$

$$\frac{d}{dt} \rho(1; t) = [W_1 + B_1 U_p(\omega_1)] \rho(0; t) - B_1 U_p(\omega_1)\rho(1; t) \tag{9}$$

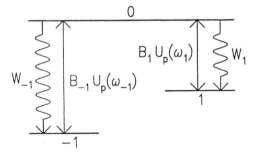

Figure 2   Driven three-level system as a minimal model for controllable charge transfer. Transitions are induced by heat bath coupling (rates $W_{\pm 1}$) and pump light field (rates $B_{\pm 1}U_p(\omega_{\pm 1})$).

$B_s U_p (\omega_s)$ ($s = \pm 1$) is the light field induced transition rate $s \leftrightarrow 0$, where B_s is the Einstein coefficient and $U_p (\omega_s)$ the spectral energy density of the pump light field at the transition energy

$$\hbar \omega_s = E_0 - E_s \qquad s = \leftrightarrow 1 \tag{10}$$

W_s is the abbreviation for the spontaneous transition rate $W_{s,0}$ from state 0 to state $s = \pm 1$. Due to Eqs. (1) and (2), only those spontaneous transitions have to be considered. Both the spontaneous and light field induced direct charge transfer $-1 \leftrightarrow 1$ is neglected because of the negligible overlap between the two states. Only transfer via the transient state 0 is possible.

To restrict to the essentials of the charge transfer dynamics, the rate equations (7)–(9) can be simplified further if we assume a weak pump light field

$$B_s U_p(\omega_s) << W_s \tag{11}$$

This hierarchy implies an immediate decay from the transient state 0 after excitation; its occupation can be neglected. Mathematically, this can be treated by an adiabatic elimination of the "fast" variable $\rho(0;t)$. Inserting the adiabatic solution of Eq. (7) (obtained from $d/dt\rho(0;t) = 0$) into Eqs. (8) and (9) leads to the effective two-level rate equations

$$\frac{d}{dt} \rho(s; t) = R^{\text{eff}}_{-s} (\rho; t) - R^{\text{eff}}_s \rho(s;t) \qquad s = \pm 1 \tag{12}$$

Here and in the following, the variable s is assumed to take only the values $s = +1$. The rate for the effective transition $s \rightarrow -s$ is (27)

$$R^{\text{eff}}_s = \frac{W_{-s}}{W_1 + W_{-1}} B_s U_p (\omega_s) \tag{13}$$

It describes a combined transition $s \rightarrow 0 \rightarrow -s$ with the induced transition rate $B_s U_p (\omega_s)$ for the transition $s \rightarrow 0$ and a subsequent spontaneous decay $0 \rightarrow -s$ with the branching ratio $W_{-s}/(W_1 + W_{-1})$. The two rates for both directions $s \leftrightarrow -s$ of the effective transition can be adjusted independently by the spectral energy density at the two transition frequencies $\omega_{\pm 1}$. The three-level model is the minimal model for controllable charge transfer. In a two-level system, where only one transition frequency exists, this independent control is not possible.

For the stationary expectation value of the dipole moment $\langle d \rangle = d \langle s \rangle$ we get from Eqs. (12) and (13)

$$\langle d \rangle = d \frac{R^{\text{eff}}_{-1} - R^{\text{eff}}_1}{R^{\text{eff}}_{-1} + R^{\text{eff}}_1} = d \frac{W_1 B_{-1} U_p(\omega_{-1}) - W_{-1} B_1 U_p(\omega_1)}{W_1 B_{-1} U_p(\omega_{-1}) + W_{-1} B_1 U_p(\varphi_1)} \tag{14}$$

The two light modes $U_p (\omega_{-1})$ and $U_p (\omega_1)$ compete with respect to the charge transfer: the first one tries to "switch" the dipole moment into direction $+d$, the second one into the reverse direction $-d$ (21). This is reminiscent of the excitatory and inhibitory forces typical for biological system control.

The rate equations describe a mathematical ensemble, i.e., an infinite ensemble of identical copies, and not a single system. $\rho(s;t)$ is the portion of ensemble members that are in state s at time t. As observed in various experiments in different fields (1,4), the dynamics of a single open quantum system cannot be described by the continuous and deterministic time development of the rate equations. On the contrary, it is a discrete stochastic process where the system jumps between its quantum levels. However, this stochastic dynamics can be derived from the ensemble rate equations if one postulates that the stochastic process defined by the transition rates is the real dynamics of the single system (28).

Figure 3 shows a computer simulation of the stochastic two-level dynamics of the single subunit for the effective transition rates of Eq. (13). Such two-level random telegraph signals are the prototype of a stochastic quantum dynamics and observed in many experiments (1,4).

Ergodic theory connects time averages over the stochastic signal $s(t)$ with corresponding ensemble averages. Stationary occupation probabilities $\rho(s)$, e.g., can be recovered from the expression (29)

$$\rho(s) = \lim_{\Delta t \to \infty} \frac{1}{\Delta t} \int_0^{\Delta t} dt' \, \delta_{s,s(t')} \tag{15}$$

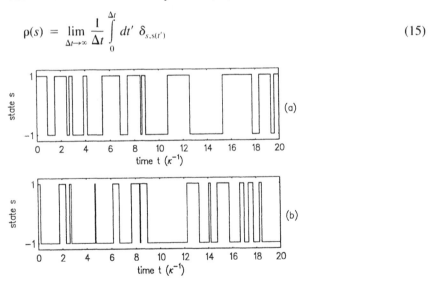

Figure 3 Effective stochastic two-level dynamics of the isolated charge transfer subunit. (a) Symmetric random telegraph signal $R_{-1}^{\mathrm{eff}} = R_1^{\mathrm{eff}} = \kappa$; (b) asymmetric random telegraph signal $R_{-1}^{\mathrm{eff}} = R_1^{\mathrm{eff}}/2 = \kappa$.

B. Two Interacting Subunits: Cooperative Random Telegraph Signals

We now examine the stochastic dynamics of two interacting subunits. As a consequence of the dipole–dipole interaction, the transition energies in one subsystem depend on the state of the other. $\hbar\omega_{s_1}$ (s_2) is the transition energy between state 0 and s_1 of subunit 1 if subunit 2 is in state s_2. From Eqs. (5) and (10) we obtain

$$\hbar\omega_{s_1}(s_2) = E_{\{0,s_2\}} - E_{\{s_1,s_2\}} = \hbar\omega_{s_1} - \hbar C_{12} \cdot s_1 s_2 \qquad s_1, s_2 = \pm 1 \qquad (16)$$

as shown in Fig. 4. The two subunits are assumed to be identical, so that interchanging the indices 1 and 2 gives the same transition energies for subunit 2. Via the neighbor dependence of the transition energies, the transition rates of Eq. (13) also become dependent on the state of the respective neighbor (27):

$$R_{s_1}^{\text{eff}}(s_2) = \frac{W_{-s_1}}{W_1 + W_{-1}} B_{s_1} U_p(\omega_{s_1}(s_2)) \qquad (17)$$

If we assume that the four transition energies (16) are all different, the four transition rates of Eq. (17) can be adjusted independently by the spectral energy density function $U_p(\omega)$. The stochastic signal $\{s_1(t), s_2(t)\}$ defined by the rates of Eq. (17) is a "random walk" in the four-dimensional state space $\{s_1, s_2\}$. For our computer simulations, we restrict ourselves to the choice

$$R_1^{\text{eff}}(-1) = R_{-1}^{\text{eff}}(1) = \kappa_1 \qquad (18)$$

$$R_1^{\text{eff}}(1) = R_{-1}^{\text{eff}}(-1) = \kappa_2$$

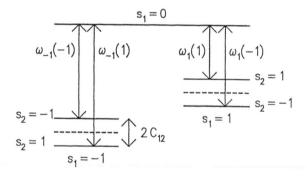

Figure 4   Dipole–dipole renormalization of the unperturbed energy levels of subsystem 1 (dashed) due to the presence of subsystem 2. $\omega_{s_1}(s_2)$ is the transition frequency between state $s_1 = \pm 1$ and the transient state 0 when subsystem 2 is in state s_2. The reverse influence is analogous.

Figure 5 shows random telegraph signals $\{s_1(t), s_2(t)\}$ for different parameters κ_1/κ_2. For $\kappa_1 = \kappa_2$ the effective transition rates do not depend on the state of the respective neighbor, and the two subsystems jump independently. Increasing the ratio κ_1/κ_2 leads to a correlation of the jumps. For $\kappa_1 \gg \kappa_2$, the two systems tend to align their dipole moments on the fast scale $1/\kappa_1$ ("ferro-order"), which on the slow scale $1/\kappa_2$ looks like entrained jumps of the two subsystems. This behavior can be easily visualized in Fig. 6, where the thickness of the arrows denotes the magnitude of the transition rates in the four-dimensional state space. This mechanism of correlation between the subsystems is generic for the applications of our quantum network model as discussed below.

C. Two-Dimensional Network: Optical Properties

We now consider a network of N subsystems, arranged on a square lattice as in Fig. 1. The stochastic dynamics now proceeds in a 2^N dimensional state space. The expressions for the transition energies and transition rates can be deduced from the results in the last section. Equation (16) is generalized to (27)

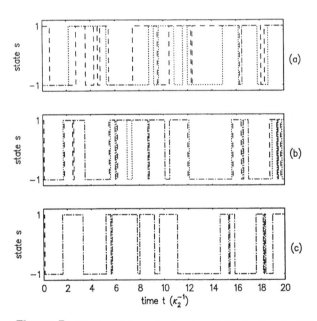

Figure 5 Random telegraph signals $s_1(t)$ (dashed) and $s_2(t)$ (dotted) in two interacting subunits. For transition rates see Eq. (18). Increasing "ferro-coupling" from (a) to (c). (a) $\kappa_1 = \kappa_2$; (b) $\kappa_1 = 10 \, \kappa_2$; (c) $\kappa_1 = 1000 \, \kappa_2$.

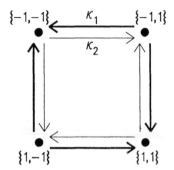

Figure 6 Visualization of the stochastic dynamics of two interacting subunits in the four-dimensional state space $\{s_1, s_2\}$. The thickness of the arrows indicates the magnitude of the transition rates. Here, "ferro-coupling" for $\kappa_1 > \kappa_2$.

$$\hbar\omega_{s_m}(\{s_n\}') = \hbar\omega_{s_m} - \hbar\sum_{\substack{n=1 \\ n \neq m}}^{N} C_{mn} s_m s_n \tag{19}$$

$\{s_n\}'$ denotes the states of all other subsystems $n \neq m$. Correspondingly, we obtain from Eq. (17) the following expression for the effective transition rate $s_m \rightarrow -s_m$ that depends on the neighborhood $\{s_n\}'$:

$$R_{s_m}^{\text{eff}}(\{s_n\}', m) = \frac{W_{-s_m}}{W_1 + W_{-1}} B_{s_m} U_p(\omega_{s_m}(\{s_n\}'), m) \tag{20}$$

The spectral energy density $U_p(\omega, m)$, now, is allowed to have a spatial dependency. Therefore, $R_{s_m}^{\text{eff}}(\{s_n\}', m)$ explicitly depends on the site m.

For a network consisting of only a few or even a single subsystem, the direct observation of the stochastic dynamics in principle should be possible, as demonstrated in the experiments referred to above. For a large network ($N \gg 1$), however, the full information $\{s_1(t), s_2(t), \ldots, s_N(t)\}$ neither will be accessible to an actual measurement nor should it be relevant for possible applications. Instead of this, we investigate in so-called "pump and probe" scenarios the information obtained by measuring the absorption of a probe field as a function of the pump field induced complex network dynamics. This information is embodied in the local absorption spectrum $\chi''(\omega, m)$. This quantity can be understood as a time average over the stochastic dynamics (23):

$$\chi''(\omega, m) = \frac{1}{\hbar\epsilon_0 v} \lim_{\Delta t \to \infty} \frac{1}{\Delta t} \int_0^{\Delta t} dt \, \frac{|d_{s_m(t)}|^2 \gamma_{s_m}(t)}{(\omega_{s_m(t)}(\{s_n(t)\}') - \omega)^2 + \gamma_{s_m(t)}^2} \tag{21}$$

v is the volume of a single subunit, d_{s_m} the transition dipole moment, and γ_{s_m} the homogeneous line width of the transition $0 \rightarrow s_m$. The integrand in Eq (21) may be interpreted as the time-dependent absorption spectrum where the central frequency of a Lorentzian evolves as a random telegraph signal. Such "spectral jumps" indeed have been observed, e.g., for single defects in a crystal (7).

D. Mean Field Approximation

The computer simulation of the stochastic dynamics defined by the transition rates of Eq. (20) is usually the only way to treat such systems. Exact analytical results can be obtained only in very limited cases (30). The simplest approximation that gives at least a glance of the qualitative network behavior is the familiar mean field approximation widely applied for different problems of statistical physics. Here, this approximation amounts to replacing the dependence of the transition rates on the actual dipole moments of the neighborhood $\{ds_n\}'$ by its mean values $\{d \langle s_n \rangle \}'$ (31):

$$R_{sm}^{\text{eff}}(\{\langle s_n \rangle\}', m) \Rightarrow R_{sm}^{\text{eff}}(\{\langle s_n \rangle\}', m) \tag{22}$$

The single subunit performs a stochastic dynamics in the "mean field" of its neighborhood. The generalization of Eq. (14) to this mean field dynamics leads to the nonlinear "self-consistent" equation

$$\langle s_m \rangle = \frac{R_{-1}^{\text{eff}}(\{\langle s_n \rangle\}', m) - R_1^{\text{eff}}(\{\langle s_n \rangle\}', m)}{F_{-1}^{\text{eff}}(\{\langle s_n \rangle, m) + R_1^{\text{eff}}(\{\langle s_n \rangle\}', m)} \tag{23}$$

We shall analyze this equation now for a special choice of the transition rates. Equation (19) defines two frequency bands centered around ω_{-1} and ω_1 with a width $2\tilde{C} = 2 \Sigma_n C_{nm}$ caused by the possible neighborhoods $\{s_n\}'$. They do not overlap if

$$2\tilde{C} < |\omega_1 - \omega_{-1}| \tag{24}$$

This inequality is fulfilled for our model system (23). For the spectral pump energy density $U_p(\omega, m)$ we take the sum of two Gaussians with (spatial varying) amplitudes $U_s(m)$, the same bandwidth Δ and detuning δ from the unperturbed transition frequencies ω_s, each confined to its respective band of transition energies:

$$U_p(\omega, m) = \sum_{s=\pm1} U_s(m) \exp\left(-\left(\frac{\omega - \omega_s - \delta}{\Delta}\right)^2 \right) \Theta(\tilde{C} - |\omega - \omega_s|) \tag{25}$$

As usual, the Θ function equals 1 for $\omega_s - \tilde{C} < \omega < \omega_s + \tilde{C}$, and 0 otherwise. With Eqs. (25) and (19), the neighborhood dependent transition rates of Eq. (20) take the form

$$R_{sm}^{\text{eff}}(\{s_n\}') = \kappa_{sm}(m) \exp\left(-\left(\frac{\delta + \sum_{\substack{n=1 \\ n \neq m}}^{N} C_{mn}s_m s_n}{\Delta}\right)^2\right)$$ (26)

with

$$\kappa_{sm}(m) = \frac{W_{-sm}}{W_{-1} + W_1} B_{sm} U_{sm}(m)$$ (27)

For the symmetric case $\kappa_{-1}(m) = \kappa_1(m)$ the mean field equation Eq. (23) for the homogeneous solution $\langle s \rangle = \langle s_m \rangle$ takes the simple form

$$\langle s \rangle = \tanh\left(\frac{2\delta\tilde{C}}{\Delta^2}\langle s \rangle\right)$$ (28)

Figure 7 shows the solutions of this equation as a function of the bandwidth Δ for fixed parameter $\tilde{C}$ and detuning $\delta = \tilde{C}$. For $\tilde{C}/\Delta < (\tilde{C}/\Delta)_c = 1/\sqrt{2}$, the only solution is $\langle s \rangle = 0$. For the critical parameter $\tilde{C}/\Delta = (\tilde{C}/\Delta)_c$ this solution becomes unstable, and two additional solutions with $\langle s \rangle \neq 0$ appear (32). The comparison with exact results (Fig. 7) obtained from the numerical simulation of the stochastic model Eq. (26) shows that the mean field approximation describes this nonequilibrium phase transition of the network at least in a qualitative manner.

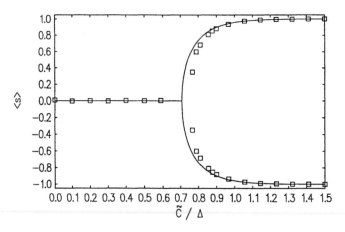

Figure 7 Expectation value $\langle s \rangle$ as a function of the inverse band width Δ^{-1} of the pump light field. Network of 40×40 subsystems, periodic boundary conditions. Fixed parameters $\delta = \tilde{C}$ and spatial homogeneous, "symmetric" pumping $\kappa_1(m) = \kappa_{-1}(m) = \kappa$ (independent of m). Squares: numerical simulation of the stochastic model; continuous line: mean field approximation.

IV. APPLICATIONS

A. Simulation of the Ising Model

The mean field equation Eq. (28) is reminiscent of the mean field theory of the Ising model for a spin system $s^m = \pm 1$ (33). Indeed, we shall show that our stochastic model Eq. (26) can be mapped onto the kinetic Ising model with the Hamiltonian (34)

$$\hat{H} = - \sum_{m=1}^{N} \mu H_m s_m - \frac{1}{2} \sum_{\substack{m,n=1 \\ n \neq m}}^{N} J_{mn} s_m s_n \tag{29}$$

μ is the magnetic moment, H_m the local magnetic field, and $J_{mn} = J_{nm}$ the exchange interaction between spin m and n. Coupling to a heat bath with temperature T induces "spin flips" $s_m \rightarrow -s_m$ with a rate $R_{s_m}(\{s_n\}')$ (30). The ratio

$$\frac{R_{s_m}(\{s_n\}')}{R_{-s_m}(\{s_n\}')} = \exp\left(-\frac{2}{k_B T} s_m \left(\mu H_m + \sum_{\substack{n=1 \\ n \neq m}}^{N} J_{mn} s_n \right) \right) \tag{30}$$

here is determined by a Boltzmann factor (34) and is independent of the specific model of the heat bath. Comparing Eq. (30) with the corresponding ratio for our optically driven model Eq. (26) yields the mapping

$$\frac{J_{mn}}{k_B T} = \frac{2 C_{mn} \delta}{\Delta^2}$$

$$\frac{\mu H_m}{k_B T} = \frac{1}{2} \ln \left(\frac{\kappa_{-1}(M)}{\kappa_1(M)} \right) \tag{31}$$

The parameter choice leading to Eq. (28), as expected, corresponds to the case of vanishing magnetic field $H_m \equiv 0$.

The mapping Eq. (31) is the starting point for an application of our artificial network in molecular electronics. It simulates another system (here, the Ising model) and thus fulfills a "computer" function. This function is even programmable, since the parameters ($J_{mn}/k_B T$ and $\mu H_m/k_B T$) of the simulated system can be adjusted by the pumping light field according to Eq. (31). The simulation of the Ising model by the quantum network has several advantages compared to the simulation on a conventional computer: firstly, quantum dynamics is intrinsically stochastic, whereas a conventional computer is deterministic and can only be made "quasi-stochastic" by special tricks. Secondly, the network dynamics is highly parallel as each spin in the Ising model has its own "processor" realized by a charge transfer subunit.

Nevertheless, simulation by the quantum network also has disadvantages: a lack of structural control may lead to a restricted control of the stochastic dynamics. The distance dependence of the exchange interaction J_{mn} is determined by the R^{-3} distance dependence of the dipole–dipole interaction C_{mn} and cannot be adjusted at will as in a conventional Monte Carlo simulation. This shows how physical constraints restrain the function of such a "quantum device." Another disadvantage concerns obtaining the results of the simulation. On a conventional computer, the state $\{s_n\}$ can be observed continuously. As outlined above, this detailed information as shown in Fig. 8 will not be observable in

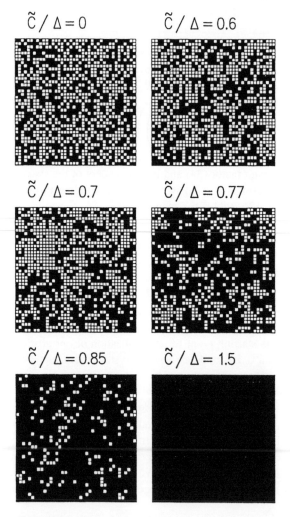

Figure 8   Typical configurations $\{s_n\}$ of the network ("snapshots of numerical simulation") for different values $\tilde{C}/\Delta$. Black squares: $s = +1$; white squares: $s = -1$. Fixed parameters same as in Fig. 7.

the quantum network: typical observational levels are not identical with the relevant level of the underlying network dynamics. This "reconstruction problem" appears to be typical for molecular electronics, where it is a fundamental problem how to connect the microscopic dynamics with the macroscopic observational level.

We illustrate a kind of "network analysis" through an absorption measurement. Figure 9 shows the absorption spectrum as measured by an additional probe light field for the same parameters as in Fig. 8. We obtained two lines centered around the resonance frequencies $\omega_{\pm 1}$ of the isolated subsystems, broadened and shifted due to the dipole–dipole interaction. The difference between the integral absorption over each line is proportional to the mean value $\langle s \rangle$ as shown in Fig. 7. However, the specific line structure contains additional information: the broad band absorption ("inhomogeneous line") for $\tilde{C}/\Delta = 0$ results from the uncorrelated neighborhood and the corresponding dynamical shift of the resonance frequencies (cf. Eq. (21)). The special line form is a consequence of the R^{-3} interaction. One would obtain the usual Gaussian line if each subsystem were coupled to any other with the same strength. The cluster formation for increasing $\tilde{C}/\Delta$ as observed in Fig. 8 shows up as a shift of the absorption

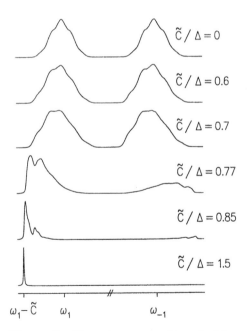

Figure 9 Absorption spectra $\chi''(\omega, m)$ of the network. Because of the spatial homogeneous pump field, the absorption is independent of m. Same parameters as in Fig. 8. Homogeneous linewidths $\gamma_{-1} = \gamma_1 = 0.01\tilde{C}$; transition dipole moments $|d_{-1}|^2 = |d_1|^2$. For convenience, the spectra are normalized to equivalent peak height.

lines to lower frequencies (see Eq. (19)). For $\bar{C}/\Delta > (\bar{C}/\Delta)_c$ one line starts to disappear due to the symmetry breaking instability (cf. Fig. 7). For large enough $\bar{C}/\Delta$, one single "homogeneous" line remains as a consequence of the complete order state.

B. Optical Image Processing

For the application as a Monte Carlo simulator in the last section, we focused on the stochastic dynamics of the network itself; the probe field merely was an instrument to get information about this dynamics. For the present application, the network dynamics as such is not relevant; it is rather the connection between the now spatial dependent pump field as "input pattern" and the spatial absorption of the probe field as "output pattern" that will be interpreted as a computer function. We shall show that our network model is complex enough that this function performs simple image processing tasks and can even be programmed, though in a restricted manner.

The input pattern $\beta(m)$ is defined by

$$\beta(m) = A^{-1} \ln\left(\frac{\kappa_{-1}(m)}{\kappa_1(m)}\right) \tag{32}$$

where

$$A^2 = \frac{1}{N} \sum_{m=1}^{N} \left[\ln\left(\frac{\kappa_{-1}(m)}{\kappa_1(m)}\right)\right]^2 \tag{33}$$

is a normalizing factor, so that

$$\frac{1}{N} \sum_{m=1}^{N} \beta(m)^2 = 1 \tag{34}$$

The comparison with Eq. (31) shows that $\beta(m)$ corresponds to the normalized spatial distribution of the magnetic field in the Ising model. We consider a spatially homogeneous probe ("test") light field with a spectral energy density $U_t(\omega)$ parameterized by a cut-off Gaussian

$$U_t(\omega) = U_t \exp\left(-\left(\frac{\omega - \omega_1 - \delta_t}{\Delta_t}\right)^2\right) \Theta(\bar{C} - |\omega - \omega_1|) \tag{35}$$

The Θ function makes sure that the probe field measures only absorption out of state $s = 1$. The total spatially dependent absorption

$$\alpha(m) = \int\limits_{0}^{\infty} d\omega \; U_t(\omega) \; \chi''(\omega, m) \tag{36}$$

is defined as the "output pattern." $\chi''(\omega, m)$ is obtained as result of our numerical simulations according to Eq. (21). The connection between input pattern $\beta(m)$ and output pattern $\alpha(m)$ is programmed by the parameters A ("absolute intensity of the pump field"), δ, Δ (frequency dependence of the pump field) and δt, Δt (frequency dependence of the probe field).

Our prototype input pattern $\beta(m)$ is a cosine in the two spatial directions with additional noise, as depicted in Fig. 10. For different program parameter sets we show computer simulations of the corresponding output patterns in Fig. 11. In Fig. 11a–d, where $\tilde{C}/\Delta_t = 0$, the output $\alpha(m)$ measures the mean occupation of state $s_m = 1$. For $\tilde{C}/\Delta = 0$ (Fig. 11a), the dynamics of the subsystems is not coupled at all, so that $\alpha(m)$ is local and, because the absolute input intensity A is low, linear with $\beta(m)$. For increasing $\tilde{C}/\Delta$ (Fig. 11 b, c), ferro-coupling between the subsystems leads to a smoothening of the output pattern. High input intensity A (Fig. 11d) results in saturation effects: the output can be characterized as making a threshold decision $\beta(m) \gtrless 0$. In Fig. 11e–i, the effect of the probe spectrum is investigated. For small probe bandwidth, the absorption is sensitive to the local environment. Local correlations (neighborhoods) are thus mapped into the response in frequency space. In fig. 11e (otherwise same parameters as in Fig. 11d) only subsystems absorb where the mean neighborhood dipole moment vanishes: this leads, in conjunction with the saturating parameter $A = 2$, to the projection of the "edges" of our prototype pattern. For smaller values of A the edges are smoothed (Fig. 11f, g). In Fig. 11h (detuning $\delta_t = -\tilde{C}$), the probe beam is in resonance only with subsystems that are in the same state $s = 1$ as their neighbors. In this way, the maximum of our input pattern is located. It is the sharper, the smaller the probe bandwidth is (Fig. 11i).

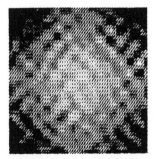

Figure 10 "Input pattern" $\beta(m)$ realized by the pumping light field. Applied to a network of 40×40 subsystems with periodic boundary conditions. Light: $\beta(m)$ large; dark: $\beta(m)$ small.

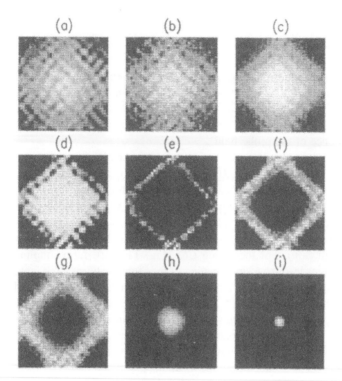

Figure 11 Spatial absorption pattern $\alpha(m)$ as "output" "program parameters." Light: $\alpha(m)$ large; dark: $\alpha(m)$ small. (a) $A = 0.1$, $\delta/C = 1$, $C/\Delta = 0$, $\delta_t/C = 0$, $C/\Delta_t = 0$. (b) $A = 0.1$, $\delta/\tilde{C} = 1$, $\tilde{C}/\Delta = 0.5$, $\delta_t/\tilde{C} = 0$, $\tilde{C}/\Delta_t = 0$. (c) $A = 0.1$, $\delta/\tilde{C} = 1$, $\tilde{C}/\Delta = 0.7$, $\delta_t/\tilde{C} = 0$, $\tilde{C}/\Delta_t = 0$. (d) $A = 2.0$, $\delta/\tilde{C} = 1$, $\tilde{C}/\Delta = 0.7$, $\delta_t/\tilde{C} = 0$, $\tilde{C}/\Delta_t = 0$. (e) $A = 2.0$, $\delta/\tilde{C} = 1$, $\tilde{C}/\Delta = 0.7$, $\delta_t/\tilde{C} = 0$, $\tilde{C}/\Delta_t = 25$. (f) $A = 0.5$, $\delta/\tilde{C} = 1$, $\tilde{C}/\Delta = 0.7$, $\delta_t/\tilde{C} = 0$, $\tilde{C}/\Delta_t = 25$. (g) $A = 0.3$, $\delta/\tilde{C} = 1$, $\tilde{C}/\Delta = 0.7$, $\delta_t/\tilde{C} = 0$, $\tilde{C}/\Delta_t = 25$. (h) $A = 2.0$, $\delta/\tilde{C} = 1$, $\tilde{C}/\Delta = 0.7$, $\delta_t/\tilde{C} = -1$, $\tilde{C}/\Delta_t = 25$. (i) $A = 2.0$, $\delta/\tilde{C} = 1$, $\tilde{C}/\Delta = 0.7$, $\delta_t/\tilde{C} = -1$, $\tilde{C}/\Delta_t = 100$.

Because of computer time limitations, we could simulate only a small network (40 × 40 subsystems). Therefore we have made the (unphysical) assumption that the light field ($\lambda \approx 10^{-6}$ m) can vary on the scale of the network lattice constant ($\alpha \approx 10^{-7}$ m). For a more realistic model we would have to take a system at least 100 times larger. However, the qualitative network properties should be independent from this scaling, as shown in Ref. 23.

C. Chemical Sensor

The optical image processor was based on the stationary network dynamics below the critical value $\bar{C}/\Delta < (\bar{C}/\Delta)_c$. For the application as chemical sensor, we exploit the transient dynamics of the first-order phase transition that can be induced between the two phases with $\langle s \rangle < 0$ and $\langle s \rangle > 0$ that exist for $\bar{C}/\Delta >$ $(\bar{C}/\Delta)_c$ (see Fig. 7). These two phases are both stable only for $\kappa_{-1} = \kappa_1$ ($H = 0$). Here, we consider a spatial homogeneous pump field, i.e., the parameters κ_s do not depend on the site m. For $\kappa_{-1} > \kappa_1$ ($H > 0$) only the state with $\langle s \rangle > 0$ is stable; for $\kappa_{-1} < \kappa_1$ ($H < 0$) it is $\langle s \rangle < 0$.

Switching the intensities of the two light modes from $\kappa_{-1} < \kappa_1$ to $\kappa_{-1} > \kappa_1$ ("turning the magnetic field") induces a first-order phase transition where the order parameter—the stable value $\langle s \rangle$—has a discontinuity. Characteristic for first order is the occurrence of metastable states (35). The initial stable states $\langle s \rangle < 0$ becomes metastable for $\kappa_{-1} > \kappa_1$, i.e., the new stable state $\langle s \rangle > 0$ possibly is reached just on a very long time scale. The new phase typically emerges via homogeneous nucleation, which requires the spontaneous formation of a critical cluster. The time scale of metastability is connected with this critical cluster size (36).

Besides homogeneous nucleation, the phase transition can be induced by heterogeneous nucleation where a critical cluster is formed by an external disturbance. This is exploited for the application of our quantum network as a chemical sensor. It works in an analogous way as a cloud or bubble chamber, where ionizing particles induce a critical cluster in a supersaturate vapor or an overheated liquid. In the chemical sensor, a molecule adsorbed on the network surface influences adjacent charge transfer subunits, which act as a nucleus for a global switching of the network dipole moment. The interaction of the molecule with the network subunits is modeled by a shifting of the resonance frequencies $\omega_{\pm 1}$. If this shifting is larger than the dipole–dipole interaction among the subunits, the dynamics of these subunits can be controlled independently by an additional light field that is tuned to the new resonance frequencies.

Figure 12 shows a simulation of this heterogeneous nucleation process. For $t < 0$, the network is in the metastable state with $\langle s \rangle < 0$ for $\kappa_{-1} = 5\kappa_1$. At $t = 0$, a molecule is adsorbed on the network surface (first picture). The additional light field that is tuned on the new resonance frequencies of the subunits neighboring the adsorbed molecule is chosen so that the spatial structure of the molecule is transferred to a spatial dipole structure on a fast time scale ($0 < t < \kappa_1^{-1}$). These switched dipoles act as nuclei for the dynamics of the other subunits, and the new stable state is reached at $t = 120\kappa^{-1}$. By this amplification process, the adsorbed molecule becomes noticeable on a macroscopic scale, e.g., by measuring the total dipole moment or absorption of the network.

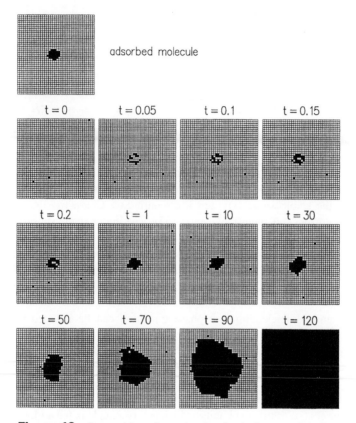

adsorbed molecule

t = 0   t = 0.05   t = 0.1   t = 0.15

t = 0.2   t = 1   t = 10   t = 30

t = 50   t = 70   t = 90   t = 120

Figure 12   Recognition of a molecule adsorbed at $t = 0$ at the network surface by a global change of the network state. For $t < 0$, the network was in a metastable state. A nucleation process is induced by the molecule.

In this application, the network is interface and information processor at once and can even be "programmed" in a certain manner: the ratio κ_{-1}/κ_1 determines the critical size from which a molecule is detected; the frequency of the additional light field makes sure that only molecules that lead to a certain frequency shift of the subunits are recognized.

V.   SUMMARY AND DISCUSSION

The challenge of molecular electronics is the preparation of synthetic nanometer-scale structures that can perform a computer function. A direct approach would try to implement abstract device concepts on this level of description. This approach has been successful in conventional electronics working on a hydrodynamical level but often fails in the molecular regime as primitive functions are here constrained and have to compete with deteriorating effects. In our approach, we try the opposite way: we consider a piece of complex structurized matter and then check whether its dynamical repertoire, controlled by external fields, can be exploited as a computer function.

Our prototype system is a network of interacting charge transfer subunits, driven by the light field. The isolated charge transfer center flips stochastically between its two dipole states. This "random telegraph signal" influences other network nodes, leading to a cooperative dynamics.

As a first application, we have demonstrated how our network model carries out highly parallel Monte Carlo calculations of the Ising model with $J(R_n - R_m) \sim |R_n - R_m|^{-3}$, and results are obtained directly as physical measurements. With a conventional computer this could only be reached by time-consuming simulations (which we actually did to obtain our results). In a second application, we have demonstrated how simple image processing tasks can be programmed and directly performed by pump and probe experiments. Finally, we exploited the first-order phase transition of the network for an application as "optical cloud chamber."

Nevertheless, there are also difficulties in this concept of "programmable matter": in our first application, the coupling constants J_{mn} of the Ising model cannot be chosen at will, as their distance dependence is restricted by the R^{-3} law of the dipole–dipole interaction: "low-level" dynamics is influenced immediately by "low-level" physics. In our second application, a set of "program parameters" can be assigned to a certain function, but the reverse problem, finding the parameters for a given function, is still unsolved. This appears to be a typical problem of many new computer concepts, e.g., for the cellular automaton, general statements about universality exist (37), but for practical applications (e.g., lattice gas model (38)), the finding of the rules is usually based on intuition rather than on a general algorithm. One may thus come to the conclusion that a molecular computer will be applied as a special-purpose rather than a general-purpose machine, but then with very high efficiency.

REFERENCES

1. Nagourney, W., Sandberg, J., and Dehmelt, H., *Phys. Rev. Lett.*, 56, 2797 (1986).
2. Moerner, W. E., and Kador, L., *Phys. Rev. Lett.*, 62, 2535 (1989).
3. Orrit, M., and Bernard, J., this volume.
4. Ralls, K. S., Skocpol, W. J., Jackel, L. D., Howard, R. E., Fetter, L. A., Epworth, K. W., and Tennant, D. M., *Phys. Rev. Lett.*, 52, 228 (1984).
5. Judd, T., Couch, N. R., Beton, P. H., Kelly, M. J., Kerr, T. M., and Pepper, M., *Appl. Phys. Lett.*, 49, 1652 (1986).
6. Sauter, Th., Blatt, R., Neuhauser, W., and Toschek, P. E., *Opt. Commun.*, 60, 287 (1986).

7. Ambrose, W. P., and Moerner, W. E., *Nature*, *349*, 225 (1991).
8. Farmer, K. R., Rogers, C. T., and Buhrman, R. A., *Phys. Rev. Lett.*, *58*, 2255 (1987).
9. Uren, M. J., Kirton, M. J., and Collins, S., *Phys. Rev.*, *B37*, 8346 (1988).
10. Ralls, K. S., and Buhrman, R. A., *Phys. Rev.*, *B44*, 5800 (1991).
11. Hopfield, J. J., *Proc. Natl. Acad. Sci. USA*, *79*, 2554 (1982).
12. Mühlenbein, H., Gorges-Schleuter, M., and Krämer, O., *Parallel Computing*, *6*, 269 (1987).
13. Wolfram, S., *Physica*, *D10*, 1 (1984).
14. Robinson, A. L., *Science*, *220*, 940 (1983).
15. Geerligs, L. J., Anderegg, V. F., Holweg, P. A. M., Mooij, J. E., Pothier, H., Esteve, D., Urbina, C., and Devoret, M. H., *Phys. Rev. Lett.*, *64*, 2691 (1990).
16. Schulz-Ekloff, G., in *Zeolith Chemistry and Catalysis* (P. A. Jacobs et al., eds.), Elsevier, Amsterdam, 1991), p. 65.
17. Ensslin, K., Hansen, W., and Kotthaus, J. P., this volume.
18. Mahler, G., Fig. 2 and Fig. 4, this volume.
19. Alexander, M. G. W., et al., *Appl. Phys. Lett.*, *55*, 885 (1989).
20. Reed, M. A., et al., *Phys. Rev. Lett.*, *60*, 535 (1988).
21. Obermayer, K., Teich, W. G., and Mahler, G., *Phys. Rev.*, *B37*, 8096 (1988).
22. Teich, W. G., Obermayer, K., and Mahler, G., *Phys. Rev.*, *B37*, 8111 (1988).
23. Körner, H., and Mahler, G., *Phys. Rev.*, *B48*, 2335 (1993).
24. Mehring, M., and Knoll, H. P., this volume.
25. Körner, H., and Mahler, G., *Phys. Rev. Lett.*, *65*, 984 (1990).
26. Peterson, I. R., this volume.
27. Körner, H., and Mahler, G., *Phys. Rev.*, *E47*, 3206 (1993).
28. Teich, W. G., and Mahler, G., *Phys. Rev.*, *A45*, 3300 (1992).
29. Reif, F., *Fundamentals of Statistical and Thermal Physics* McGraw-Hill, New York, 1965.
30. Glauber, R. J., *J. Math. Phys.*, *4*, 294 (1963).
31. Kawasaki, K., in *Phase Transitions and Critical Phenomena*, *Vol.* 2 (C. Domb and M. S. Green eds.), Academic Press, London, 1972.
32. Müller, B., and Reinhardt, J., *Neural Networks*, Springer, Berlin, 1990.
33. Brout, R., *Phase Transitions*, Benjamin, New York, 1965.
34. Suzuki, M., and Kubo, R., *J. Phys. Soc. Jap.*, *24*, 51 (1968).
35. Binder, K., *Phys. Rev.*, *B8*, 3423 (1973).
36. Binder, K., in *Fluctuations, Instabilities and Phase Transitions*, (T. Riste, ed.), Plenum Press, New York, 1975.
37. Albert, J., Culik, II, K., *Complex Systems*, *1*, 1 (1987).
38. Frisch, U., Hasslacher, B., and Pomeau, Y., *Phys. Rev. Lett.*, *56*, 1505 (1986).

11

Ensemble Properties and Applications

Günter Mahler

University of Stuttgart
Stuttgart, Germany

Volkhard May

Humboldt University of Berlin
Berlin, Germany

Michael Schreiber

Technical University
Chemnitz, Germany

A number of ideas and concepts towards the long-term goal of molecular electronics have been discussed in the foregoing sections. As explained earlier, molecular electronic devices should be based on a definite structure on the molecular level and a network of molecular entities communicating one to another, thus promising a large information processing power. It seems that we are still far away from any such prototype device. Nevertheless, simple information processing systems utilizing the properties of noninteracting single molecules do exist already. These systems have reached the level on which the mapping of "one bit onto one molecular state" may indeed become possible.

One promising concept developed in this direction will be presented in the first three contributions of this section. It deals with dye molecules embedded in an amorphous polymer matrix. These systems are appropriate for the application of the spectral hole burning technique. The use of a biological molecule as an optical processor unit is discussed in the last two contributions.

The spectral hole burning technique has been studied for many years and is based on the existence of zero-phonon lines of various types of impurity molecules within inhomogeneously broadened absorption bands. These broad absorption bands allow us to burn spectral holes, which in the helium temperature range can become as sharp as the homogeneous linewidth of the molecules. In such a manner it is possible to address between 10^3 and 10^7 sharp resonances in an inhomogeneously broadened absorption. The long lifetime of these holes results from an alternation of its chemical properties or the properties of the surrounding medium. Within this technique, nearly molecular selectivity can be reached, as the frequency space helps overcome the severe limitations in real space addressing.

The influence of an external electric field is discussed in the contribution by Bogner. This applied field offers the possibility of supplementing the frequency domain by a further dimension and of making the whole technique more flexible. The operation principle of a related two-dimensional electrooptical processor is also discussed.

Based on the concept of holography, information storage can be realized in a nonlocal manner, so that a single bit would be stored over a considerable area of the sample. Furthermore, a new type of massive parallel data processing becomes possible. The article by Rebane and Rebane introduces the application of this holographic technique in the field of persistent spectral hole burning. Among various applications an error-correcting neural network is discussed. The contribution by de Caro et al. discusses a number of questions connected with the utilization of holography in molecular electronics.

The use of the natural optical processor molecule bacteriorhodopsin is presented in the last contributions by Birge et al. and Bräuchle et al. Due to the pronounced photochromism of bacteriorhodopsin and its stable structure, various applications have been proposed, e.g., spatial light modulation, holographic recording, and pattern recognition.

Bacteriorhodopsin represents a native protein that acts as a photosynthetic proton pump in the bacterium *Halobacterium salinarium*. Natural selection resulted in a protein complex that works near the optimum for biological needs. A fascinating issue, however, is concerned with a further optimization with respect to the technical needs by genetic engineering, as discussed especially in the contribution of Bräuchle et al.

These investigations may also be considered as appealing examples for the idea of biomolecular electronics. Here, one tries to use native biomolecules, mainly protein complexes, which are modified according to the respective needs by genetic engineering or conventional chemical techniques. The advantage of this approach is that one can avoid complex chemical synthesis insofar as this work is done by the living cell; and one can benefit from the natural selection process, which results in systems with optimized properties. Probably this concept will become more important for molecular electronics in the years to come.

12

Electric Field Effects on Persistent Spectral Holes: Applications in Photonics

U. Bogner

University of Regensburg
Regensburg, Germany

I. INTRODUCTION

Persistent spectral hole burning (PSHB) is a new method in the laser spectroscopy of solids (1). This method is based on the existence of zero-phonon lines (2) in the spectra of light-absorbing species, such as impurities (molecules, ions, etc.) or color centers in crystalline or amorphous materials. The spectral holes produced by narrow-band laser excitation are detected in the absorption or fluorescence excitation spectra. At helium temperatures, hole widths can be observed that are equal to the lifetime width (3). Due to its high spectral resolution, PSHB was used from the beginning as a sensitive probe to study, for example, phonon processes (4–7), site relaxation (8), and the effects of external magnetic and electric fields (9–17) or hydrostatic pressure (18,19) in solids.

In this chapter we describe the electric field effects on persistent spectral holes and in particular special applications (20–25) in electrooptics and photonics (concerning the use of laser light for information processing). In other chapters of this book, the applications of the PSHB technique concern only digital processing, e.g., in the chapters by Rebane and Rebane and by U. Wild's group (de Caro et al.) dealing with the combination of PSHB with different types of holographic techniques (26,27). The results presented in this chapter demonstrate that PSHB is useful not only for digital processes, such as data storage in the electric field domain (21), but also for analog processes concerning, in particular, the modulation and pulse forming of laser beams (22–24) and hybrid optical bistability (25). The use of the laser beam modulator and pulse former in a light-guiding configuration (23,24) is essential to demonstrate the possible applications in integrated optics; and the measurement of subnanosecond rise time (24) is useful in order to test the possibility of gigabit information processing.

An interesting extension of possible technical applications is based on the fact that all applications can be realized as special functions in a two-dimensional electrooptic processor, operating in the wavelength and electric field dimension (20). In this universally applicable processor, the electrooptic characteristics not only can be adjusted to the requirements of the desired application but also in particular can be adjusted *separately* for *separate* laser wavelengths.

Until now, all demonstrations of applications, including the simultaneous use of two dimensions, were performed with dye molecules embedded in low concentrations in polymers. Therefore the section concerning the electric field induced level shifts (which can be termed in a general sense as Stark effects) is restricted to these materials, with special emphasis on nonpolar dyes in amorphous polymers, because these systems provide the best conditions for the use of both dimensions.

II. EFFECTS OF AN EXTERNAL ELECTRIC FIELD

Because PSHB is described in other chapters (e.g., in the chapter of Rebane and Rebane concerning hole burning time- and space-domain holography), I can begin the discussion on basic principles with electric field induced level shifts, which form the basis of electric field effects on persistent spectral holes.

A. General Remarks on Electric Field Induced Level Shifts

In the PSHB studies of dye/matrix systems, the analysis of the electric field effects (see, e.g., (10–15) and (27–36) demonstrates that the dominant effect is a frequency shift Δv of the pure electronic S_0–S_1 transition, which is proportional to the external electric field F_{ext} given by U/d, where U is the voltage applied to the sample with thickness d. Usually Δv is estimated by

$$\Delta v = -\frac{f}{h} \Delta\vec{\mu} \cdot \vec{F}_{ext} \tag{1}$$

The dipole moment difference $\Delta\vec{\mu} = \vec{\mu}_e - \vec{\mu}_g$ is the vector describing the change of the electric dipole moment between the excited and the ground state. Equation (1) is based on the assumption that the change of the energy of the optical transition is given by

$$h \, \Delta v = - \Delta\vec{\mu} \cdot \vec{F}_{loc} \tag{2}$$

The local field F_{loc} is the field at the molecule, which can be calculated in the simplest case by $\vec{F}_{loc} = f \cdot \vec{F}_{ext}$, using the Lorentz field factor f, given by $f = (\epsilon + 2)/3$, where ϵ is the dielectric constant.

In Eq. (2), describing the first-order Stark shift, the effect caused by the difference $\Delta\alpha$ of the polarizabilities (between the excited and the ground state) is neglected, because the second-order Stark shift is very small in general for the small field strength obtained for $\vec{F}_{loc}$. But in the case of polar matrices it is popular (see, e.g., Refs. 13, 28, and 30) to use the tensor difference $\Delta\alpha$ for the introduction of an *induced* dipole moment difference $\Delta\vec{\mu}_i$ by

$$\Delta\vec{\mu}_i = \Delta\alpha \cdot \vec{F}_{int} \tag{3}$$

where $\vec{F}_{int}$ is the internal field. This internal field generated by the polar groups of the matrix (surrounding the dye molecule) can be orders of magnitude larger than F_{loc}. The matrix induced dipole moment difference $\Delta\vec{\mu}_i$ is often used (instead of $\Delta\vec{\mu}$) in Eq. (1) in order to present the simplest explanation for the linear electric field effect observed in PSHB studies of dye molecules with inversion symmetry, embedded in polar matrices. Molecules with inversion symmetry do not have a permanent dipole moment in the ground state and in the excited state ($\vec{\mu}_g = \vec{\mu}_e = 0$), and therefore there is no *permanent* dipole moment difference ($\Delta\vec{\mu}_p = 0$).

The first PSHB study of electric field effects of guest molecules with $\Delta\vec{\mu}_p = 0$ was performed (12) with the centrosymmetric dye molecule perylene (see Fig. 1) embedded in the polar polymer polyvinylbutyral (PVB), shown in Fig. 2. The first results and also additional measurements with a detailed analysis (28) provided the proof that, within experimental accuracy, only a linear electric field effect can be observed in PSHB studies. Many dyes (see Fig. 1) are polar molecules like 9-amino acridine (an example of a neutral molecule) and cresyl violet (an example of an ionic dye). These molecules have a permanent dipole moment, changing its value and direction in the S_0-S_1 transition so that $\Delta\vec{\mu}_p \neq 0$. Usually for polar dye molecules in polar matrices the resulting vector $\Delta\vec{\mu}$ used in Eq. (1) is attributed to a superposition of $\Delta\vec{\mu}_p$ and $\Delta\vec{\mu}_i$ (31,32).

A linear electric field effect is also observed for perylene in nonpolar matrices like polyethylene (33) or in the Shpol'skii matrix *n*-heptane (34) and also for the dye octatetraene (which is a linear chain) in *n*-hexane (35). In the case of octatetraene the effect was discussed in terms of a dipole moment difference induced by the internal fields of the CH dipoles (of the *n*-hexane molecules of the matrix) at short distances. In the case of perylene, it is assumed that the effect can be attributed to electric field induced variations of the gas-to-matrix shift, which is caused by the dispersion interaction (33,34).

This assumption was confirmed recently by the observation that in perylene/ n-heptane, the electric field effect in the anti-hole lines (created by PSHB in certain original Shpol'skii lines) can change sign and can be up to a factor of 5

perylene   9-aminoacridine

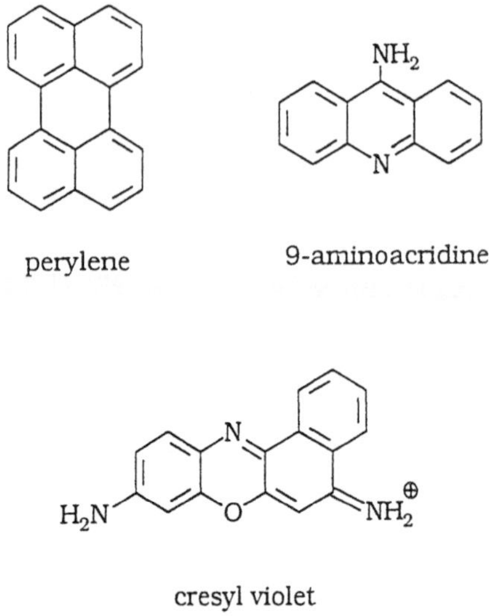

cresyl violet

Figure 1   Molecular structures of various dye molecules: perylene, 9-aminoacridine, cresyl violet.

larger as compared to the original line (36). The observation of a correlation between hole burning efficiency and electric field effect of perylene in different nonpolar matrices produced the same conclusion. On the basis of these observations a model is suggested in which the contact regions between the π electron system of the dye molecule and the surrounding matrix play a decisive part (36). In this model, the coefficient β is introduced for the description of the linear electric field effect:

$$\Delta\nu = \beta F_{loc}$$

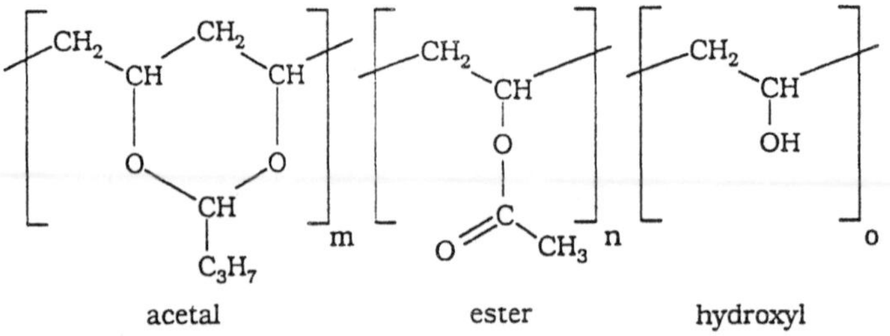

acetal   ester   hydroxyl

Figure 2   Molecular structure of the polymer polyvinylbutyral (PVB) with acetal, ester, and hydroxyl side groups.

Because of the polarization by deformation, the application of an external field changes in all dye/matrix systems the local details of the charge density distribution, and due to the short range of the dispersion interaction the influence of these changes is especially effective at the contact regions. For a distinct dye molecule, each contact with the matrix cage provides a *separate* contribution to the electric field induced level shift, and this contribution can be positive or negative. The coefficient β of the linear electric field effect is dependent on the direction $\vec{e}$ of $\vec{F}_{loc}$

$$\Delta \nu = \beta(\vec{e}) \mid \vec{F}_{loc} \mid \tag{5}$$

and for a distinct dye molecule, $\beta(\vec{e})$ describes the net effect of the separate contributions $\beta_i(\vec{e})$ of its N contact regions:

$$\beta(\vec{e}) = \sum_{i=1}^{N} \beta_i (\vec{e}) \qquad \beta_i (\vec{e}) \gtrless 0 \tag{6}$$

In general the complex dependence of each β_i on the direction $\vec{e}$ (of $\vec{F}_{loc}$) is different for various contact regions.

B. Experimental Results in Crystalline and Amorphous Systems

Comparing the electric field effect in different materials with respect to possible applications, it is decisive to distinguish between applications in which a device is used only at one fixed wavelength (e.g., light modulation or pulse shaping) and applications concerning the use of two dimensions (e.g., data storage in the electric field and wavelength dimension or more generally electrooptic applications using wavelength multiplexing techniques). For the case of an application with fixed wavelength it is in principle sufficient to study the effect of an external electric field in the *center* of the spectral hole, whereas for the simultaneous use of both dimensions it is interesting to investigate the spectroscopic details, including in particular the question whether there is electric field induced splitting or broadening of the spectral hole. From the point of view of designing a device for wavelength multiplexing, the hole broadening corresponds to a voltage controlled reversible vanishing of the two-dimensional hole.

The effects of external electric fields on persistent spectral holes and the applications were studied in a large number of different types of crystalline and amorphous dye/matrix systems. The amorphous materials concern in particular thin layers or films of polymers in which the dye molecules are embedded in low concentrations (10^{-3} mol).

In some experiments, Langmuir–Blodgett films (i.e., ultrathin monomolecular layer systems prepared by a dipping technique) have been used (37,17). In many materials, the study of the electric field effect can be performed in the convenient voltage range of less than 50 V if the thickness of the layer is less than about 10 μ. Polymers like PVB layers with this thickness can be easily prepared from a solution by evaporation of the solvent (e.g., ethanol). The sample is prepared between two electrodes. In some investigations, in which the wave vector $\vec{k}$ of the incident laser light is in the direction of the electric field $\vec{F}_{ext}$ at least one transparent electrode (e.g., an indium tin oxide layer on top of a glass slide) is used.

For all dye/matrix systems, low temperatures are necessary; the highest temperature reported until now for an organic PSHB material is about 100K (38) (PSHB at room temperature observed in inorganic systems will be handled in the paragraph concerning future prospects). In all of the following studies, the measurements were performed at temperatures less than 2K. At helium temperatures the width of the spectral holes is on the order of GHz. In general, the holes were burned and detected by using single frequency dye lasers with a linewidth of less than 50 MHz (including frequency jitter). The detection can be performed by absorption or fluorescence excitation spectroscopy. If holographic techniques are used, then the signal-to-noise ratio is substantially improved (39).

In Fig. 3, the electric field induced splitting of the spectral hole burned in

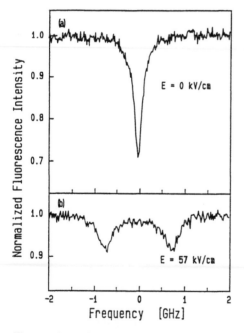

Figure 3 Electric field induced splitting of a persistent spectral hole burned in site II of perylene in crystalline *n*-heptane. (a) Without electric field, (b) with electric field strength $F = 57$ kV/cm. (From Ref. 34.)

site II of the nonpolar dye perylene, embedded in crystalline *n*-heptane, is shown. The splitting is linear with respect to the electric field strength (34). The investigated polycrystalline *n*-heptane sample consisted of crystalline lamellas that were mainly oriented parallel to the sample cell. The removal of the twofold orientational degeneracy of the perylene molecules with respect to the crystal lattice is responsible for the appearance of a splitting in two components. For this *arbitrary* orientation the splitting (or shifting) coefficient $\beta_{s,a}$ is defined as

$$\beta_{s,a} = \frac{1}{f} \frac{\Delta \nu_s}{\Delta F_{ext}}$$

The spectral distance of the split components is $2 \cdot \Delta \nu_s$ hr. The change of the external electric field strength ΔF_{ext} is required for this splitting; f is the Lorentz factor.

Figure 4a shows the electric field induced broadening of a spectral hole in the amorphous system perylene/PVB (40). In Fig. 4b, it was confirmed experimentally that the hole area remains constant when an electric field is applied. The hole broadening is the net effect of the statistically distributed shifting coefficients β_i (with Gaussian distribution function) of the individual molecules with an average value β. According to the formula

$$\beta = \frac{1}{f} \frac{\delta \nu_s}{\delta F_{ext}}$$

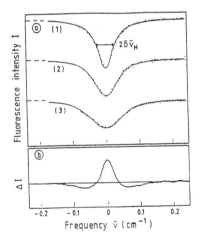
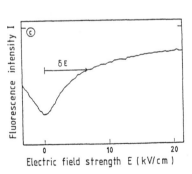

Figure 4 Electric field induced broadening of the persistent spectral holes in the amorphous system perylene/PVB. (a) Spectral hole for different electric field strengths *E*: (1) 0 kV/cm, (2) 2.9 kV/cm, (3) 4.4 kV/cm. (b) Difference between the spectral holes without and with an electric field (*E* = 8.3 kV/cm). (c) Fluorescence intensity in the center of the spectral hole versus electric field strength. (From Ref. 40.)

β can be determined by the measurement of $\delta\nu_s$ (i.e., the half-width of the hole in the frequency domain), and δF_{ext} (i.e., the half-width of the hole in the electric field domain). An example of a field hole is shown in Fig. 4c.

With polar dye molecules, even in amorphous systems, an electric field induced hole splitting can be observed for certain geometrical configurations. In Fig. 5 this splitting is shown for the system cresyl violet/PVB with the polarization of the light parallel to the electric field (41); the holes were detected by a holographic method. Dye molecules with optimal orientation are selected by the direction of the polarized laser excitation (with respect to the direction of the external electric field). Generally the splitting is stronger than the superimposed broadening only for certain geometrical conditions, which are different for different dye molecules, indicating that intramolecular effects are dominant. The geometrical conditions of splitting provide information concerning the angle between the transition dipole moment and the direction within the dye molecule for which the corresponding shifting coefficient has its *maximum* value $\beta_{s,m}$

$$\beta_{s,m} = \frac{1}{f}\frac{\Delta\nu_s}{\Delta F_{ext}}$$

where $\Delta\nu_s$ is the distance between the two splitting components and F_{ext} is the external field strength; f is the Lorentz factor.

In Table 1 the experimental results are presented for different examples of dye/matrix systems. The introduction of the wavelength λ in this table is necessary because there can be a significant wavelength dependence (30,32,41,42) of the electric field induced level shifts in particular for the case of polar dyes in polar matrices (32,41). The presentation of the values of the various coefficients

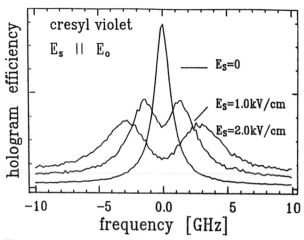

Figure 5   Electric field induced hole splitting of cresyl violet/PVB obtained for selected geometrical conditions with polarization of the light parallel to the electric field. The holes were detected by a holographic method. (From Ref. 41.)

Table 1   Examples of Experimental Results for the Coefficient ß of Linear Electric Field Induced Level Shifts[a]

Dye molecule	Matrix	λ (nm)	$\beta\left(\dfrac{MHz}{kV/cm}\right)$		Ref.
Perylene	*n*-Heptane:				
	Site I	440.8		1.5	(34)
	Site II	445.2	$\beta_{s,a} =$	9.2	(34)
	Anti-hole (A_6)				
	of site II	444.5		48	(36)
	Polyethylene	441		13	(33)
	PVB	441	$\bar{\beta} =$	72	(40)
	Cellulose	441		130	(40)
	nitrate				
Cresyl violet	PVB	616		850	(32)
Nile red	PVB	600	$\beta_{s,m} =$	2100	(32)

[a]For the nonpolar perylene in crystalline *n*-heptane, $\beta_{s,a}$ is taken from hole *splitting* (with *arbitrary* geometrical conditions), and in amorphous matrices the average value $\bar{\beta}$ is determined from hole broadening. For the two polar dyes (cresyl violet and nile red), $\beta_{s,m}$ is obtained from hole *splitting* (with an orientation of the electric field and the laser polarization selected for a *maximum* value).

β is more appropriate with respect to applications, and in addition it offers the possibility of avoiding problems (see, e.g., Ref. 32 and Refs. 41–45) connected with the description of the electric field effects in terms of the various types of dipole moment differences $\Delta\mu$.

III. APPLICATIONS IN PHOTONICS

In principle there are two different types of applications of electric field effects in photonics and more specifically in the field of electrooptics. In the first type, the laser is used only at a fixed wavelength, and in this case it does not matter whether the hole is split upon application of an electric field or not. Such applications are, for example, data storage in the electric field dimension, light modulation and pulse shaping, and hybrid optical bistability. The second type of application also uses the tunability of lasers, and in general it requires the fulfillment of the condition that the hole is only broadened and not split upon application of an electric field. This provides a universally applicable electrooptic processor with the unique feature of simultaneous operation in the electric field and frequency dimensions, which will be handled in Sec. III.D for the examples of optical data storage and light modulation with wavelength multiplexing and demultiplexing.

A. Optical Data Storage in the Electric Field Dimension

A multiplicity of persistent spectral holes can be used for optical data storage in the wavelength dimension. Instead of the hole in the electric field dimension, which is shown in Fig. 4b, a multiplicity of holes can also be burnt in the electric field domain. This was first demonstrated in the case of 9–aminoacridine in PVB films (21). Here a fixed frequency He–Cd laser was used to burn and detect the spectral holes, which, by mere chance, is suited to excite this dye in its pure electronic transition (46). The experimental result is shown in Fig. 6, demonstrating the electric field readout on the basis of a measurement of the normalized fluorescence intensity I/I_0 versus the electric field strength E (lower scale) and voltage U (upper scale). The rectangular hole shown in Fig. 6a was burned before demonstration of data storage shown in Fig. 6b. This rectangular hole was burned while a periodically linearly ramped voltage (-50 to $+50$ V) was applied to the sample. The purpose of this preparation procedure was to remove those dye molecules that could be very easily burned.

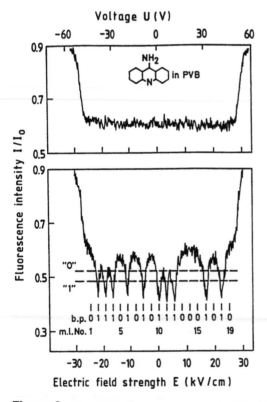

Figure 6 Normalized fluorescence intensity I/I_0 of 9-aminoacridine in polyvinylbutyral demonstrating data storage in the electric field domain. (From Ref. 21.)

This preparation procedure provides the advantage that the 19-bit memory scheme shown in Fig. 6b can be read very often with moderate laser intensity without being affected by the readout process. It is worth mentioning that this preparation procedure is something like a substitute for a photon gated PSHB process. In Fig. 6b the fluorescence intensities at memory locations 1–19, which are below the broken line "1" and above the broken line "0", represent a logical 1 and 0 respectively. The letters "b.p." in the bottom of Fig. 6b mean bit pattern, and the letters "m.l. No." represent the memory location number.

There are several advantages in this method of voltage-tunable data storage. The electric field dimension is distinguished by its simplicity and speed in addressing the memory locations by selecting the voltage applied to the sample. When a two dimensional spatial array of spots is used, it is possible to address the spots individually by applying the appropriate voltage to the electrodes of each spot. This provides a simple way of parallel writing the information, which increases the effective writing speed compared to serial addressing. The information at a definite memory location in the electric field domain may be written simultaneously at the spatial spots by illuminating the whole sample with one laser beam with sufficient intensity for hole burning (logical 1). At the spots where a logical 0 (no hole) is written, a common voltage (e.g., $U = 0$) is applied that acts as a trap picking up the undesired holes. In addition, it should be mentioned that it is possible to erase the information "holes" in each spot separately using one of the electrodes of the spot to generate a short heat pulse (4).

B. Light Modulation and Pulse Shaping

Further possible applications (22–24) are based on the fact that the transmission (or absorption) of a solid can be changed by voltage induced changes of the population in the center of a persistent spectral hole. The most simple use was the demonstration (22) of intensity modulation of laser light by electroabsorption. In this publication it was also shown that this novel technique provides a versatile electrooptic device with unique features, because the electrooptic characteristic (i.e., the curve that describes the dependence of the transmission T of the sample on the applied voltage U) can be adjusted to match the requirements of the specific desired applications. The preparation and the use of tailor-made electrooptic characteristics $T(U)$ was demonstrated for two examples of laser pulse shaping. The adjustment of the electrooptic characteristic is carried out by applying a time-dependent voltage $U_B(t)$ of selected sign, magnitude, and temporal course during spectral hole burning with full laser intensity.

The sample used in the experiments consisted of a PVB film doped with perylene molecules at a concentration of about 2×10^{-3} mol/L. The film was contained between two electrodes made of transparent indium tin oxide and indium. An additional metal reflecting layer was used to multipass the laser beam, resulting in a total interaction length of about 400 μm. The spectral holes were burned with He–Cd laser light (441.6 nm) which is absorbed in the 0–0' band of the $S_1;lwS_0$ transition of perylene. The voltage modulated transmission of the sample was probed using an attenuated He–Cd laser beam.

In the first experiment with this device it was demonstrated that even the T (U) characteristic, obtained after burning while a dc voltage was applied to the sample, contains a part of sufficient linearity that is suited for voltage-activated intensity modulation of a laser beam.

In the second experiment, shown in Fig. 7, tailor-made electrooptic characteristics (a,b) were prepared in the device, and subsequently their use for laser pulse shaping was demonstrated (c, d). The (normalized) transmission vs. voltage characteristic curve $T(U)$, shown in Fig. 7a, represents a steplike change of the transmission. It was adjusted by burning the spectral hole while simultaneously applying a voltage U_B, which was periodically linearly ramped in a triangular manner between -90 and -40 V. The problem of undesired spectral hole burning during repeated probing of the characteristic curve has already been mentioned. In the case of Fig. 7a, it was reduced by removing the easily burnable

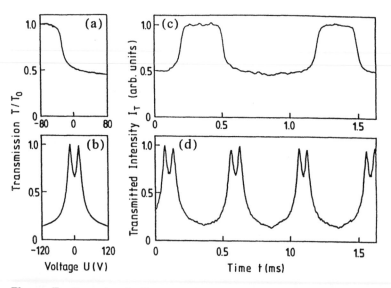

Figure 7 Examples of tailor-made electrooptic characteristics (a and b) and demonstration of their use for laser pulse shaping (c and d). (a,b) Normalized transmission vs. voltage characteristic curves $T(U)$ that have been prepared in the sample. (c,d) Resulting time-dependent laser intensity I_T transmitted through sample, if a sinusoidal (c or d) triangular time-dependent voltage is applied to samples with electrooptic characteristics like a and b, respectively. (From Ref. 22.)

molecules by additional hole burning, while a periodic triangular voltage (-100 to $+100$ V) was applied to the sample.

Another example of a characteristic curve is shown in Fig. 7b. The $T-U$ characteristic with two peaks was formed by spectral hole burning while alternately the dc voltages $U_B = 15$ and -15 V were applied to the sample. It should be noted that all of the characteristic curves are erasable, for example by increasing the temperature of the sample or by heat pulses, both of which refill the spectral hole.

The steplike function shown in Fig. 7a is a characteristic curve which may be of technical interest. When an ac voltage (amplitude 80 V, frequency 1 kHz) was applied to the sample with this characteristic, nearly rectangular pulses were generated (see Fig 7c). Possible applications include the generation of rectangular laser pulses of high repetition rate (e.g., for pulse code modulation) by using a high-frequency voltage generator.

The described electrooptic device can also be used for pulse forming of laser light. The essential point of the technique is that the characteristic curve (i.e., the dependence of the sample transmission T on the applied voltage U) is adjusted to the desired pulse shape. When a voltage $U(t)$ with a periodic triangular time dependence is applied to the sample, the shape of the $T-U$ characteristic is impressed on the transmitted laser intensity I_r as a function of time t. This is clearly seen from the transmitted intensity $I_r(t)$ in Fig. 7d, which represents the exact replica of the double peak $T-U$ characteristic in Fig. 7b.

The applications of the device can be extended to the field of image processing by using a two-dimensional spatial array of spots with separate electrodes. The transmission of each spot can be modulated individually by varying the applied voltage. The technique presented in this publication is not restricted to the use of organic materials. In this respect the electric field effects recently observed (11) in BaCIF:Sm^{2+} are interesting in particular because of the method of photon-gated (47) hole burning. Other inorganic materials may also be useful for special applications.

Concerning the comparison between organic and inorganic materials it is interesting to note that in the case of organic materials, it is easy to prepare an experimental setup for the demonstration of electroabsorption modulation in a simple light guiding configuration suitable for integrated optics (23). A schematic diagram of the experimental setup is shown in Fig. 8.

The light beam L_m of an He–Cd laser is focused on the core of the optical fiber F_1 and propagates via the fiber F_2 to the perylene-doped polyvinylbutyral (PVB) film, where it is modulated.

The entrance surface of the long fiber F_1 (length 5 m) is outside the helium cryostat at room temperature. Its exit surface is coupled to the short fiber F_2 (length 8 cm) in the helium cryostat. The exit part of fiber F_2 is embedded in the thin PVB film. The PVB film is prepared on top of a transparent indium tin

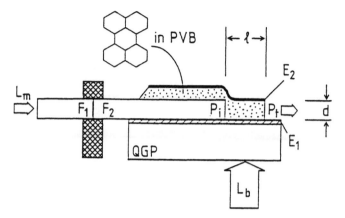

Figure 8 Schematic diagram of the experimental setup for the demonstration of electroabsorption light modulation in a light-guiding configuration suitable for integrated optics. (From Ref. 23.)

oxide layer E_1 on a quartz glass plate QGP. It is covered by an indium layer E_2. The doped PVB film is the center of a light guiding structure. The thickness d and the active length l of the waveguide modulator are 38 μm and 0.42 mm, respectively.

The spectral hole in perylene is burned (without a voltage) with a second light beam L_b of the He–Cd laser, which propagates vertically to the first beam L_m (through the quartz glass plate QGP and the transparent layer E_1) to the perylene-doped PVB film. After hole burning, the second light beam L_b is turned off. When a voltage is applied to the electrodes E_1 and E_2, the hole center is filled, and the laser light L_m propagating in the light guiding structure is strongly absorbed.

We measured P_t without and with an applied voltage. P_i and P_t are the incident and transmitted laser powers, respectively. The measured extinction ratio or maximum modulation depth of 0.94 is in strong agreement with the estimated value [23].

In another investigation, the same light-guiding configuration was used to provide the experimental proof for a response time on the subnanosecond scale [24]. The experimental result is shown in Fig. 9, where (a) is the voltage pulse applied to the sample versus time and (b) is the transmitted power versus time t. In Fig. 9b, the dots show the measurement, and the solid line demonstrates the comparison with a computer simulation, taking into account all experimental details, including the overall rise time of the detecting system (which contains an avalanche silicon photodiode).

Within experimental accuracy, there is strong agreement between the mea-

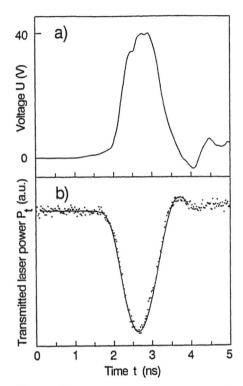

Figure 9 Subnanosecond response time of a voltage-controlled laser beam modulator. (a) Voltage pulse applied to the sample versus time; (b) transmitted laser power versus time. (From Ref. 24.)

sured (points) and the calculated transmitted laser power P_t. This result demonstrates that the response time of the electric field effect on the investigated spectral hole is less than 300 ps. This fast response provides modulation rates exceeding several gigabits per second.

The principal limitation of the response time of the electric field induced changes of a spectral hole is due to the finite hole width $\delta\nu_H$. The Fourier transform yields a characteristic time τ on the order of $1/(2\pi\delta\nu_H)$. Using the measured hole width $\delta\nu_H = 6*10^9$ s^{-1}, we obtain $\tau \approx 27$ ps. Shorter response times can be obtained when broader spectral holes are used.

C. Hybrid Optical Bistability

Because of the nonlinear electrooptic characteristics, the effect of an external electric field on persistent spectral holes can be used for the realization of *hybrid*

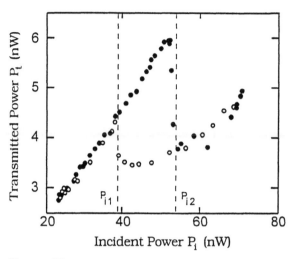

Figure 10   Hybrid optical bistability based on voltage-controlled changes of persistent spectral holes in perylene/PVB. (From Ref. 25.)

optical bistability. The original concept of optical bistability handles the controlling of light with light in an all-optical device (48). In the case of hybrid optical bistability, the transmission dependence of the power level of the transmitted light is caused by a combination of the optical elements with an *external* feedback circuit. This external feedback can be realized for example by a Pockels cell or by other electrooptic elements, such as a liquid-crystal electrooptic modulator (49). Recently, it was demonstrated (25) that the electric field induced changes of the absorption in the center of a persistent spectral hole can also provide a hybrid optical bistable device. The dye/matrix system used in this demonstration was perylene/PVB. The external feedback was performed through measurement of the fluorescence of the first vibronic zero-phonon line with a photomultiplier, and by applying the amplified voltage signal to the sample. In Fig. 10, the transmitted laser power P_t vs. incident laser power P_i shows the region of hybrid optical bistability between P_{i1} and P_{i2}. The full and open circles represent the increasing and decreasing incident laser powers, respectively. The use of the zero-phonon line intensity—instead of the transmitted laser power P_i—for the external feedback is an essential point for successful experiment operation. The intensity of the zero-phonon lines is strongly influenced by the PSHB process and by the applied voltage, whereas the transmitted laser power P_t is determined mainly by the contribution of the phonon sidebands to the absorption. In perylene/ PVB, which has a Debye–Waller factor of about 0.5 (50), the contribution of the phonon sidebands to the absorption is relatively large. Therefore changes in the transmitted laser power P_t with the applied voltage are relatively small. The operating conditions could be improved by using a dye/matrix system with a Debye–Waller factor close to 1.

D. Two-Dimensional Electrooptic Processor

In all of the application cases described in the preceding sections, a fixed laser wavelength was used. These applications are based on the electric field induced changes in the *center* of the *spectral* hole, and it does not matter whether the hole is shifted or split in the frequency dimension. In the base Sec. II.B it was demonstrated that electric field induced hole splitting can occur not only in crystalline materials but even in amorphous polymers if *polar* dye molecules are used. In the case of nonpolar dye molecules like perylene, one can be sure that *only* electric field induced hole broadening is possible, because there is a Gaussian statistical distribution of the corresponding level shifts. It is this Gaussian distribution of the electric field coefficients that provides the unique possibility of using both dimensions simultaneously (i.e., the electric field dimension and the wavelength dimension). The most simple use as a two-dimensional processor is data storage in the electric field and wavelength dimension. This two-dimensional data storage is shown in Fig. 11 for the example of perylene in PVB (51).

The experimental conditions for the production of Fig. 11 were quite similar to those of Fig. 6. Instead of the fixed wavelength of the He–Cd laser, a single-frequency dye laser was used in a scan range of 30 GHz. In order to perform PSHB according to the bit pattern shown in the bottom panel of Fig. 11, the laser was tuned in steps of about 5.5 GHz, and the voltage applied to the sample was changed stepwise, so that the corresponding changes of the electric field dimension were about 12 kV/cm. Like holes in the wavelength dimension, such two-dimensional patterns can be detected also by the holographic method (52) with high accuracy. The details of this method are described in the chapter from U. Wild's group.

Data storage in the electric field and wavelength dimension is the simplest application involving the simultaneous use of both dimensions because it is based on *stepwise* changes in both dimensions. More generally, it is possible to combine the two-dimensional operation with continous changes in at least one of the two dimensions. It has been demonstrated in Sec. III.B that the electric field dimension can be scanned continously, even with extremely high speed. Therefore it appears reasonable to assume that possible applications employing real two-dimensional operation will be based on the simultaneous use of different discrete wavelengths (i.e., in the case of wavelength multiplexing), whereas a continous scan is used in the electric field dimension. In such a continuous scan, the optical properties of the solid can be changed according to an electrooptic characteristic that is separately adjustable for separate wavelengths.

The schematic drawing of Fig. 12 demonstrates how PSHB can be used for

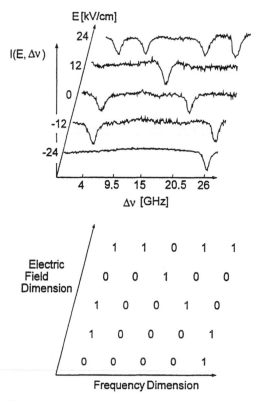

Figure 11 Two-dimensional data storage in the electric field dimension and in the wavelength dimension or dimension of the light frequency. (From Ref. 51.)

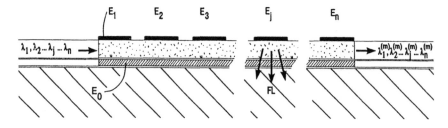

Figure 12 Schematic device configuration for wavelength multiplexing (and demultiplexing).

the preparation of a device capable of wavelength multiplexing and demultiplexing. In the first step, the part of the solid between electrode E_j and the ground electrode E_0 is prepared (according to Sec. III. A) so that an electrooptic characteristic is produced that is suited for the modulation of the laser wavelength λ_j with an operating voltage scanned between 0 and V. The further preparation steps for additional $(n-1)$ discrete laser wavelengths $(\lambda_i \neq \lambda_j)$ are performed so that for all λ_i, a completely flat hole (or ''groove'') is produced for the voltage dimension in the range between the values 0 and V. The specific part, located below electrode E_j, will operate as a modulator only for λj, whereas all λ_i will pass without modulation. The preparation of the device is complete if each separate part of the array below one of the electrodes E_i is prepared as a modulator for *one* of the separate discrete wavelengths λ_i. At the end of the electrode array (i.e., after electrode E_n) the signals of n information channels are transformed into the n modulated light signals $\lambda_1^{(m)}; \lambda_2^{(m)}; \ldots \lambda_j^{(m)}, \lambda_n^{(m)}$ emitted from the device, and they can be transferred, for example, into an additional waveguiding device. It is worth noting that the same device can be also used for wavelength demultiplexing. In the case of the application of a voltage V there is an increased absorption of an irradiated light signal with wavelength λ_j in that part of the array that is below electrode E_j. The increased absorption results in an increased fluorescence FL (which can be radiated out of the region below electrode E_j; see Fig. 12).

IV. FUTURE PROSPECTS

Independent of the further developments in the possible applications of PSHB and electric field effects, the material science concerning basic solid-state physics and technical optimization problems will play the decisive part. The role of material science can be completely different for different applications, such as information processing in large-scale, high-technical systems, or in equipment used in everyday life. For future possible applications involving PSHB techniques in everyday appliances, it is important to provide room-temperature operation. Recently, room-temperature PSHB has been performed in different Sm^{2+} doped inorganic materials (53–57); however, the electric field induced level shifts for these materials (measured at helium temperatures) were very small (11) compared to organic systems. Therefore it is worth mentioning that in neutron-irradiated IaB-type diamond—recently used as a material for room-temperature PHSB (58)—we observed the largest electric field effect reported until now (59). This large electric field effect, exceeding even that of nile red in PVB, was obtained for one of the new color centers with a zero-phonon line in the near infrared spectral range, which is accessible by diode lasers. It appears reasonable to assume that these color centers, generated by neutron irradiation, will be generated in the future also by irradiation of other particles, for example in industrial accelerator (or ion implantation) equipment, if thin diamond-like layers in the μm range are used as a target.

If possible applications concern large, high-technical information processing systems (e.g., communications network centers or computer centers), then the use of low temperatures, which is necessary for organic PSHB materials, will not be a crucial limitation. This is because in such centers, cooling of the active elements is often required, and temperatures in the range of about 100K are easily accessible with closed-cycle cryogenic engineering.

Future prospects regarding the necessary background studies concern in particular the physical origin of electric field induced level shifts. These studies should be based on the microscopic details of the theory of dielectrics, applied, for example, to the electric field induced changes of the local charge density distribution of the dye molecule π electron system and of the surrounding matrix. The main object of research in material science will involve not only a large coefficient of the electric field effect but also in particular the combination of different features such as photon-gated PSHB (57) and operation in the spectral range accessible with diode lasers. In the further development of PSHB studies, semiconducting materials will probably be used as well. The use of semiconductors may include a combination of electric field effects on persistent spectral holes and photovoltaic processes, so that light can be controlled *directly* by light. This reasoning meets the original goal of bistability (45) concerning an all-optical device (without *external* feedback), or more generally this reasoning may lead to the creation of an optical transistor.

In the future, the simultaneous use of the electric field and frequency dimensions may provide optical implementations for new computing methods (requiring in particular parallel processing). Cellular automata (60) exemplify an appropriate *digital* method, but regarding the special features of the electric field effects on persistent spectral holes, it is suggested that we investigate the potential applicability for optical computing also in the case of present (and future) *analog* methods.

ACKNOWLEDGMENTS

This research has been supported by the Deutsche Forschungsgemeinschaft (Grant: Bo 743/1–3), by the Stiftung Volkswagenwerk (photonic project), and by the Fraunhofer-Gesellschaft (patent office for the German research). The author wishes to thank M. Maier and U. Wild for useful discussions and for permission to reprint figures.

REFERENCES

1. Moerner, W. E., ed., *Persistent Spectral Hole-Burning: Science and Applications*, Springer-Verlag, New York, 1988.
2. See, e.g., K. K. Rebane, *Impurity Spectra of Solids*, Plenum Press, New York, 1970.

3. Levenson, M. D., Macfarlane, R. M., and Shelby, R. M., *Phys. Rev.*, *B22*, 4915 (1980).
4. Bogner, U., *Phys. Rev. Lett.*, *37*; 909 (1976); Bogner, U., and Schwarz, R., *Phys. Rev.*, *B24*, 2846 (1981).
5. Bogner, U., Low temperature properties probed by selective laser-excitation, in *Nonmetallic Materials at Low Temperatures* (G. Hartwig and D. Evans, eds.), Plenum Press, New York, (1986), p. 79.
6. Hizhnyakov, V. V., and Reineker P., *Chem. Phys.*, *135*; 203 (1989).
7. Beck, K., Röska, G., Bogner, U., and Maier, M., *Solid State Commun.*, *57*; 703 (1986); Beck, K., Bogner, U., and Maier, M., *Solid State Commun.*, *69*, 73 (1989); Attenberger, T., Beck, K., and Bogner, U., Persistent spectral holes in disordered systems: a probe of barrier crossing in double-well potentials, in *Phonon Physics* (S. Hunklinger and W. Ludwig, eds.), World Scientific, 1990, p. 555.
8. Jankowiak, R., and Bässler, H., A comparative study of site relaxation and hole-burning for tetracene in aromatic organic glasses, *J. Molecular Electronics*, *1*, 73 (1985).
9. Personov, R., Site selection spectroscopy of complex molecules in solutions and its applications, in *Spectroscopy and Excitation Dynamics of Condensed Molecular Systems* (V. M. Agranovich and R. M. Hochstrasser, eds.), North-Holland, Amsterdam, 1983, p. 555 and references therein.
10. Marchetti, A. P., Scozzafava, M., and Young, R. H., *Chem. Phys. Lett.*, *51*; 424 (1977).
11. Macfarlane, R. M., and Shelby, R. M., Persistent spectral hole-burning in inorganic materials, in *Persistent Spectral Hole-Burning: Science and Applications* (W. E. Maerner, ed.), Springer-Verlag, New York, 1988, p. 127.
12. Bogner, U., Seel, R., and Graf, F., *Appl. Phys.*, *B29*; 152 (1982).
13. Maier, Max, *Appl. Phys.*, *B41*, 73 (1986).
14. Haarer, D., Photochemical hole-burning in electronic transitions, in *Persistent Spectral Hole-Burning: Science and Applications* (W. E. Maerner, ed.), Springer-Verlag, New York, 1988, p. 79.
15. Wild, U. P., Renn, A., Molecular computing: a review, *J. Molecular Electronics*, *7*; 1 (1991).
16. Jankowiak, R., Hayes, J. M., and Small, G. J., *Chemical Review*, *93*; 1471 (1993).
17. Orrit, M., Bernard, J., and Personov, R. I., *J. Chem. Phys.*, *97*; 10256 (1993).
18. Zollfrank, J., and Friedrich, J., *J. Phys. Chem.*, *96*; 7889 (1992).
19. Ellervee, A., Jaaniso, R., Kikas, J., Laisaar, A., Suisalu, A., and Scherbakov, V., *Chem. Phys. Lett.*, *176*; 472 (1991).
20. Bogner, U., Method for providing separately adjustable voltage dependent optical properties, U. S. pat. 4 733 369.
21. Bogner, U., Beck, K., and Maier, M., *Appl. Phys. Lett.*, *46*; 534 (1985).
22. Schätz, P., Bogner, U., and Maier, M., *Appl. Phys. Lett.*, *49*; 1132 (1986).
23. Hartmannsgruber, N., Bogner, U., and Maier, Max, *J. Molecular Electronics*, *5*; 193 (1989).
24. Hartmannsgruber, N., Bogner, U., and Maier, Max, *Opt. Quant. Electronics*, *23*; 361 (1991).
25. Hartmannsgruber, N., and Maier, Max, *Appl. Phys. Lett.*, *58*; 1585 (1991).

26. Ollikainen, O., Rebane, A., and Rebane, K., *J. Optical and Quantum Electronics*, in press.
27. Wild, U. P., Renn, A., DeCaro, C., and Bernet, S., Spectral hole-burning and molecular computing, *Applied Optics*, *29*; 4329 (1990).
28. Bogner, U., Schätz, P., Seel, R., and Maier, M., *Chem. Phys. Lett.*, *102*; 267 (1983).
29. Johnson, L., Murphy, M., Pope, C., Foresti, M., and Lombardi, J., *J. Chem. Phys.*, *86*; 4335 (1987).
30. Kador, L., Jahn, S., Haarer, D., and Silbey, R., *Phys. Rev.*, *B41*, 12215 (1990).
31. Renn, A., Bucher, S., Meixner, A., Meister, E., and Wild, U., *J. Luminescence*, *39*, 181 (1988).
32. Vauthey, E., Holliday, K., Wei, C., Renn, A., and Wild, U., *Chem. Phys.*, *171*, 253 (1993).
33. Gerblinger, J., Bogner, U., and Maier, M., *Chem. Phys. Lett.*, *141*, 31 (1987).
34. Attenberger, T., Bogner, U., and Maier, M., *Chem. Phys. Lett.*, *180*, 207 (1991).
35. Gradl, G., Kohler, B. E., and Westerfield, C., *J. Chem. Phys.*, *87*, 6064 (1992).
36. Bogner, U., Attenberger, T., and Bauer, R., Correlations between dynamical processes and electric-field effects in perylene/*n*-heptane probed by persistent spectral holes, in *Spectral Hole-Burning and Luminescence Line-Narrowing: Science and Applications*, Technical Digest, Optical Society of America, Washington D.C., 1992, Vol. 22, p. 22.
37. Bogner, U., Röska, G., and Graf, F., *Thin Solid Films*, *99*, 257 (1982).
38. Lin, W. I., Tada, T., Saikan, S., Kushida, T., and Tani, T., The study of weak linear electron–phonon coupling in iron-free hemeproteins, in *Persistent Spectral Hole-Burning: Science and Applications*, Technical Digest Series, Optical Society of America, Washington D.C., 1991, Vol. 16, p. 210.
39. Holliday, K., Wei, C., Meixner, A., and Wild, U., *J. Luminescence*, *48/49*, 329 (1991).
40. Kanaan, Y., Attenberger, T., Bogner, U., and Maier, Max, *Appl. Phys.*, *B51*, 336 (1990).
41. Renn, A., Bucher, E., Meixner, A., Meister, E., and Wild, U., *J. Luminescence*, *39*, 181 (1988).
42. Hartmannsgruber, N., and Maier, Max, *J. Chem. Phys.*, *96*, 7279 (1992).
43. Dicker, A., Johnson, L., Noort, M., and van der Waals, J., *Chem. Phys. Lett.*, *94*, 14 (1983).
44. Orrit, M., Bernard, J., and Zumbusch, A., *Chem. Phys. Lett.*, *196*, 595 (1992); Erratum, *199*, 408 (1992).
45. Yoshimura, M., Nishimura, T., Yaggu, E., and Tsukada, N., *Polymer*, *33*, 5143 (1992).
46. Kharlamov, B., Personov, R., and Bykovskaya, L., *Opt. Commun.*, *12*, 191 (1974).
47. Winnacker, A., Shelby, R. M., and Macfarlane, R. M., *Opt. Lett.*, *10*, 350 (1985).
48. Gibbs, M., *Optical Bistability: Controlling Light with Light*, Academic Press, Orlando, FL, 1985.
49. Hong-Jun, Z., Jian-Hua, D., Jun-Hur, Y., and Cun-Xiu, G., *Opt. Commun.*, *38*, 21 (1981).

50. Bogner, U., Beck, K., Schätz, P., and Maier, M., *Chem. Phys., Lett., 110*, 528 (1984).
51. Hartmannsgruber, N., Bogner, U., and Maier, M., unpublished.
52. Meixner, A., Renn, A., and Wild, U., *J. Chem. Phys., 91*, 6728 (1989) and references therein.
53. Jaaniso, R., and Bill, H., *Europhys. Lett., 16*, 569 (1991).
54. Holliday, K., Wei, C., Croci, M., and Wild, Urs. P., *J. Lumin., 53*, 227 (1992).
55. Zhang, J., Huang, S., and Yu, J., *Chin. J. Luminescence, 12*, 181 (1991).
56. M. Nogami, Y. Abe, K. Hirao, and A. H. Cho, *Appl. Phys. Lett., 66*, 2952 (1995), and references therein.
57. Kurita, A., Kushida, T., Izumitani, T., and Matsukawa, M., *Opt. Lett., 19*, 314 (1994).
58. Bauer, R., Osvet, A., Sildos, I., and Bogner, U., *J. Luminescence, 56*, 57 (1993).
59. Weber, A., Bauer, R., and Bogner, U., to be published.
60. McAulay, A., *Optical Computer Architectures: The Application of Optical Concepts to Next Generation Computers*. John Wiley, New York, 1991, Chapter 15 (Optical cellular automata), pp. 409–430.

13

Persistent Spectral Hole Burning: Time- and Space-Domain Holography

Karl K. Rebane

Institute of Physics
Tartu, Estonia

Alexander Rebane

Swiss Federal Institute of Technology
Zürich, Switzerland

I. INTRODUCTION

Persistent spectral hole burning (PSHB) in *inhomogeneously* broadened impurity absorption bands of solids was found twenty years ago (1–3). For a brief historical overview of spectral hole burning, including the *transient* version of it, see Ref. 4.

The fact that the spectral shape of absorption can be changed with high spectral resolution by illumination, and the changes preserved for long times, opened up the new fields of photophysics and photochemistry (3,5–10 and references therein). Spectral selectivity of the modification can be very high; it is limited by the *homogeneous* width of the absorption line $\Gamma(T)$, which for a purely electronic zero-phonon line (ZPL) in low temperature (liquid helium temperatures, $T \leq 4.2K$) impurity-doped solids is $\Gamma(T) = 10^{-2} \div 10^{-4}\mathrm{cm}^{-1}$ (8–10). The inhomogeneous bandwidth Γ_i depends largely on the host solid and the interaction of it with the optical transition in the impurity; it varies from $0.1–1$ cm^{-1} for rare-earth-ion-doped single crystals to $10^2–10^3$ cm^{-1} for molecular impurities in polymeric and glassy organic matrices. Thus a low-temperature inhomogeneous ZPL spectral line represents a broad band comprising $10^3–10^7$ sharp resonances.

These resonances can be addressed by illumination, which creates photochemical or other changes in that frequency-selected body of impurities, whose ZPLs happen to be in resonance with the excitation. The changed impurities have their absorption frequency shifted far (compared to the ZPLs' width) away from the initial ones, whereby absorption at the illuminated frequencies decreases. The changes are fixed for long periods of time by chemical changes of the impurity molecule or rearrangements of the constituents of the solid in its vicinity. If the illumination is a narrow-line laser light, the change results in a sharp hole of transparency in the inhomogeneous absorption band. Illumination by a light of more complicated spectral structure creats a profile of transparency, which corresponds to the spectral distribution of intensity (more precisely, dose) of the illumination. Thus the distribution of the ZPLs over frequencies is modified. Note that the change of absorption is not the only consequence. The modifications can be displayed in various ways in the responses to various excitations of the "information storage medium"—the impurity-doped solid with the written-in profile of transparency.

One possibility is to burn a sharp single hole, consider it as an ("negative") image of the homogeneous ZPL, and measure its shape in the luminescence excitation spectrum, which reflects absorption well but has a much better signal-to-noise ratio than the direct measurement of the latter. In this way, very high spectral resolution data about the *homogeneous* shape of a ZPL may be obtained (11,9–10 and references therein). Note that it is impossible to get these high spectral resolution data by trying to study absorption or luminescence directly, by conventional methods, because of the strong masking influence of *inhomogeneous* broadening. PSHB shows itself here as an effective tool to get rid of the tremendous inhomogeneous broadening and address the very sharp ZPLs indirectly, but with the accuracy up to the homogeneous shape of them. Remarkable progress in the matrix spectroscopy of molecules has been achieved (3,5–7,9–14) and references therein), including chlorophyll and its relatives (13). For instance, in studies of chlorophyll the resolution compared to that of conventional spectroscopy has been improved a thousandfold. Naturally, the sharp holes in absorption can be utilized also to store information coded in bits.

The other option is to store the information carried by a light pulse as broadband structured profile of the decrease of the inhomogeneous absorption. Note that the spectral accuracy of the write-in of the profile may be very high—up to the homogeneous width $\Gamma(T)$. A simple but impressive example is utilizing PSHB material directly as a spectrometer with a data storage device. A piece of material whose spectrum of luminescence we are interested in is excited, and the luminescence falls onto a block of PSHB substance and burns a hole profile into it. A rather fine structured spectrum (e.g., Shpolsky spectrum) may be detected and stored in such a simple way (Fig. 1, J.Kikas (14). The obvious version of PSHB data storage is to interpret the presence of a hole at a certain $x,y;\omega$ position as "one" and the absence of a hole as "zero." Thus the frequency dimension enlarges the spatial density of storage by a factor of $\Gamma_i : \Gamma(T)$, i.e., about 10^4 times. The number of bits available in principle becomes very large,

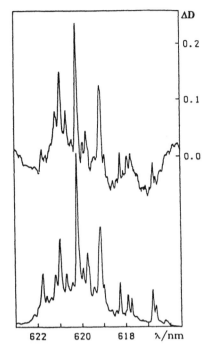

Figure 1 The fluorescence spectrum of octaethylporphin molecules embedded in *n*-hexane (Shpolsky system) at $T = 5K$ measured in a conventional way by means of a DFS-24 spectrometer (below) and the same spectrum recorded by illuminating a persistent spectral hole burning material (the same octaethylporphin doped *n*-hexane but with large inhomogeneous broadening) directly with a photoexcited fluorescence flux from the sample under study. (From Ref. 14.)

and parallel processing becomes a must. Considerably more advanced is space- and frequency-domain holography (15,6,7 and references therein), which is a coherent write in and read out procedure.

The real capability of broad-band information storage opens up when frequency is utilized (via Fourier and Hilbert transforms) to store the data coded *in time*. Time- and space-domain PSHB holography of pico- and femtosecond pulses, which performs *coherent* data storage and processing, has been proposed and experimentally realized (16,17,5–7 and references therein). In addition to the spectacular features provided by coherence itself, the holographic approaches are based on *distributed* (over spatial coordinates and frequencies) data storage and fully parallel processing, which make the storage considerably more resistant to errors and the processing faster by several orders of magnitude. A recent application is the PSHB optical modelling of error-corrective neural networks

(18,19). One approach of this simultaneous frequency and time PSHB indicates the potential of PSHB in wavelet studies and applications (20).

Up to now, PSHB has been demonstrated and studied for hundreds of impurity-doped organic and inorganic materials. The main requirement is the presence of narrow and intense *homogeneous* lines, constituting a broad *inhomogeneous* absorption band. The inhomogeneous distribution of impurities is always the case for a real solids. The first requirement is usually satisfied if there are ZPLs in the spectra. Fortunately, there are hundreds of host solids (single crystals and polycrystalline, polymeric, and glassy materials) and thousands of impurities as dopants, which have nice ZPLs at liquid helium temperatures and thus can serve as good PSHB materials. The condition for the presence of ZPLs is the absence of strong interaction between the impurity (more precisely, between the given electronic transition) and the lattice modes. First of all, the Stokes losses must be small and temperatures low. A large variety of the potential PSHB substances matching these conditions makes the search for the really suitable ones for various applications quite prospective.

Actually, persistent spectral hole burning provides a unique possibility of controlling, by means of illumination with high spectral resolution, optical properties, the coefficient of absorption and, correlated to it (via Kramers–Kronig relations), the index of refraction in the impurity absorption band of doped solid materials. The foundation stone is the zero-phonon line.

II. ZERO-PHONON LINE. PEAK VALUE OF ABSORPTION CROSS-SECTION

The cornerstone of PHSB is the purely electronic zero-phonon line (ZPL) in the impurity spectra of solids (Fig. 2), the optical analog of the Mössbauer resonance γ line (8,5,10 and references therein). ZPLs are not Doppler-broadened. For molecular impurities, the rotational structure is also almost always eliminated. At liquid helium temperatures, the ZPL of a single molecule (impurity center) is often very narrow and has high peak intensity.

The homogeneous line width (i.e., that for a single impurity) $\Gamma(T)$ is given by

$$\Gamma(T) = \frac{1}{\pi c \tau_2(T)} = \frac{1}{\pi c}\left[\frac{1}{2\tau_2^*(T)} + \frac{1}{\tau_1}\right] \quad \mathrm{cm}^{-1} \tag{2.1}$$

where τ_1 is the excited electronic state lifetime, τ_2 is the full dephasing time, and τ_2^* is pure dephasing time. At the temperature $T = 0$, dephasing is absent ($\tau_2^* \to \infty$), $\Gamma(0)$ is determined solely by τ_1, and (in the absence of the quenching of luminescence) equals the radiative line width of the free molecule. For a temperature around 2K, and for typical excited electronic state lifetimes of 10^{-8}

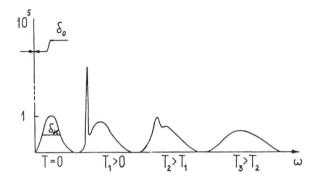

Figure 2 Schematic of the purely electronic zero-phonon line (ZPL) and phonon sideband in the impurity homogeneous absorption spectrum and its dependence on temperature; $T_3 \approx 40K$ for organic impurities in organic hosts. (From Ref. 10.)

s, we can estimate, for a well pronounced ZPL, $\Gamma(2K) \sim 2\Gamma(0) \sim 10^{-3} - 10^{-4} cm^{-1}$. The absorption at the maximum of a ZPL $\sigma(T)$ is by a factor of $\alpha(T)$ smaller than the universal value for the radiative peak absorption cross-section σ_τ of a free atom (21 and references therein)

$$\sigma(T) = \alpha(T)\sigma_\tau, \qquad \sigma_\tau = \frac{\lambda^2}{2\pi} \qquad (2.2)$$

Here σ_τ is given for nonpolarized excitation light (22). To take account of polarization, a factor $\beta(\theta)$ has to be introduced. In case of linear polarization and dipole transition, $\beta(\theta) = 3 \cos^2 \theta$, where θ is the angle between the electric vector of the excitation light at the impurity site and the direction of the transition dipole moment. Note that the force caused by the first vector depends on the local situation at the impurity site and thus, in principle, also additionally on the angle θ. λ is the excitation wavelength at the maximum of absorption. A simple but reasonable consideration gives for ZPL absorbing nonpolarized light (23,21),

$$\alpha(T) = F_{DW}(T) \frac{\Gamma(0)}{\Gamma(T)} \qquad (2.3)$$

where n_0 and n are the refractive indices in vacuum and in solid, and thus for polarized excitation,

$$\sigma(T) = \frac{\lambda_n^2}{2\pi} F_{DW}(T) \frac{\Gamma(0)}{\Gamma(T)} \beta(\theta) \qquad (2.4)$$

Here $\lambda_n = \lambda/n$ is the wavelength in the sample; $F_{DW}(T)$ is the Debye–Waller factor, defined as the ratio of the integrated intensity of the ZPL, S_0, and the sum of S_0 and the integrated intensity of the phonon sideband, S_κ, accompanying the ZPL (Fig. 2) (8,5,10),

$$F_{DW}(T) = \frac{S_0}{S_0 + S_\kappa} \qquad (2.5)$$

F_{DW} shows how much of the oscillatory strength of the electron transition is located in ZPL and is the smaller the stronger is the electron–phonon interaction and the higher the temperature. $F_{DW}(T)$, consequently, has different values for different impurity systems, the common feature being that it decreases rather rapidly with the increasing temperature, and for most molecular systems it drops to zero at temperatures above $20K$. As rough estimates, the values $F_{DW}(T = 2K) = 0.1$, and $\Gamma(0) : \Gamma(T) = 0.5$, may be taken. Then, $\alpha(T = 2K)\ 0.05$, $\beta(\theta) = 1$, and we have

$$\sigma(T) = \alpha(T)\frac{\lambda^2}{2\pi} \qquad \sigma(2K) \approx 10^5 d^2 \approx 10^{-11} cm^2 \qquad (2.6)$$

where d^2 is the geometrical size of the molecule, and for estimate $\lambda = 2500d$.

One of the finest uses of this precious feature of low temperature ZPL is the recent development of the spectroscopy of a single impurity molecule (24–25,6,7). It shows that the peak cross-section is large enough for selective excitation of a *single* molecule in a *dense matter* environment, i.e., in a situation where about 10^{10} molecules are under exciting illumination. This is a new level of matrix isolation spectroscopy of molecules, solid state spectroscopy, and also for the experiments reaching back to the fundamentals of the quantum theory of interaction of light with matter. Whether that development could be used for data storage on a single-molecule level ("one bit per one molecule") is an open question, and the current answer seems to be rather negative: no (or not yet).

III. INHOMOGENEOUS BROADENING OF ZERO-PHONON LINES

Because of its narrowness, ZPL is extremely sensitive to the slightest fluctuations of electrical, deformation, etc. fields at the site of the impurity. Even a weak perturbation can shift the ZPL's frequency out from its initial position and rather far away compared to its line width. An estimate for dipole point defects at concentrations 10^{-3}–10^{-4} mol/mol gives $\Gamma_i \approx 1$ cm^{-1} (8).

Thus, for a body of molecules, ZPL as a noble spectral feature is strongly distorted by large inhomogeneous broadening: in the absorption and conventionally measured luminescence spectra of an impurity-doped medium, ZPLs form a broad band.

The experimental inhomogeneous bandwidth Γ_i is about 1 cm^{-1} for single

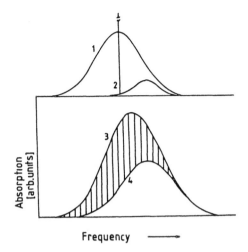

Figure 3   1, the inhomogeneous distribution function; 2, the homogeneous ZPL with its accompanying phonon sideband; 3, the inhomogeneous absorption band as the sum of two parts; 4, a really continuous spectral band formed by the overlapping phonon sidebands of spectrum 2 and a pseudocontinuous broad band (between curves 3 and 4) of very narrow homogeneous ZPLs distributed as shown by curve 1. The calculation is performed for the Debye–Waller factor of spectrum 2, being $F_{DW} = 0.5$ (10).

crystal hosts, $1-10$ cm^{-1} for polycrystalline structures (e.g., Shpolsky systems), and 10^2-10^3 cm^{-1} for disordered solids—glasses and polymers. Actually inhomogeneity of the sample creates a distribution of the impurities over the frequencies $g(\vec{r}, \omega)$, which may depend on the site in the sample. The profile of the inhomogeneous distribution function $g(\vec{r}, \omega)$ is often rather complicated and depends on the way the sample was prepared; in the presence of slow relaxation processes in the sample, it may depend on time, but it does not depend directly on temperature.

We can see that $\Gamma_i = (10^3-10^6)\Gamma(2K)$, i.e., the inhomogeneous ZPL band comprises thousands of distinguishable very sharp resonances (Fig 3). This feature is a precious gift of nature; it enables us to perform fine optics, spectroscopy, and photochemistry. New possibilities for applications open up, including various versions of data storage and processing. But an effective tool to address, control, modify, and fix modifications of these resonances is necessary.

IV. PERSISTENT SPECTRAL HOLE BURNING

Persistent spectral hole burning (PSHB) is the tool that enables one to control by illumination with high spectral resolution the absorption coefficient and the (bound to it via Kramers–Kronig relations) index of refraction of the impurity-

doped materials. The possibility of burning holes in low temperature inhomogeneous ZPL band is indicated in Ref. 8, p. 150. The new and essential feature
is that in this case the holes can be persistent (1,2). Illumination with a narrow
laser line brings along spectrally selective interaction with that subset of impurity
molecules whose ZPLs are in resonance with the laser frequency. These impurities are, because of their very high absorption cross-section $\sigma(T)$, under "heavy
bombardment" by photons, and they can undergo photostimulated changes, i.e.,
photochemical reaction in the molecule itself or any other process in the impurity
center or in its surrounding matrix. These changes shift the ZPL frequency to a
new value, typically rather far away from the initial one. The number of molecules able to absorb resonantly at the excitation laser frequency decreases. This
leads to the appearance of a hole (dip) in the inhomogeneous band of the impurity
absorption. The hole width in the inhomogeneous absorption band (more precisely in the inhomogeneous distribution function $g(\vec{r}, \omega)$) is given by the convolution of the shapes of the laser line and the ZPL's homogeneous line; the hole
area is determined by the dose of illumination. The hole width detected in a
spectrum is broader, since the readout procedure brings along another convolution
with the ZPL's homogeneous line shape function. Holes in $g(\vec{r} \omega)$ may be studied
by using various techniques, the most popular one being the measuring of the
luminescence excitation spectrum, which is closely correlated to absorption. If
the hole burning and the readout laser width are negligibly small compared
to the ZPL's homogeneous width $\Gamma(T)$, the hole width in the absorption (or
luminescence excitation) spectrum is twice $\Gamma(T)$. Note that the spectral and the
spatial resolution of a PSHB material is, in principle, several orders of magnitude
higher that that of a photographic one (ten thousand different frequencies as
"colors"); one wavelength is the limit for spatial resolution only in conventional
optical addressing; the PSHB capability of feeling and fixing the action has the
spatial resolution down to the distance between impurities, at which there is still
no energy transfer between them. Thus utilizing the frequency as an additional
coordinate in optical addressing brings the spatial resolution to the nanometer
scale. For other fine features of the hole burning art see e.g., Refs. 26,27,10.
Lifetime τ_H of the holes depends on the rate of the reverse chemical reactions
or on the time of reverse rearrangements of the constituents in the impurity
center and its environmental host. τ_H is different for different systems and shows
a large variety of temperature dependencies. We could accept to call *persistent*
the holes whose τ_H exceeds 100 s, considering the faster decaying ones to be
transient. Actually for the most of today's popular PSHB systems τ_H exceeds
this roughly put boundary value by several orders of magnitude and may reach
quite a number of years, especially if the sample is stored at low temperatures.
In parallel with the deterioration of the holes because of filling (the area of the
holes decays), the broadening of the holes' shape (the area being approximately
constant) is also present. The PSHB materials mostly used today show no essen-

tial changes of the hole structure during the experiments, i.e., during several hours until the liquid helium evaporates (see Ref. 28 and references therein). In all PSHB applications, the light–matter interaction may be, and actually mostly is, rather weak. The conventional linear description of this interaction holds well. No intense excitation light to create nonlinear light–matter interaction is needed. The space-and frequency-domain hole pattern is written in PSHB material in step-by-step accumulation of small changes of the coefficient of absorption and the index of refraction under modest illumination. Coherent times domain optical responses like photochemically accumulated stimulated photon echo (PASPE) (16,29,5,10) and other PSHB effects arise from the circumstance that after hole burning the piece of matter has its optical characteristics changed, and a new frequency or space-and-frequency pattern of permittivity is formed. This pattern stores for long times (determined by the lifetime of the spectral holes) the information transferred by the illumination. This difference between the optical properties of the material before and after PSHB is caused by gradually collecting small changes of the number of impurity molecules, absorbing via their ZPLs, at a given frequency ω, at a given spatial spot. In other words, the light–matter interaction is well described by conventional linear Maxwell equations, but the equations describing the properties of matter are essentially different before and after hole burning. It should also be mentioned that in PSHB, because of the possibility of keeping excitation intensities low, the *power* broadening of ZPLs and, consequently, the power broadening of spectral holes caused by saturation effects can be avoided. *Hole* broadening arising quite easily with the growing dose of illumination and also from spectral diffusion is always possible. Additional effects arise in optically dense samples. They include hole narrowing (26,27) and dependence of the hole profiles on the geometry of burning (one- or two-way burning) and on temporal ordering of the irradiation at bichromatic burning (27).

Let us note a remarkable possibility: PSHB enables one to create and study exotic photochemical reactions. It allows one to give a small fraction of impurity molecules a really high excitation load, initiate the reaction, and study its results with high spectral selectivity, without producing a prohibitive amount of heat in the sample.

V. PSHB MEMORY

A broad absorption band with a sharp spectral hole in it is a narrow-band wide-aperture transparency filter. PSHB can be used to produce filters with various spectral and spatial transparency profiles (10,26,27). An important case is a filter as an optical memory: bits can be coded along the frequency axis ω, by interpreting the presence of a hole at the frequency w_i as a bit value 1 and the absence of a hole as a bit value 0. An inhomogeneous band of a typical PSHB material,

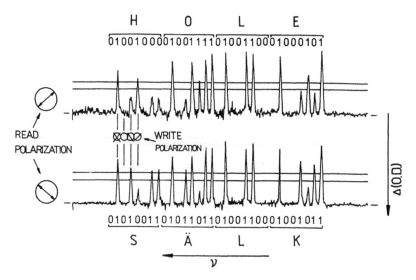

Figure 4 Words "HOLE" and "SÄLK" (the Estonian equivalent of "dip") are recorded and played back by using two mutually perpendicular polarizations at the same 0.5 cm² spot of PSHB material (polystyrene block doped with octaethylporphin molecules at $T \sim 5K$, R. Jaaniso, 1989 (30)).

e.g., octaethylporphyrin molecules in polystyrene matrix (concentration about 10^{-4} mol/liter), can comprise up to 10,000 bits on the frequency axis at $T \approx$ 2K. An example of frequency-coded data is shown in Fig. 4 (30); see also (3). The smallest size of the spatial spot at which the frequency bits can be written in is limited by the diffraction, i.e., the area per one focused laser spot can ultimately be of λ^2 (λ^3 for volume storage), where λ is the hole-burning wavelength. Note that this limit is put by the conventional optical addressing and not by the spatial resolution capabilities of PSHB material.

The theoretical upper limits for an optical PSHB storage density are high. Optimistic estimates place them to 10^{12} bits per cm² (i.e., 10^4 spectral holes per each diffraction-limited spatial pixel of the size $\lambda^2 \sim 10^{-8}$cm²) or 10^{16} bits per cm³ for a volume storage memory (cell size $\lambda^3 \sim 10^{-12}$cm³ (3,31). If the impurity concentration is too high, energy transfer between them may occur. This results not only in some broadening of homogeneous ZPLs but also, decisively, in addition to the direct and spectrally selective hole burning, spreading the hole burning all over the inhomogeneous band of ZPL frequencies. After an impurity is selectively excited and *before* the act of burning-out is fixed, the excitation energy may be transferred to a neighbor impurity and the burning-out act performed there. The neighbor's ZPL frequency may be quite different (the effective resonant energy transfer can be provided by the overlap with phonon sidebands), and thus the contribution to the hole may go to a frequency quite different from the excitation one. This is one of quite a number of situations when it is obligatory to remember about phonon sidebands. The erasing of the hole structure arises also from the changes of the positions or orientations of the molecules in the sample after the hole structure has been implemented, i.e., spectral diffusion is initiated by spatial diffusion. On the other hand, spectral diffusion opens a window to study even very slow diffusion processes in low temperature solids (12). The narrow width of ZPLs makes them sensitive to even the smallest changes of the kind. In disordered matrices (glasses, polymers), two-level systems contribute strongly to spectral diffusion.

Interactions between impurities can be suppressed by taking lower concentrations, but this will result in a critical decrease of the average number of impurity molecules available for burning a hole at a fixed wavelength at a fixed spatial spot. If we assume that the concentration of impurities in the absence of energy transfer can be 10^{18} molecules per cm^3, and that the storage density is 10^{16} bits per cm^3, there will be on the average only 100 molecules to store one bit of information. That number is obviously too small, because its fluctuations from one pixel $\delta x\ \delta y\ \delta z\ \delta\omega$ to another are too large compared to the number of burnt-out impurities. Note that this estimate holds for both three-dimensional $\delta x\ \delta y\ \delta\omega$ and four-dimensional $\delta x\ \delta y\ \delta z\ \delta\omega$ storage: in the first case the depth of focusing brings along an additional limitation, δz, for the storage volume.

To increase the number of molecules per bit and to observe at the same time the concentration limitations, imposed by the requirement not to have energy transfer between impurities, a PSHB memory composed of impurity-activated optical fibers has been proposed (32). This design guarantees also a good control of the propagation of light beams in the memory. A 1 cm long and λ thick piece of fiber comprises at a given concentration 10^4 times more impurity molecules than a λ^3 cell. Estimates show that electromagnetic cross talk between the illuminated adjacent fibers of 1 cm is negligible if they are separated by a distance of about $1.5\ 10^{-4}$ cm, i.e., 3λ, if $\lambda = 500$ nm (33). On the other hand, to avoid cross talk during the optical addressing caused by the overlapping of the light beams focused at neighboring positions at the memory surface, about the same separation 3λ between the centers of the adjacent illuminated spots is needed (34). Compared to other versions of bit by bit storage, the PSHB fiber memory seems prospective, especially for tasks where the use of all of the spectral capacity is instrumental.

Bit by bit (digital) storage is not a full-scale utilization of the capabilities of PSHB materials' very high spectral selectivity and very high spatial resolution. The practical limit for the latter is the storage wavelength $\lambda \approx 10^{-4}$ cm, the principal limit that makes sense when speaking about single impurity molecule data storage—the distance between the adjacent impurities, i.e., approximately 10^{-6}cm. This limit may be reached in addressing if shorter than visible wavelengths or sharper than λ^2 (e.g., nonlinear or via sharp pointed fibers) focusing are used. But, as has already been mentioned above, the number of molecules puts the principal limit. The most important advantage is the physical dimensionality of the frequency, the reciprocal of the time. It is not used in digital storage. The practical problem is the enormous numbers of bits, which make parallel processing a must. No effective schemes for large scale bit by bit PSHB storage have been proposed up to now.

Fourier transform turns frequency functions into time-dependent ones, and this feature has to be exploited. Time-dependent storage and playback for every diffraction-limited spot means time and space-domain holography of the entire time-dependent picture, i.e., the scene or event (16,35–40 and references therein).

Two versions of PSHB distributed data storage and processing have been realized up to now—holography and models of neural networks. In PSHB holography, the number of space-frequency memory cells involved and addressed in parallel is really large. Potentials of PSHB materials were described above on the example of digital storage. (See Chapter 14). Note that in the case of digital storage only the presence or absence of holes matters. In holography, depths of holes are also important, and the storage capacity increases approximately n-fold, where n is the number of different hole depths. Further, in parallel writing, the hole burning light is *simultaneously* on all over the space and frequency domain of the PSHB sample, and it compensates to some extent for the errors arising in sequential bit by bit storage, where the products of burning the latter holes can fill partly the holes burnt earlier.

VI. HOLOGRAPHY

New versions of holography are the most successful application of PSHB data storage and processing up to now. It utilizes both distributed storage and parallel processing and demonstrates the advantages and potentials of PSHB in this field.

PSHB has enriched holography with the frequency domain and the time domain as additional "degrees of freedom," which, owing to their physical dimensionalities, ω and t, not only enlarge the number of available bits but also open up principally new possibilities. Two main new fields of holography have been born: space and frequency-domain stationary holography (15,41,42) (see also Chapter 14 by De Caro et al. in this book) and space and time-domain holography (16,35–40,43, see also 5–7). The former utilizes ω directly as an additional coordinate to enlarge the storage capacity several thousand times. The latter one exploits the high-frequency selectivity of the PSHB materials to realize the holography of time-dependent pictures, i.e., events. It is shown that PSHB holography can also include the polarization of light pulses (39). It means that all the features carried on by a classical light pulse—three-dimensional time-dependent color picture, including also polarization of light—can be holographically stored and played back. A first attempt to store several picosecond events at different intervals of the frequency axis, i.e., to realize space, time, and frequency-domain holography, has been used in one of the versions of PSHB modelling of neural networks (19,44,45, see also Sec. 8).

Introducing time and frequency brings PSHB holography under the reign of causality: Kramers–Kronig relations between the coefficient of absorption and the index of refraction are in power. This fundamental bound, which naturally is an obvious must for time sequence of events (causality clearly preserves and plays back the sequence of events in time), holds also for time-independent pictures stored in the frequency domain, and the awareness of this connection helps one to minimize, in stationary holography, cross talk between the holograms stored at different frequencies in the same PSHB sample (42). Causality teaches one to perform spectacular experiments on the time arrow and its reversal (PSHB conjugation of wave fronts) in time and space-domain holography, exploiting associative memory properties (e.g., correction of distorted signals (46), imaging through light scattering media (47), and storage and study of femtosecond domain light pulses (48).

An interesting property of time and space-domain holograms occurs in connection with external electric fields (43): the pseudo-Stark splitting of narrow spectral holes due to the linear Stark effect in an external electric field manifests itself as an additional modulation of holographically stored signals in the time domain. At suitable material parameters (the value and orientation of the intrinsic and induced dipole moments' difference in the ground and in the excited electric state of the impurity molecules), it is possible to change the phase of the holographic images by $\pm\pi$ applying an external electric field.

The possibility of fully parallel optical access to the stored data, even if the information is partially (or completely) coded in the frequency and/or time dimensions, makes the application of hole burning materials for associative memories (49) and parallel processing especially attractive. This possibility has been demonstrated by the reconstruction of full holographic spatial and also space and time-domain images, as the readout pulse comprises only a part of the image in space and/or in time (46,50).

A remarkable property of associative memories is that the input information may comprise errors (e.g., probe images with erroneous elements or missing parts) but can be corrected in the readout without any loss of essential information: the recall automatically corrects for the errors towards the closest matching amongst the memorized information units (51,52). This possibility is realized in PSHB modelling of neural networks (19,44,45). We consider below the theory of the time and space-domain holography (Secs. VII. A and VII. B) and some examples of PSHB distributed data storage and parallel processing (Secs. VIIC and VIID).

VII. PERSISTENT SPECTRAL HOLE BURNING TIME- AND SPACE-DOMAIN HOLOGRAPHY

A. Hole Burning by Picosecond Pulses

PSHB by light pulses opens up a new and promising field for optical data storage and signal processing on the nano-, pico-, and femtosecond time scales. These applications bring PHSB into correlation with the transient photon–echo phenomena (53), but here the storage is *permanent,* and the playback may be performed many times during the lifetime of the persistent holes, i.e., after hours, days, or more time has passed from the time of the storage.

The conventional method of performing high-resolution spectroscopy and optical data storage is first to burn spectral holes and then to study them by means of tunable narrow-line cw lasers. When dealing with large numbers of bits (10^{10}–10^{12} bits·cm^{-2}), parallel processing becomes a must. From the basic principles of the physics of wave processes it follows that the detection of a spectral hole with accuracy $\delta\omega$ requires a measurement time at least as long as $1/\delta\omega$. This means that the probing of a hole of width $\delta_H = 10^{-3}$ cm^{-1} takes at least 10^{-7} s to establish the probing quasi-monochromatic frequency. The burning time of such a narrow single hole is at least as long or longer, depending on the quantum efficiency of the process.[1] Therefore, single-beam readout of 10^{12} bits stored on a 1×1 cm^2 area takes at least 10^5 seconds, i.e., one day. Of course, working with poorer spectral resolution enables one to read faster. The loss of spectral storage density has to be compensated by increasing the number of spatial storage spots via making the spatial storage area larger.

A complementary technique to conventional quasi-monochromatic data processing is the excitation and reading of the PSHB medium by means of short light pulses. A pulse of 2–3 ps duration covers about 5 cm^{-1} of frequency space comprising tens of thousands of homogeneous ZPL lines. The problem is then how these sharp resonances will respond when excited by a picosecond pulse. Can we expect that the unique narrowness of the ZPLs determined by the excited electronic state lifetime τ_1 will be displayed at all, since τ_1 ($\simeq 10^{-8}$ s) is 10^4-fold longer than the excitation time? The answer is yes, by virtue of well-known Fourier-transform considerations, which certainly will be valid if the interaction processes can be considered as linear. This is the case in the experiments described below. Actually a proper understanding of the situation needs a bit of theory of time-dependent spectra. It shows (54) that the Fourier components of the pulse can, in fact, be stored with the accuracy of the homogeneous line width

---

[1] Hole burning and reading time of 30 ns (corresponding to a hole width of 10^{-3} cm^{-1}) have actually been achieved already by means of semiconductor heterojunction lasers (see W. E. Moerner et al. in Ref. 53).

provided the light-matter interaction is linear, and that the results of PSHB are observed not earlier than τ_1 after the excitation.[2] There is another useful feature of this approach: one can utilize the very high spectral resolution provided by ZPLs without having to use very narrow line lasers. Thus one can write information in tens of thousands of spectral channels at each spatial spot with the dimension of about 10^{-8} cm^2 over an area of several cm^2 simultaneously. In principle it can be done in pico- or femtoseconds, but it requires a strong exciting light, which may bring along nonlinearities or power broadening of holes. Thus in reality to burn persistent holes takes a longer time: how long depends on the efficiency of the PSHB and how deep holes must be. Actually sequences of picosecond pulses (pairs of pulses in holography) of equal shape are applied to obtain the desired hole profile. The whole write-in procedure takes from fractions of a second up to minutes.

To store all the intensities of the Fourier components of a pulse lasting t_p seconds, an inhomogeneous bandwidth Γ_i broader than t_p^{-1} is needed, i.e., 100 cm^{-1} is a sufficient bandwidth for picosecond pulses. This width is rather typical for organics. For a 10 femtosecond pulse $\Gamma_i \geq 500$ cm^{-1} is required. Storage of fast events is clearly an application where large inhomogeneous broadening turns out to be a useful feature of the materials. Doping with two (or more) different PSHB active impurities helps to get broad bands.

The homogeneous line width $\Gamma(T)$ determines the ultimate spectral resolution in the frequency domain. As we know, when $T \to 0$, $\Gamma(T) \to \Gamma(0) = (\tau_1)^{-1}$, where τ_1 is the excited electronic state decay time. The hole width $\delta_H(T) = 2\Gamma(T)$, and as the limit $\delta_H(O) \geq 2\tau_1^{-1}$. The inverse hole width $\delta_H^{-1} \sim \tau_{max}$ places an upper limit on the temporal duration of the signal pulse to be stored in the main scheme described below. New possibilities utilizing frequency chirping have been recently proposed (55).

B. Theory of Space- and Time-Domain Holographic Recording and Playback

The idea of PSHB time- and space-domain holography is simple. When a PSHB sample is illuminated by a pulse of light, the intensity (more precisely, the doses of illumination), in dependence of the frequency, is stored in the hole profile. In other words, the squares of the Fourier amplitudes are stored. The phase shifts between the Fourier components are lost, and the stored information is in general not adequate to restore the time profile and spatial structure of the initial pulse.

---

[2]Communication by Inna Rebane. The theory of time-dependent ZPL spectra and hole burning tells us also that under two and more pulse burning and detection hole widths narrower than $2\Gamma(T)$ are in principle possible (54).

If, turning to the idea of holography, two pulses separated in time (e.g., a signal pulse and a reference pulse) are applied, then in addition to the intensities of each of the pulses, also the contribution to the burning intensity from the interference between the pulses is stored and preserved for long periods of time as a hologram in the hole space–frequency profile. A frequency distribution of the phase differences is stored, and it proves to be enough to serve as a substitute for knowledge of all the phases. (More precisely, because of causality the amplitudes and phases are not completely independent but bound via the Kramers–Kronig relations. If the event is limited in time and space, and if certain conditions of storage are observed, the restoration of phases from the stored structure of intensities is possible (56).)

Applying an interrogating readout pulse of a proper space–frequency structure and incident angle (e.g., the reference pulse) to the PSHB hologram recalls the signal pulse in its full initial structure. Note that there is no need to have the pulses overlap in space and time, i.e., to be at the PSHB sample at the same moment of time. Interference occurs if the pulses are separated in time by not more than the impurity centers' phase relaxation time τ_2, which is in the nanosecond range. The excited electronic state of the impurity keeps the phases of the first pulse memorized for a time interval τ_2, and if the second pulse arrives before the phase memory is erased, interference takes place.

Below we follow a version of theory (17,36,37) (see also (38–40)) that considers the problem as linear filtering and shows very clearly that nonlinear light–matter interaction is not necessary to build PSHB time-and space-domain holography. This approach provides also some additional understanding of photon–echo phenomena.

We consider a PSHB sensitive plate with dimensions $2x_0$, $2y_0$, d, positioned in the plane $z = 0$ (Fig. 5), illuminated with a signal light pulse $S(\vec{r}, t)$ propagating in the positive direction of the z axis:

$$S(\vec{r}, t) = s(x, y, t - \frac{z}{c}) \exp[i\omega_0(t - \frac{z}{c})] \tag{7.1}$$

Let the bandwidth $\Delta\omega_s$ of the signal be narrow, i.e., $\Delta\omega_s \ll \omega_0$. The leading edge of the signal reaches the plate at $t = 0$, and the trailing edge leaves the plate at $t = t_s$.

Let us suppose further that the plate is illuminated also by a second pulse $R(\vec{r}, t)$, propagating at a small angle θ to the z axis and delayed by t_R from the first pulse (Fig. 5). $R(\vec{r}, t)$ has a flat wave front, is short compared to the duration of the signal pulse, and serves as a δ-function like a reference pulse in the conventional holographic write-in procedure:

$$R(\vec{r}, t) = R_0\delta(t - \frac{\vec{n}_R \cdot \vec{r}}{c} - t_R) \exp[i\omega_0 (t - \frac{\vec{n}_R\vec{r}}{c} - t_R)] \tag{7.2}$$

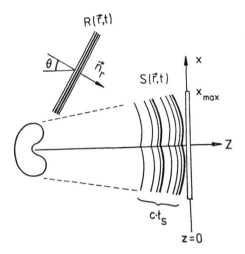

Figure 5 Time- and space-domain holographic recording. The object pulse $S(\vec{r}, t)$ reaches the recording PSHB plate at the moment $t = 0$; the reference pulse $R(\vec{r}, t)$ is shown to be delayed. Thus here the signal is put into the future relative to the reference pulse, and it may be recalled in the wave front conjugated form (17,38–40).

where $\vec{n}_R - (-\sin\theta, 0, \cos\theta) \approx (-\theta, 0, 1)$ shows the pulses' direction of propagation.

The delay t_R and the duration of both pulses are chosen to be within the phase memory time τ_2 of the excited electronic state of the impurity. Thus the interference between the signal and reference pulses takes place, and the distribution of the joint field, which forms around ω_0 via PSHB the space and frequency-dependent profile of transparency, is

$$I(x, y; \omega) = R_0^2 + |s(x, y, \omega - \omega_0)|^2 + R_0 s_F(x, y, \omega - \omega_0)$$

$$\times \exp\left[i\omega\left(-\frac{x\theta}{c} + t_R\right)\right] + R_0 s_F(x, y, \omega - \omega_0)\exp\left[-i\omega\left(-\frac{x\theta}{c} + t_R\right)\right] \quad (7.3)$$

where the index F denotes the Fourier transform.

The PSHB medium is characterized by its function of dielectric permittivity:

$$\epsilon(\vec{r}, \omega) = \epsilon_0 + \frac{\sigma c}{2\pi\omega} \int_{-\infty}^{+\infty} \frac{g(\vec{r}, \omega')d\omega'}{\omega' - \omega + i\tau_2^{-1}} \quad (7.4)$$

where σ is the integrated (over frequencies) value of the ZPL absorption cross-section; τ_2^{-1} is homogeneous width of ZPL; $g(\vec{r}, \omega)$ is the inhomogeneous distribution function, $g(\vec{r}\ \omega)d\omega$ giving the number of impurities whose ZPL frequency is in the interval $\omega + d\omega$; ϵ_0 is the part of dielectric permittivity that does not depend on the PSHB transition, and we can put $\epsilon_0 = 1$. Note that we neglect here the contribution from the phonon sidebands.

If the width of the inhomogeneous band Γ_i is considerably broader than the spectral width of the reference pulse Δ_r (here we speak about the pulse realized in experiment, not about the δ-approximation in the simple version of the theory), and the reference pulse is broader than the signal's spectral width Δ_s, and the ZPL homogeneous width $\Gamma(T)$ is much smaller than Δ_s (Fig. 7a), then the dielectric permittivity may be written as.

$$\epsilon(\vec{r}, \omega) = 1 - \frac{\sigma_c}{2\omega}i(1 + i\hat{H})\,\{g(\vec{r}, \omega)\} \qquad (7.5)$$

Here $(1 + i\hat{H})\,\{g(\vec{r}, \omega)\} \equiv g(\vec{r}, \omega) + i\hat{H}\{g(\vec{r}, \omega)\}$, where $\hat{H}\{g(\vec{r}, \omega)\}$ is the Hilbert transform of $g(\vec{r}, \omega)$:

$$\hat{H}\,\{f(\omega)\} \equiv \frac{1}{\pi} \int \frac{d\omega' f(\omega')}{\omega' - \omega} \qquad (7.6)$$

Formula (7.5) is obtained and works well for $\Gamma_i >> \Delta_r >> \Delta_s >> \Gamma(T)$. Let us note that the decisive requirement is $\Gamma_i >> \Gamma(T)$, i.e., the inhomogeneous band has to comprise a large body of homogeneous ZPLs, which is really the case for PSHB materials at liquid helium temperatures. The inequality $\Delta_r >> \Delta_s$ is actually irrelevant: two signal pulses of the durations Δ_s comprised in the $\tau_2 \sim \Gamma(T)^{-1}$ time interval may be holographically stored and played back without applying any reference pulses. This is just the interesting case of a spectacular display of associative memory properties of time and space domain holograms (50 and references therein).

The depth of PSHB-created profile of transparency grows with the dose of hole burning illumination. If the intensity is modest (which is mostly the case in experiments) and no saturation effects of hole burning are present (they appear after a long burning time), the dose is proportional to the illumination intensity $I(x, y; \omega)$ and we can write for the modified inhomogeneous distribution function,

$$g(\vec{r}, \omega) = g_0\{1 - \kappa(\omega)I(x, y; \omega)\exp[-g_0\sigma d]\} \qquad (7.7)$$

Where g_0 is the inhomogeneous distribution function of the ZPL frequencies before hole burning and κ shows the efficiency of PSHB. The approximations made above suggest the taking of $g_0(\vec{r}, \omega) = \kappa(\omega) = const$.

In PSHB holography the sample is optically thick, i.e., $g_0\sigma d >> 1$, the *spectrally selective* PSHB bleaching cannot be strong, and usually the PSHB changes are small compared to the initial absorption, i.e., the second term in Eq. (7.7) is small. Then the transparency of the hologram for the frequency ω may be written as:

$$K(x, y; \omega) = \exp\left[-\frac{1}{2}g_0\sigma d - i\frac{\omega d}{c}\right]$$

$$\times [1 + \frac{1}{2}\kappa(1 + i\hat{H}\{I(x, y; \omega)\}]$$

$$(7.8)$$

Here the exponential factor shows the obvious changes of the amplitude and the phase of the wave transmitted through the absorbing sample of thickness d.

Let us now turn to the recall procedure. We choose for the readout pulse a copy of the reference pulse and consider for the beginning also only one of its Fourier components, i.e., a plane wave of the frequency ω propagating in the $\vec{n}_R$ direction. This component, transmitting the sample with burnt-in $\varepsilon(\vec{r}, \omega)$ profile, creates the wave at the back surface of the plate (we drop here the trivial exponential in Eq. (7.8)):

$$E^{\text{out}}(\omega_0) = \exp[i\omega_0(t + \frac{\theta x}{c} - \frac{z}{c})]K(x, y; \omega) \qquad (7.9)$$

Substituting $\omega - \omega_0 \equiv \Omega$ and taking account of Eqs. (7.3) and (7.8), we have for the response of the hologram

$$E^{\text{out}}(\omega_0, \Omega) = [1 + \frac{1}{2}\kappa(1 + i\hat{H})\{R_0^2 + |s_F(x, y; \Omega)|^2\}]\exp[i\omega_0(t + \frac{\theta x}{c} - \frac{z}{c})]$$

$$+ [\frac{1}{2}\kappa R_0(1 + i\hat{H})\{s(x, y; \Omega)\exp[-i\Omega(\frac{\theta x}{c} - t_R)]\}]\exp[i\omega_0(t + t_R - \frac{z}{c})]$$

$$+ [\frac{1}{2}\kappa R_0(1 + i\hat{H})\{s_F^*(x, y; \Omega)\exp[i\Omega(\frac{\theta x}{c} - t_R)]\}]\exp[i\omega_0(t - t_R + \frac{2\theta x}{c} -$$

$$- \frac{z}{c})]$$

$$(7.10)$$

We can see that the response to one Fourier component is very much the same as with a conventional hologram: the response comprises three waves given by the three terms in Eq. (7.10). The first one describes the attenuated readout wave continuing along $\vec{n}_R$; the second wave propagates along the z axis and restores the Ω Fourier component of the signal $s(\vec{r}, t)$, i.e., represents this signal component, created by the readout pulse, transmitted through the hologram as the linear filter; the third one propagates in the direction $(-2\theta, 0, 1)$, and its wave front is conjugated to the signal's wave.

Let now the readout pulse be a linear superposition of the monochromatic waves considered above that forms a δ-like readout pulse $R(\vec{n}_R, t)$ propagating along the direction $\vec{n}_R$. Each component is transformed according to Eq. (7.10) and, as far as the light–matter interaction is linear, the response of $R(\vec{n}_R, t)$ is the waves constructed as the linear superposition of the responses (7.10) for the individual monochromatic components. The summation over frequencies results in the Fourier transformation of (7.10) from frequencies to the time domain. Taking into account that the Fourier transform of the operator $(1 + i\hat{H})$ is the Heaviside step function $Y(\tau)$, we have for the response

$$
\begin{aligned}
E^{out}(x, y; t) &= \left[R_0(1 + \frac{1}{2}\kappa R_0^2)\delta(t + \frac{\theta x}{c}) + R_0\kappa Y(t + \frac{\theta x}{c}) \right. \\
&\left. \times \int_{-\infty}^{+\infty} s(x, y; \tau)s^*(x, y; \tau - t - \frac{\theta x}{c})dr \right] \exp[i\omega_0(t + \frac{\theta x}{c})] \\
&+ R_0^2\kappa Y(t + \frac{\theta x}{c})s(x, y; t + t_R)\exp[i\omega(t + t_R)] \\
&+ R_0^2\kappa Y(t + \frac{\theta x}{c})s^*(x, y; t_R - t - \frac{\theta x}{c})\exp[i\omega(t - t_R + \frac{2\theta x}{c})] \\
&\equiv f_0(\vec{r}, t)\exp[i\omega_0(t + \frac{\theta x}{c})] + f_s(\vec{r}, t)\exp[i\omega_0(t + t_R)] \\
&+ f_s(\vec{r}, t)\exp[i\omega(t - t_R + \frac{2\theta x}{c})]
\end{aligned}
$$

(7.11)

where

$$
Y(\tau) = \begin{cases} 0 & \text{if } \tau < 0 \\ \\ 1 & \text{if } \tau > 0 \end{cases}
$$

(7.12)

$Y(\tau)$ represents the time arrow, i.e., guarantees, firstly, the absence of any response before the readout pulse arrives and, secondly, the absence of any part of the response that contradicts the causality of the events in the write-in procedure.

Equations (7.11) and (7.12) show that in the output of the spectral hologram there will appear three different light waves (pulses) (Fig. 6). The first term with f_0 $(\vec{r}, t)$ propagates in the direction $\vec{n}_R$ and describes a transmitted readout pulse with a distorted shape determined by the autocorrelation function of the object pulse. The second term f_1 $(\vec{r}, t)$ propagates along the z axis and represents the playback of the stored object signal as a virtual image of the event recorded. The last term indicates the possibility of recalling the conjugated image of the object with reversed time behavior.

If the reference pulse in the recording procedure arrives at the PSHB plate before the signal, we have for the delay

$$
t_R \leq -\frac{\theta x_{max}}{c}
$$

(7.13)

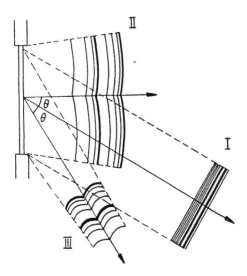

Figure 6 Diagram showing three different pulses appearing in the output of the hologram when the readout pulse is applied in the same direction as the reference pulse in recording. I, readout pulse transmitted and modified by absorption; II, reproduced object pulse; III, conjugated object pulse. The latter occurs in the case of thick holograms only if the readout is applied in the direction opposite to that of the recording reference pulse (17,38–40).

and in readout the replica of the signal pulse (Fig. 6) (the second term in Eq. (7.11)) appears behind the hologram and carries the imaginary image of the event.

In the opposite case, when

$$t_R \geq -\frac{|\theta x_{max}|}{c} + t_s \tag{7.14}$$

i.e., the reference pulse visits the sample after the signal pulse has already left, the response is a real image of the conjugated signal pulse, i.e., the wave fronts are reversed in time. In the intermediate situation,

$$\frac{|\theta x_{max}|}{c} + t_s > t_R - \frac{\theta x_{max}}{c} \tag{7.15}$$

when the reference and signal pulses partly overlap in the sample, the response comprises both waves: the normal copy of the signal and also the conjugated one. But the copies are truncated: the parts allowed by causality appear, and those in contradiction with it are cut off. In experiments these consequencies of causality are present as a matter of fact. If a theory pretends to be reasonable, it has to reflect them, too. Here it is guaranteed by the Heaviside step functions.

Our consideration resulting in the presence of three response pulses holds for thin (two-dimensional) holograms. If the optical path in the sample comprise many wavelengths (thick sample, three-dimensional hologram), the structure of the light field is written in the hologram in the same way as in the thin one, but in the recall additional interference between the partial waves takes place, and only those waves propagate out from the thick sample whose direction obeys the spatial synchronism (Bragg condition). Only one pulse (besides the attenuated continuation of the readout pulse) emerges from a thick hologram. Whether it is pulse I or II depends on the order in time of the signal and the reference pulses in the recording procedure and on the direction ($\vec{n}_R$ or $-\vec{n}_R$) of the readout pulse. To recall in readout the "future" event, i.e., the event or a part of it, that in recording arrived at the PSHB plate ahead of the reference pulse, the readout probe pulse should be applied in the opposite direction to the reference write-in pulse. This dependence, a fine reflection of the causality, is illustrated in Fig. 10.

In conclusion of the theoretical part, two remarks:

1. In experiments, one pair of pulses is usually not enough to burn in the PSHB profile of required contrast. Therefore actually a rather long series of identical pairs is applied. The total duration of a pair of two pulses (i.e., the duration of the signal, plus the duration of the reference, plus the time delay between them) has to be shorter than the phase memory τ_2 of the excited electronic state of the impurity. On the other hand, the temporal separation between two successive pairs has to be longer than τ_2 (Fig. 7b). This condition guarantees the absence of interference between pulses in different pairs. The formulae given above hold for n identical pairs of pulses, if, starting with (7.7), k is replaced by nk, i.e., the second term is multiplied by n.

2. Light–matter interactions may be kept linear and described well by the conventional Maxwell equations before, during, and after hole burning. The point is that the dielectrical permittivity before the procedure $\varepsilon_0(\vec{r},\omega)$ and after it $\varepsilon(\vec{r},\omega)$ are essentially *different* $\varepsilon_0(\vec{r}, \omega) \neq \varepsilon(\vec{r}, \omega)$.

The response of the matter to the light pulses before and after hole burning is, naturally, also essentially different. But the difference $\varepsilon(\vec{r}, \omega) - \varepsilon_0(\vec{r}, \omega)$ is collected little by little in many identical acts of modest and linear light–matter interaction. Our consideration shows also that the conventional photon–echo phenomena are to be considered as nonlinear only because the changes of $\varepsilon_0(\vec{r}, \omega)$, created here via excitation of short-lived transient excited electronic states, require strong light pulses and, consequently, a nonlinear interpretation.

C. Experimental

The experimental equipment, PSHB materials, and temperature for time-and space-domain holography are chosen to meet the principal requirements put by

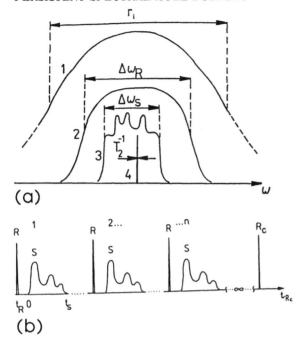

Figure 7 Relations between the bandwidths of the spectra required for time- and space domain holography. (a) 1, inhomogeneous absorption band; 2, reference pulse; 3, signal; 4, homogeneous zero-phonon line. The main (and actually the only strict) requirement is $\Gamma_i >> \tau_x^{-1} = \Gamma(T)$ (17,38–40).

To collect the depth of the PSHB profile, many identical pairs of reference and signal pulses may be applied. (b) Duration of each pair has to be shorter than the phase relaxation time τ_2; the delay between two sequential pairs has to be much longer than τ_2, $t_{R_{i+1}} - t_{R_i} >> \tau_2 >> t_s - t_R$. Depending on PSHB material and on the temperature at which the hologram is preserved, the recall moment is performed by applying the pulse R_C at an arbitrary moment, which may be chosen after hours, months, or years have passed from the storage time.

the spectral bandwidths and by the corresponding relaxation times (Fig. 7). The following is typical for the experiments.

An actively mode-locked Ar-ion laser (e.g., Spectra-Physics 171) is utilized to pump a Rhodamine 6G picosecond dye laser, which provides 2–3 ps duration pulses (5–6 cm^{-1} spectral FWHM) at 82 MHz repetition rate with an average output power of 100 mW.

As a PSHB medium for recording time-and-space holograms by PSHB, polystyrene doped with octaethylporphin (OEP-PS) at concentrations of 10^{-4} to 10^{-3} M is used. The inhomogeneously broadened ZPL impurity absorption band is 200 cm^{-1} (FWHM) centered at the wavelength of 617 nm. Homogeneous ZPL

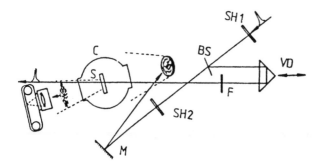

Figure 8 Experimental setup for recording time- and space-domain holograms. Beam splitter BS divides the expanded picosecond laser beam between the reference and the object channel. The light from the object scene (a picosecond pulse scattered by a coin) travels through the windows of the cryostat C to the sample S; the reference pulse passes the delay VD and strikes the sample at $10°$ with respect to the object beam. To reproduce the holographic image of the scene, the reference beam is attenuated by the filter F, and the input beam illuminating the object is blocked by a shutter SH2 (17,38–40), see also (10).

widths are less than 0.05 cm^{-1} at 1.8K. Samples are prepared in blocks with thickness $0.3–1.0$ cm and optical density $1–3$ cm^{-1} and are placed in a liquid He cryostat with optical windows. To record high-contrast spectral holograms in OEP-PS, PSHB exposures of at least 0.1 mJ·cm^{-2} are needed. Depending on the average intensity of the incident light, this exposure contains $10^9–10^{11}$ identical pairs in the sequence of writing pulses.

In Fig. 8, a typical experimental setup (36–38) is presented. An object—a ten kopeck coin—is positioned in front of the entrance window of the cryostat containing an OEP-PS block at 2K. One beam is directed at the coin to provide a scattered object beam. The other beam, carrying plane-wave reference pulses, is passed through the sample at a $10°$ angle to the direction of the first beam and 20 ps ahead of the object pulse. Ten minutes of PSHB exposure is enough to store a high-contrast spatial frequency-dependent holographic image of the coin. The spatial image is clearly visible from behind the cryostat when the illumination of the original coin is blocked and only the plane-wave beam is passed through the sample to act as a readout reference pulse (Fig. 9). The holographic images, which stay ready to be recalled for at least several hours (until the liquid helium boils away), are either photographed through the output window of the cryostat or temporally analyzed with the help of a synchroscan streak camera system. As usual, the photographs and streak camera images cannot reproduce, to a full extent, the rich information contained in the holographic images of the picosecond

Figure 9 Image of a coin photographed from the time- and space-domain hologram. The time dependence lost in the photo is here the picosecond pulse of illumination sliding over the coin. Interference between the holographic image and the image formed from the light scattered directly by the coin proves the coherent nature of the holographic image (36–39,10).

events. For example, the time-dependent course of the illumination of the coin with a picosecond pulse and the spatial depth of the image are lost. Display of causality in time and space-domain holography is illustrated in Fig. 10. Experimental results are shown in Fig. 11.

A slightly different setup serves to show the wave front conjugation capabilities of time and space-domain holography (17,38,39); the original object image incident on the PSHB recording sample is first spoiled by a distorter (a fragment of a broken glass vial) and the distorted image recorded (Fig. 10). During the writing of the hologram the reference delay is adjusted so that the reference pulse arrives at the sample some tens of picoseconds *later* than the distorted image's wave front. The recorded hologram is illuminated by plane wave readout pulses travelling in the direction opposite to the original reference pulses (compare with the last term of Eq. (7.11)). This readout recalls the fully conjugated (i.e., time and space-reversed) object pulse. The pulse passes through the same spot of the piece of glass in the opposite direction, the changes of the opposite sign compensate the initial temporal and spatial distortions, and the nondistorted object image emerges (Fig. 12).

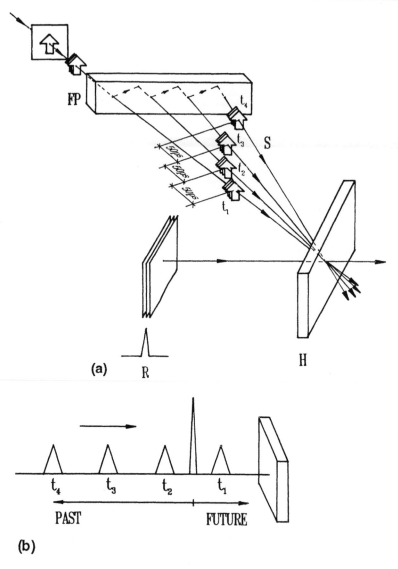

(b)

Figure 10 Time arrow in the time- and space-domain holography. Writing (a, b) and reading (c, d) of the holograms. (a) The object scene S (event) is four arrow-shaped picosecond pulses separated in time. The reference pulse R, timed to arrive at the PSHB recording medium with delay t_R, divides the signal into two parts, "past" and "future" (b). In a thick hologram, the reading out in the direction of the reference pulse (c) plays back the "past"—the three pulses that arrived *after* the pulse R at t_R. The reading out in the opposite direction (d) reproduces conjugated "future" part of the pulse reversed in time, which arrived before t_R (38,39). For the experiment see Fig. 11.

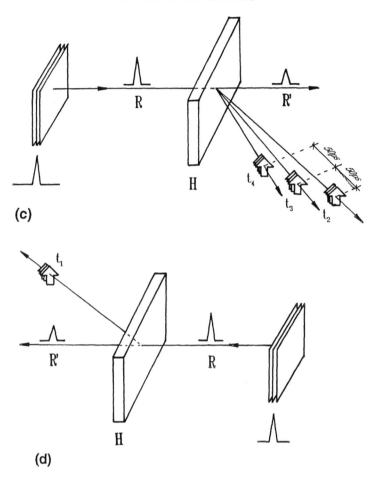

(c)

(d)

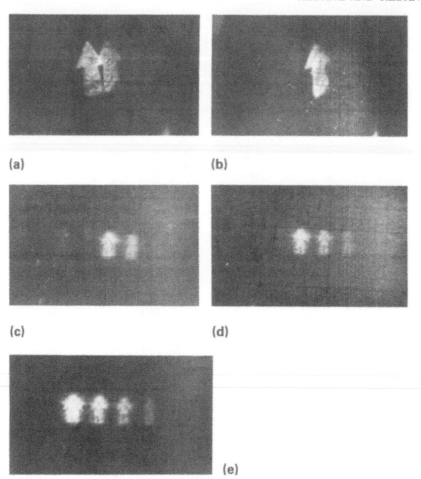

Figure 11 Experimental results obtained in realization of the scheme given in Fig. 10; *a*, *b*, conjugated images, photographed in the scheme 10d; *c*, *d*, *e*, direct images (scheme 10c). For a and c the reference pulse was placed between the second and third pulses; for b and d, after the first and before the second pulse; e, reference arrived at the recording PSHB plate ahead of all four pulses.

D. Associative Space-and-Time Memory

The role of the reference signal in recording and playback of space- and time-domain holograms can not only be accomplished by a δ-pulse but also be performed in an associative manner by two separate signals (space and time fragments (episodes) of the object light field), each serving as the reference to the other (50). The recall of this kind of associative hologram is accomplished,

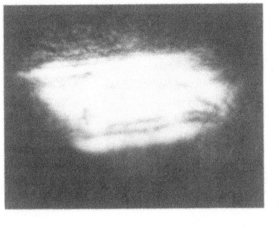

(a)

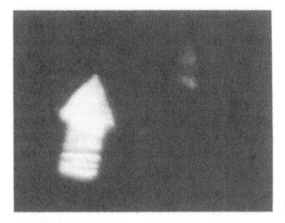

(b)

Figure 12 The object, an arrow cut in a screen, as seen and holographically recorded through a light-scattering piece of a glass vial (a). Restored image of the arrow photographed at the output of the cryostat obtained by recalling the conjugated readout pulse and letting it pass the same distorter in the opposite direction (b) (36–39,10).

analogously to an associative recall of ordinary space-domain holograms, by illuminating the hologram with one or several fragments of the stored-in signal. In space- and time-domain holography there occurs not only the reproduction of the spatial image but also the recall of the temporal features of the original signal missing in the readout signal. In other words, one can reproduce from the hologram the whole event, provided an episode of this event is available to be used as a readout pulse.

The output beam of the laser was expanded in a telescope and passed through

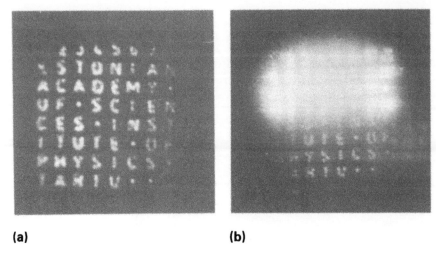

(a)   **(b)**

Figure 13 Photographs of the full image of a picosecond space–time signal (an eight-line text, each line delayed by 34 ps from the previous one) recorded in the hologram (a) and an image recalled from the recorded hologram by applying images of the first three lines of the text as the readout pulse (b) (50).

a Michelson echelon (Fig. 13). The optical path through the neighboring segments of the echelon differed by 34 ps. At the output of the echelon the laser pulse appeared to be divided up into eight pulses. Each of the eight pulses corresponded to a different spatial fraction of the input pulse and travelled through a certain segment of the echelon so that the relative delays of the segments were correspondingly 34 ps, 68 ps, 102 ps, etc.

The total duration of the pulse train did not exceed, as required in space-time holography, the phase relaxation time τ_2 of the excited electronic state of the impurity molecules, which was $\tau_2 \simeq 500$ ps for the octaethylporphin molecules in polystyrene used here, at $T = 2$K.

By inserting various transparencies into the laser beam, the spatial outlines of the light pulses could also be tailored, say, in the form of stripes, lines of printed text, etc. Each pulse separated in time carries its own spatial image (line). For the details of the experiment see (50).

In order to write a hologram, mutual interference of various parts of the incident object light field had to be arranged by making all the spatial parts of the object pulse meet each other all over the hologram. To do this a mat glass plate was inserted into the laser beam some 15 cm from the incident window of the cryostat. As a result, the laser beam, which at the output of the echelon comprised a train of copropagating pulses with spatially nonoverlapping wave fronts, turned at the incident plane of the hologram into a train of pulses with each pulse illuminating almost the whole area of the sample. It should be noted that the possibility of holographic temporal recall, using as readout pulses fragments of the recorded signals, has been indicated earlier by Gabor (57), and the analogy between holographic recall in the time domain and some properties of human memory was also pointed out.

E. Imaging Through a Light-Scattering Medium

In (47), the features of time- and space-domain holography described in the previous section are used to select one image out of the two (or a sequence of images) propagating in space, separated from each other by a picosecond domain delay. This forms the basis for a new holographic method that allows one to improve considerably the visibility of an object positioned behind a semiopaque light-scattering medium.

Consider two optical pulses incident on a PSHB recording medium as shown in the inset of Fig. 14. Let the pulse F arrive τ_0 seconds before the pulse G.

In the experiments (47), the recording medium was a 3-mm-thick block of polystyrene doped with protoporphyrin at a concentration of 10^{-3} mol^{-1}. The spectral characteristics (homogeneous and inhomogeneous width) of the sample at temperature $T = 2K$ were similar to those described in the previous section. The useful sample area was 4 cm^2. The peak of the broad inhomogeneous absorption band was at $\lambda = 621$ nm, at which the optical density was 1.6. The light source was a synchronously pumped tunable dye laser producing not-transform-limited pulses, with a coherence time of 0.5 ps. A beam splitter divided the beam from the picosecond dye laser into a reference beam and a signal beam to illuminate various objects in the scene, as shown in Fig. 14. The reference beam was expanded by a telescope to illuminate the entire polystyrene block. The scattered light from the illuminated objects simply propagated to the polystyrene block with no intervening lens. The angle between the reference beam and the image-bearing one was $\sim 14°$.

The recorded scene consisted of two objects. The nearby object was a 1.0-mm-thick glass slide with the letters 'HOLO' attached to its front surface. The distant object was a white paper screen carrying the letters 'GRAM', which was pressed against the back of the transparent slide, so the separation between the two objects was ~ 1 mm. To increase the amount of the light scattered by the slide, its front surface was coated with a frosting aerosol spray. Of the scattered light from the slide reaching the polystyrene block, $\sim 80\%$ came from the front sprayed surface of the slide, and only $\sim 20\%$ came from the rear surface. Viewing the laser-illuminated slide by eye from the position of the polystyrene block, one could clearly read the HOLO, but the intense glare from the front surface almost completely obscured the GRAM written in the back of the slide.

The light from the laser reached the front of the slide 5 ps before it reached

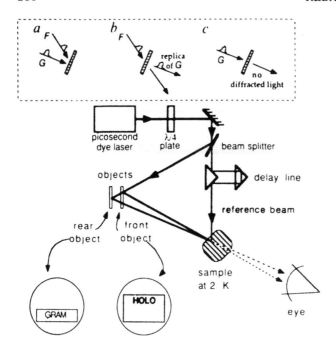

Figure 14 Main figure: experimental setup. The front object is a frosted slide with the letters 'HOLO' pasted on the front. The rear object is the letters 'GRAM' pasted in the back of the same slide. Inset: (a) Writing the hologram with two light pulses separated in time. (b) Reading with the earlier pulse F recreates the later pulse G. (c) Reading with the later pulse G does not produce any diffracted pulse (47).

the back of the slide. Consequently, the light from the front of the slide reached the storage medium 10 ps (twice the glass travel-through time) before the light from the back of the slide.

If the reference beam was timed to arrive at the PSHB storage block before both of these object waves, then both images (events) F and G were put relative to the reference pulse into the "past," and the image reconstructed from the hologram (Fig. 15a) comprised both F and C. The letters HOLO on the front of the slide are plainly visible. The image of the latter event is also present, but the glare stored in the hologram from the front of the slide again almost obscures the letters GRAM near the back of the slide. However, if the hologram was written in with the reference beam carefully timed to arrive within the 10 ps interval between the two object waves, then only the second event G belongs to the past; the first one F is a "future" event and would be recalled only if the time arrow were reversed, i.e., the probe pulse sent in the opposite direction to the reference pulse. Consequently, the reconstructed image from the second

(a)

(b)

Figure 15   (a) Holographic reconstruction of both objects. The readout reference beam of pulses is applied in the same direction as in writing in the two holograms. In recording of this hologram the reference pulse was set to arrive before light from either of the objects. The glare from the frosted glass in front with the word HOLO obscures the object GRAM behind it. (b) This hologram was recorded with the reference pulse delayed to arrive after the light from the front object (frosted glass) but before the light from the rear object. The front object is no longer reconstructed, whereas the rear object (GRAM) is now plainly visible (47).

hologram (Fig. 15b) shows only the distant object that belonged to past, the nearby (future) object being completely eliminated. We note that the maximum time delay permitted between the object and the reference beam was 10^3 ps and was set by the phase decay time τ_2 of the sample.

VIII.   MODELLING OF ERROR-CORRECTIVE NEURAL NETWORKS

The motivation of PSHB modelling of neural networks (NN) is to implement one more version of distributed data storage and parallel processing and to show that PSHB provides frequency as a new dimension, which allows one to increase considerably the number of interconnects in NN models, in principle, up to 10^3–10^5 times.

Optical realizations of NN as associative memories (51,52,44,45,18,19) exploit the additional possibilities that optics has in comparison with electronic devices, particularly those in massive matrix–matrix multiplication procedures. In optical processors the matrices can be coded as spatial transparencies or holograms, and the multiplication procedure can be carried out simultaneously in many parallel channels. Conventionally optics uses only the spatial (and polarization) degrees of freedom of the storage media. However, a larger throughput capacity of NN than is provided by conventional optical storage is needed, especially for image processing. Also, for direct input–output of images in the two-dimensional format, the optical multiplication process itself needs at least one more (in addition to three spatial coordinates) degree of freedom. PSHB, introducing frequency domain and also time domain, can provide a four-dimensional format (x,y,ω,t) for transformation (memory) matrix (19,44,45).

A.   Mathematical Model

In (18,19,44,45), four different models are implemented and their capability to correct errors shown. In all four models the data storage is fully distributed, the readout is, differently from holography, only partly parallel. The latter means that even after the preparation of masks for implementation of a part of the basic input information via the spatial structure, the write-in of the frequency-carried component of the input information is not parallel: the masks are consequently changed and the frequency is scanned to predetermined values for each mask(19).

For example, the memory was coded in three dimensions (spatial coordinates x,y, and frequency ω) and materialized $32 \times 32 \times 32 = 32768$ optical interconnects. The input (probe) vector, carrying 32 bits, was transformed into a 32×32 matrix (coded in x and y); the output was one-dimensional, consisted of 32 bits coded in frequency, and corrected 4 erroneous bits of the probe vector.

Simultaneous read out for the spatial locations and also for the frequency-domain bits is performed experimentally in (58). It serves as an example of an almost full-scale realization of PSHB NN modeling. It may mean that improvements are still possible in the write-in procedure and also in arranging the output to be directed back into input to obtain multistep corrections.

Below we consider briefly the mathematical model used in a parallel space, time, and frequency storage experiment (18). The elements of the four-dimensional associative memory matrix (interconnection matrix) T_{ijkl} (i, k = 1, ... ,4; j, l = 1, ... ,3) are calculated as (52)

$$T_{ijkl} = \frac{T''_{ijkl}}{\sum_{k,l}^{M,N} T''_{ijkl}} \tag{8.1}$$

where

$$T''_{ijkl} = \begin{cases} T'_{ijkl} & \text{if } T'_{ijkl} < 0 \\ \\ = 0 & \text{otherwise} \end{cases} \tag{8.2}$$

Here

$$T'_{ijkl} = \sum_{s=1}^{S} (2V^{(s)}_{ij} - 1)(2V^{(s)}_{kl} - 1) \tag{8.3a}$$

$$T'_{ijij} = 0 \tag{8.3b}$$

$V^{(s)}_{ij}$ is the ijth element of the sth stored binary image: $V^{(s)}_{ij} = 1$ or 0.

The readout of a memory by an interrogating image $\tilde{V}^{(in)}_{kl}$ is mathematically equivalent to a thresholded scalar product

$$\tilde{V}^{(out)}_{ij} = TRH\left\{ \sum_{k,l}^{M,N} \tilde{V}^{(in)}_{kl} T_{ijkl} \right\} \tag{8.4}$$

where $TRH\{\}$ stands for a threshold function

$$TRH\{x\} = \begin{cases} 1 & \text{if } x < 0.5 \\ \\ 0 & \text{otherwise.} \end{cases} \tag{8.5}$$

B. Experimental

In the following experiments PSHB samples with enlarged inhomogeneous bandwidth were used. This was achieved by introducing two different organic impurity molecules (octaethylporphyrin and protoporphyrin) in a polymeric matrix (poly-styrene) at approximately equal concentrations of 10^{-3}–10^{-4} mol/L. The par-tially overlapping inhomogeneous absorption bands of the two impurities give in the 615–623 nm region of wavelengths a nearly flat absorption frequency interval of a width of about 150 cm^{-1}, where the transmission of the sample (measured before the write-in exposures) varies between 1% and 2%. The spatial dimensions of the sample (block polystyrene) are 3×3 cm across and 4 mm in thickness. The sample is positioned inside an optical cryostat and immersed in liquid helium. At the working temperature of 2K, twice the ZPL homogeneous width is about 0.05 cm^{-1} for both types of impurities.

As a laser source a synchronously pumped picosecond Rhodamine 6G dye laser was used. The frequency of the dye laser was tuned in steps within the 150 cm^{-1} spectral interval.

To materialize the four-dimensional memory matrix (8.1–8.3), four different physical variables were involved. Two variables, corresponding to the indices k and l, are both accommodated by the spectral dimension ω of the PSHB storage media. k corresponds to three different delays of the time-domain holographic photochemically accumulated stimulated photon echo (PASPE) (17,36–40); re-sponses τ_1, τ_2, τ_3 and l correspond to four different spectral ranges, ω_1, ω_2, ω_3, ω_4. At each ω_1 three PASPE signals are stored, each consuming a frequency diapason $\Delta\omega_l \approx 10$ cm^{-1}. The two remaining variables, indices, i and j, label the two orthogonal spatial coordinates x and y in the plane of the PSHB plate. In accordance with the transition theorem of the Fourier transform, the time delay between the signal and the reference pulse, τ_κ, corresponds to multiplying the transmission spectrum of the time–space hologram by the function $\exp(i\omega\tau_\kappa)$, where ω is the light frequency. The corresponding time response of the hologram has the same time delay, τ_κ, with respect to the readout pulse.

The frequency-domain storage range (150 cm^{-1} in the flat region of the inhomogeneous band) was divided into four nonoverlapping subintervals, each interval corresponding to a different laser frequency mode of a width of about 2 cm^{-1}. In each of the frequency subintervals, three picosecond time-domain (PASPE) holograms are stored. These frequency intervals, compared to the ZPL homogeneous width, are quite broad: one PASPE hologram of a 3 ps pulse consumes the frequency space for a thousand of ZPL limited holes. Thus this approach seems not to give any considerable increase of storage capacity. On the other hand, it is an interesting version of simultaneous time-and-frequency description, to which the wavelet concept (20) may be applied.

Concerning the spatial coordinates, the PSHB storage plate was divided into 12 square elements arranged in the same way as the storaged images to be recorded, i.e., into a 4×3 element pattern of squares (Fig. 16). At each of the 12 spatial locations an additional two-coordinate (frequency and time) storage area was available. The first coordinate corresponded to four carrier-wave fre-

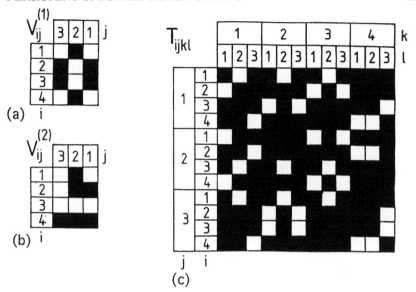

Figure 16 Graphic representation of the two binary images assumed as storage basic data vectors (matrices) (a, b) and of the memory matrix (c) calculated according to Eqs. (8.1–8.4) (45,19).

quencies ($\omega = \omega_1, \omega_2, \omega_3, \omega_4$), while the second one corresponded to the temporal delays ($\tau = \tau_1, \tau_2, \tau_3$) of a holographic time-domain (PASPE) signal. In other words, the 4-D associative memory matrix was physically coded in the strength of the optical response—the transparency $T(x, y; \omega, \tau)$—of the PSHB plate in dependence on the spatial coordinates x, y, frequency ω, and time delay τ.

A scheme of the experimental arrangement is presented in Fig. 17. The incoming expanded picosecond dye laser beam was divided into two parts with a beam splitter. At the position of the PSHB plate the two beams (signal and reference) cross at an angle of 6°. The time delay between the arrival of the pulses of the intercrossing beams had three different values (40, 104, and 180 ps) depending on the thickness of the glass block positioned in the reference beam. The signal beam, which had a shorter optical path, contained a holder with an interchangable mask slide. The signal laser beam projected the image of a slide upon the PSHB sample so that every spatial element of the slide coincides with a corresponding spatial element of the storage plate. The reference beam possessed a plain wave front and illuminates uniformly the whole storage area of the PSHB plate.

The memory matrix was written in by tuning the dye laser wavelength sequentially to four fixed storage frequencies. At every frequency three different write-

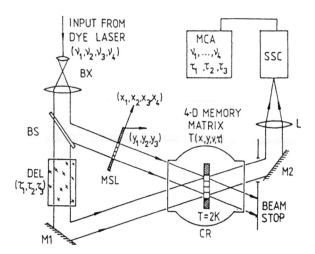

Figure 17   Scheme of the experimental arrangement used for the storage and interrogation of 4-D associative memory matrix. BX, beam expander; BS, 50% beam splitter; DEL, glass block delay; M1, M2, mirrors; MSL, mask slide; CR, cryostat; L, collecting lens; SSC, synchrotron streak camera; MCA, multichannel analyser (45,19).

in PASPE exposures were carried out, one for each of the three different delay values. Each exposure (the total number of write-in exposures is 12) was performed through its own spatial mask slide, which guaranteed the *xy* structure of the elements of the memory matrix.

The multiplication procedure described by formula (8.4) was performed optically as follows. In the slide holder was fixed a mask $\bar{V}_{kl}^{(\mathrm{in})}$ that resembles one of the two storaged images but differs from it by two error bits (Fig. 18, left). The write-in reference beam was blocked and the sample was illuminated through the slide only with the signal beam of frequency ω_1. Its intensity was attenuated by a factor of 100 to avoid erasing the memory by further hole burning during the readout. In this experiment, the readout was not fully parallel; it is performed sequentially at four frequencies ω_1, ω_2, ω_3, ω_4. At the output of the memory, a focusing lens collected the light diffracted from different spatial elements of the PSHB plate onto a single focus spot at the entrance slit of the streak camera, and the latter measured the output intensity integrated over the whole spatial coordinates area, in dependence on the time delay for one of the readout frequencies ω_1. The frequency was sequentially scanned to the four values, and thus the outcoming signal intensity arranged into a 3×4 pattern along time delays and frequencies (Fig. 18). The result clearly reproduces the correct basic vector in Fig. 15a. Note that the holographic approach has well suppressed the light

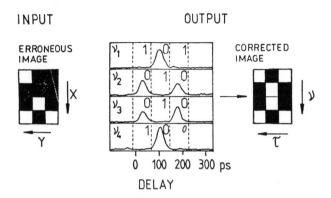

Figure 18 Readout of the 4-D memory by interrogation with an erroneous 2-D input image. Output signal measurement is carried out by monitoring the time delay of the diffracted signal at four different readout laser frequencies. The input information (coded in spatial coordinates x, y) comprises two error bits that are corrected in the procedure of optical multiplication, and the output (coded in frequency and time delay) is correct, i.e., it reproduces the stored basic information matrix closest to the erroneous one. The interchange of ones and zeros (white and black squares), caused by the mathematical scheme, is irrelevant and easy to avoid if necessary (45,19).

scattering background and thus contributed to making the thresholding procedure trivial (Fig. 18).

Feedback and multiple pass-through of the processed signals was not needed in these experiments, since an error-free recollection appears from the memory after the first pass-through of the interrogating signal.

This experiment actually performs a scalar multiplication of two- and four-dimensional matrices. It is also noteworthy that the input image is coded in two spatial coordinates and the output information in the frequency and time domain of an optical signal, which makes it, in principle, convenient to transmit along optical waveguides.

Simultaneous parallel readout not only for the spatial and temporal locations, as described above, but also for all the frequency-domain bits is also feasible (58) (see also (45)). In this experiment, a quadratic 3-D associative memory is constructed (52) on four 10-bit vectors, and the capability of correcting two erroneous bits is demonstrated. The procedure is in principle simple: the read-out spectrum is broad enough to comprise all the 10 write-in frequencies, and a monochromator performs the discrimination of the output intensity by frequencies.

Compared to holography, where pictures and events from the outer world provide the write-in information and modulate the input signal beam both spatially and in frequencies (in time), in the schemes of NN considered here the input information has to be first arranged into bits (digitized), the (x, y) masks have to be prepared, and the write-in frequency has to be scanned. This means an additional and rather time-consuming operation. However, once the memory matrix is ready, it may be used repeatedly to check and correct various input information, each probe taking fractions of a second. If we take, in place of the digital input information, a direct "picture of the outer world," we actually come close to PSHB holography, where the associative memory aspects have been shown to provide a remarkable error-corrective capability.

IX. CONCLUSION

The prospects for continued progress in PSHB science and applications are promising (see also Chapter 12 by Bogner and Chapter 14 by De Caro et al. in this book). Several hundreds of PSHB materials have been used ranging from organic glasses doped with various impurities to inorganic single crystals, the latter including neutron-irradiated sapphire and diamond (59).

ZPL is a rather common feature in the spectra of low-temperature impurity-doped solids. Thousands of dopants in hundreds of matrices should show ZPLs, and many of them may display also quite effective hole burning. Among the great variety of potentially promising systems, suitable ones for new spectacular experiments and for various technological applications are to be found.

The requirement of liquid helium temperatures is certainly a limiting factor, but actually only for large-scale commercial applications. The first high-temperature PSHB materials have been found (3,59–61) (see also (6,7)), but the possible number of holes given by the ratio of the inhomogeneous to homogeneous linewidths $\Gamma_i : \Gamma(T)$ is small. For instance, at room temperature only about ten widths of holes burnt in an Sm^{2+}-doped mixed crystal $Sr_yBa_{1-y}FCl_xBr_{1-x}$ can be implemented in the inhomogeneous band (61). Note that a mixed host crystal is taken to enlarge the inhomogeneous band width. Theoretical considerations indicate that it is really difficult, if not completely impossible, to have for optical transitions at room temperature *simultaneously* narrow homogeneous and broad inhomogeneous absorption bands, i.e., to have really large ratios $\Gamma_i : \Gamma(T)$. The point is that the conditions for having narrow ZPL widths $\Gamma(T)$ at high temperatures are largely (but not exactly!) the same that make inhomogeneous broadening small.

On the other hand, PSHB materials, which preserve the written-in hole structure at high temperatures for long periods of time, are possible, and several have already been found. For instance, a neutron-irradiated sample of sapphire with holes burnt in it at liquid helium temperatures may be kept for about ten minutes at $T \approx 400$–$600K$ (depending on the absorption band in which the hole is located), and when it is cooled down to helium temperatures the holes are there again, only broadened to some extent.

As a storage medium for holograms, PSHB has the interesting property that the coherent linear time response of the material to light can be changed in broad intervals between femtoseconds and nanoseconds. Because the local time response function of the PSHB material obeys causality strictly, the Kramers–Kronig dispersion relations play a central role in evaluating the diffraction properties of the holograms. The comparison of distributed and digital PSHB storage is overwhelmingly in favor of the former. Space and frequency-domain and space and time-domain holographies, both processing really large amounts of data, and in fact representing pictures and time-dependent pictures, have been successfully implemented. Models of error-corrective neural networks have also been realized. The feasibility of both PSHB distributed data storage applications is well shown on the level of basic research. Realizations of the digital approach are limited up to now to examples with only a few hundred bits involved, but prospects for future development do exist also here.

ACKNOWLEDGMENTS

The author would like to thank Olavi Ollikainen for providing figures from his publications and many fruitful discussions, and Liivia Juhansoo for helping to prepare the manuscript.

REFERENCES

1. Gorokhovskii, A. A., Kaarli, R. K., and Rebane, L. A., Hole burning in the contour of purely electronic line in the spectra of Shpol'skii system (in Russian), *Pis'ma JETP Lett.*, *20*, 216 (1974).
2. Kharlamov, B. M., Personov, R. I., and Bykovskaya, L. A., Stable "gap" in absorption spectra of solid solutions of organic molecules by laser irradiation, *Optics Commun.*, *12*, 191 (1974).
3. Moerner, W. E., ed., *Persistent Spectral Hole Burning: Science and Applications*, Springer-Verlag, Berlin, Heidelberg, 1988.
4. Moerner, W. E., Introduction. 1.3. Historical overview and survey of mechanisms, in (3), p. 7.
5. Sild, O., and Haller, K., eds., *Zero-Phonon Lines and Spectral Hole Burning in Spectroscopy and Photochemistry*, Springer-Verlag, Berlin, Heidelberg, 1988.
6. Technical Digest on Persistent Spectral Hole-Burning: Science and Applications, Conference 26–28 Sept., 1991, Monterey, California, Optical Society of America, Washington, D.C., Technical Digest Series, *16* (1991).
7. Technical Digest on Spectral Hole-Burning and Luminescence Line Narrowing: Science and Applications, Conference 14–18 Sept., 1992, Monte Verita, Ascona,

Switzerland; Optical Society of America, Washington, D.C., Technical Digest Series, 22 (1992).

8.   Rebane, K. K., *Impurity Spectra of Solids*, Plenum Press, New York, 1970.

9.   Rebane, L. A., Gorokhovskii, A. A., and Kikas, J. V., Low-temperature spectroscopy of organic molecules in solids by photochemical hole burning, *Applied Physics*, *B29*, 235 (1982); Horie, K., and Furusawa, A., Photochemical hole burning, *Progress in Photochemistry and Photophysics* (J. Rabek, ed.), CRC Press, Boca Raton, Ann Arbor, Boston, 1992 V, p. 50.

10. Rebane, K. K., and Rebane, L. A., Basic principles and methods of persistent spectral hole-burning, in (3), p. 17.

11. Gorokhovskii, A. A., Kaarli, R. K., and Rebane, L. A., The homogeneous pure electronic line width in the spectrum of a H_2-phthalocyanine solution in *n*-octane at 5K, *Opt. Commun.*, *16*, 282 (1976); Gorokhovskii, A. A., and Rebane, L. A., Temperature broadening of pure electronic line by hole burning technique, *Opt. Commun.*, *20*, 144 (1977); Gorokhovskii, A. A., and Rebane, L. A., Temperature broadening of zero-phonon lines in the spectra of impurity molecules by hole burning methods (in Russian), *Fizika Tverdogo Tela*, *19*, 3417 (1977); Voelker, S., Macfarlane, R. M., Genack, A. Z., Trommsdorf, H. P., and van der Waals, J. H., Homogeneous line width of the S_1-S_0 transition of free-base porphyrin in an *n*-octane crystal as studied by photochemical hole-burning, *J. Chem. Phys.*, *67*, 1759 (1977); Voelker, S., and Macfarlane, R. M., Photochemical hole burning in free-base porphyrin and chlorin in *n*-alkane matrices, *IBM J. Res. Dev.*, *23*, 547 (1979). Volker, S., Macfarlane, R. M., and van der Waals, J. H., *Chem. Phys. Lett.*, *110*, 7 (1984).

12. Friedrich J., and Haarer, D., Photochemical hole-burning: spectroscopic study of relaxation processes in polymers and glasses, *Angew. Chem. Int. Ed. Engl.*, *12*, 113 (1984). Friedrich, J., and Haarer, D., Structural relaxation processes in polymers and glasses as studied by high resolution optical spectroscopy, *Optical Spectroscopy of Glasses* (I. Zschonke, ed.), D. Reidel, 1986, p. 149.

13. Avarmaa, R. A., and Rebane, K. K., Zero-phonon lines in the spectra of chlorophyll-type molecules embedded in low-temperature solid-state matrices (in Russian), *Uspekhi Fiz. Nauk*, *154*, 433 (1988); English translation, *Sov. Phys. Usp.*, *31*(3), 225 (1988).

14. Kikas, J., Spectral hole burning (SHB): scientific and practical applications, in (5), p. 89; Fearey, B. L., and Small, G. J., New studies of non-photochemical chemical holes of dyes and rare-earth ions in polymers I. Spontaneous hole filling, *Chem. Phys.*, *101*, 269 (1986): Fearey, B. L., Carter, T. P., and Small,G. J., New studies of non-photochemical holes of dyes and rare-earth ions in polymers II. Laser-induced hole filling, *Chem. Phys.*, *101*, 279 (1986); Trommsdorf, H. P., Zeigler, J. M., and Hochstrasser, R. M., Spectral hole burning in polysilanes, *J. Chem. Phys.*, *89*, 4440 (1988).

15. Renn, A., Meixner, A. J., Wild, U. P., and Burkhalter, F. A., *Chem. Phys.*, *93*, 157 (1985); Wild, U. Renn, A., Caro, C., and Bernet, S., Spectral hole burning and molecular computing, *Appl. Opt.*, *24*, 1526 (1986); Meixner, A. J., Renn, A.,

and Wild, U. P., *Appl. Opt.*, *26*, 4040 (1987); *J. Chem. Phys.*, *93*, 6728 (1989); Wild, U. P., and Renn, A., *J. Mol. Electron.*, *7*, 1 (1991); Bernet, S., Renn, A., Kohler, B., and Wild, U., Molecular computing: parallel binary additions, in (7), p. 218.

16. Rebane, A. K., Kaarli, R. K., and Saari, P. M., Burning out a complex-shaped hole by a coherent series of picosecond pulses, *Opt. Spectrosc. (USSR)*, *55*, No.3, 238 (1983); Rebane, A., Kaarli, R., Saari, P., Anijalg, A., and Timpmann, K., Photochemical time-domain holography of weak picosecond pulses, *Optics Commun.*, *47*, 173 (1983); Rebane, A., Kaarli, R., Picosecond pulse shaping by photochemical time-domain holography, *Chem. Phys. Lett.*, *101*, 317 (1983); Saari, P. M., Kaarli, R. K., and Rebane, A. K., Holography of spatial-temporal events (in Russian), *Kvantovaya Elektron.*, *12*, 672 (1985); English translation, *Sov. J. Quantum Electronics*, *15*, 443 (1985).

17. Mossberg, T. W., Time-domain frequency-selective optical data storage, *Opt. Lett.*, *7*, 77 (1982).

18. Rebane, A., and Ollikainen, O., Error-corrective optical recall of digital images by photoburning of persistent spectral holes, *Optics Commun.*, *83*(3/4), 246 (1991).

19. Ollikainen, O., Rebane, A., and Rebane, K., Error-corrective optical neural networks modelled by persistent spectral hole-burning, *Optical and Quantum Electronics*, *25*, S569 (1993).

20. Combes, J. M., Grossmann, A., and Tchamitchian, Ph., eds., *Wavelets, Time-Frequency Methods and Phase Space*, Springer-Verlag, Berlin, Heidelberg, New York, 2nd ed., 1990, 331 pp.

21. Rebane, K., and Rebane, I., Peak value of zero-phonon line absorption cross-section and single impurity molecule spectroscopy, *J. Lumin.*, *56*, 36 (1993).

22. Heitler, W., *The Quantum Theory of Radiation*, Oxford, 1954.

23. Hizhnyakov, V. V., Fluctuations of impurity absorption and fluorescence at low temperatures (in Russian), *Proc. Estonian Acad. Sci., Phys. Math.*, *38*, 113 (1989).

24. Moerner, W. E., and Kador, L., Optical detection and spectroscopy of single molecules, *Phys. Rev. Lett.*, *62*, 2535 (1989); Ambrose, W. P., and Moerner, W. E., *Nature*, *349*, 225 (1991); Ambrose, W. P., Basche, Th., and Moerner, W. E., Detection and spectroscopy of single pentacene molecules in a *p*-terphenyl crystal by means of fluorescence excitation, *J. Chem. Phys.*, *95*, 7150 (1991).

25. Orrit M., and Bernard, J., Single pentacene molecules detected by fluorescence excitation in a *p*-terphenyl crystal, *Phys. Rev. Lett.*, *65*, 2716 (1990); Orrit, M., Bernard, J., and Zumbusch, A., Stark effect on single molecules in a polymer matrix, *Chem. Phys. Lett.*, *196*, 595 (1992); Wild, U., Güttler, F., Pirotta, M., and Renn, A., Single molecule spectroscopy: Stark effect on pentacene in *p*-terphenyl, *Chem. Phys. Lett.*, *193*, 451 (1992); Bräuchle, Chr., *Angewandte Chemie*, *104*, 431 (1992); Basche, Th., Ambrose, W. P., and Moerner, W. E., Optical spectra and kinetics of single impurity molecules in a polymer: spectral diffusion and persistent spectral hole burning, *JOSA,B9*, 829 (1992); Palm, V., Rebane, K., and Suisalu, A., *J. Phys. Chem.*, *98*, 2219 (1994).

26. Kikas, J., and Malkin, J., Spectral hole burning in optically dense sample (in

Russian), *Proc. Estonian SSR Acad. Sci. Phys. Math.*, *36*, 62 (1987); Malkin, E., and Kikas, J., Spectral hole burning and pulse filtering in optically dense samples, *Opt. Commun.*, *73*, 295 (1989).

27. Jaaniso, R., Kikas, J., Malkin, E., and Truusalu, P., Temporal memory in spectral hole burning (SHB), *Opt. Commun.*, *75*, 397 (1990); Kikas, J., and Malkin, J., Spectral hole burning with two-way irradiation in an optically dense sample (in Russian), *Proc. Estonian Acad. Sci. Phys. Math.*, *39*, 163 (1990).

28. Breinl, W., Friedrich, J., and Haarer, D., *J. Chem. Phys.*, *81*, 3815 (1984).

29. Hesselink, W., and Wiersma, D. A., *J. Chem. Phys.*, *75*, 4192 (1981); Wiersma, D. A., in *Advances in Chemical Physics XLVII* (Jortner, J., Levine, R. D., and Rice, S. A., eds.), John Wiley, New York, 1981, p. 421.

30. Rebane, Karl, Useful properties and problems of spectral hole burning memories (in Estonian), *Estonian Physical Society, Annual Report*, p. 32 (1991); Rebane, K. K., and Jaaniso, R. V., Means to manufacture optical persistent spectral hole burning memories (in Russian), USSR Patent Application No. 4836210/10 (030409), 1989.

31. Rebane, K. K., Four dimensional optical memory, based on persistent spectral hole burning (in Russian), USSR Patent Application, 1990.

32. Kikas, J. V., and Rebane, K. K., Persistent spectal hole-burning optical fiber memory (in Russian), USSR Patent Authors Certificate No.1105942, 1984.

33. Rebane, K. K., and Fedoseyev, V. G., to be published elsewhere.

34. Kikas, J., and Leiger, K., Effect of geometry on storage density in spectral hole burning memories, *Opt. Commun.*, *94*, 557 (1992); Rebane, I., and Kikas, J., On spatial resolution of light-gated spectral hole burning, submitted to *Opt. Commun.*

35. Rebane, A., Coherent response and time- and space-domain holography using impurity systems with photochemical hole burning (in Russian), Thesis Cand. Sci. (Ph.D.), Tartu University, 1984.

36. Rebane, A., and Kaarli, R., Picosecond pulse shaping by photochemical time-domain holography, *Chem. Phys. Lett.*, *101*, 317 (1983).

37. Saari, P. M., Rebane, A. K., and Kaarli, R. K., Writing of space and time-domain holograms in spectrally highly selective media (in Russian), *Optical Holography in Three-Dimensional Media* (Yu. N. Denisyuk, ed.), Leningrad, Nauka, 1986, p. 30.

38. Saari, P., Kaarli, R., and Rebane, A., Picosecond time-and-space domain holography by photochemical hole burning, *J. Opt. Soc. Am.*, *B3*, 527 (1986).

39. Saari, P., Zero-phonon lines and time-and-space-domain holography of ultrafast events, in (5), p. 123; Rebane, K. K., and Saari, P. M., Spectral hole burning and its applications for picosecond optical data processing, *Physica Scripta*, *T19B*, 604 (1987); Saari, P. M., Kaarli, R. K., Sarapuu, R. V., and Sõnajalg, H. R., Polarization-preserving phase conjugation and temporal reversal of an arbitrary-polarized pulsed optical signal by means of time and space-domain holography, *IEEE J. of Quant. Electronics*, *25*, 339 (1989); Kaarli, R. K., Saari, P. M., and Sõnajalg, H. R., Storage and reproduction of an ultrafast optical signal with arbitrary time dependent wavefront and polarization, *Opt.Commun.*, *65*, 170 (1988).

40. Rebane, A., Time domain holography, in (7), p. 196; Mitsunaga, M., Time-domain optical data storage by photon echo, *Opt. and Quantum Electronics*, *24*, 1137 (1992).

41. Bernet, S., Kohler, B., Rebane, A., Renn, A., and Wild, U., Holography in frequency selective media II: Controlling the diffraction efficiency, *J. Lumin.*, *53*, 215 (1992); Bernet, S., Kohler, B., Rebane, A., Renn, A., and Wild, U., Spectral hole burning and holography V: Asymmetric diffraction from thin holograms, *JOSA*, *B9*, 987 (1992).

42. Rebane, A., Bernet, S., Renn, A., and Wild, U., Holography in frequency selective media: hologram phase and causality, *Opt. Commun.*, *86*, 7 (1991).

43. Gygax, H., Görlach, E., Rebane, A., and Wild, U., Photochemically accumulated photon echoes and Stark effect, *J. Lumin.*, *53*, 59 (1992); Gygax, H., Rebane, A., and Wild, U. P., Stark effect in dye-doped polymers studied by photochemically accumulated photon echo, *JOSA*, *B10*, 1149 (1993).

44. Rebane, A., and Ollikainen, O., Error-corrective optical recall of digital images by photoburning of persistent spectral holes, *Opt. Commun.*, *83*, 246 (1991); Ollikainen, O., and Rebane, A., Optical implementation of a Hopfield-type neural network by use of persistent spectral hole burning media, *Optical memory and neural networks*, *SPIE*, *1621*, 351 (1991); Ollikainen, O., Optical implementation of quadratic associative memory by the use of persistent spectral hole burning media, *Appl. Opt.*, *32*, 1943 (1993).

45. Rebane, K., Ollikainen, O., and Rebane, A., Error-corrective recall of digital optical images in neural network models by photoburning of spectral holes, in (6), p. 24; Rebane K., Optical modelling of neural networks as a realization of spectral hole burning distributed data storage, in (7), p. 214.

46. Rebane, A., Associative recall of time- and space-domain holograms in spectrally selective photoactive media (in Russian), *Proc. Estonian Acad. Sci. Phys. Math.*, *37*, 89 (1988); Rebane, A., Compression and recovery of temporal profiles of picosecond light signals by persistent spectral hole-burning holograms, *Opt. Commun.*, *67*, 301 (1988).

47. Rebane, A., Feinberg, J., Time resolved holography, *Nature*, *351*, 378 (1991).

48. Rebane, A., Aaviksoo, J., Kuhl, J., Storage and time reversal of femtosecond light signals via persistent spectral hole burning holography, *Appl. Phys. Lett.*, *54*, 93 (1989); Schwoerer, H., Erni, D., Rebane, A., and Wild, U. P., Subpicosecond pulse shaping via spectral hole-burning, *Opt. Commun*, *107*, 123 (1994).

49. Kohonen, T., *Self-Organization and Associative Memory*, Springer-Verlag, Berlin, Heidelberg, 1987.

50. Rebane, A., Associative space-and-time-domain recall of picosecond light signals via photochemical hole burning holography, *Opt. Commun.*, *65*, 175 (1988).

51. Hopfield, J. J., Neural networks and physical systems with emergent collective computational abilities, *Proc. Natl. Acad. Sci.*, *USA*, *79*, 2554 (1982).

52. Farhat, N., Psaltis, D., Prata, A., and Paek, E., *Applied Optics*, *24*, 1469 (1985); Shariv, I., and Friesem, A. A., All optical neural network with inhibitory neurons, *Opt. Lett.*, *14*, 485 (1989); Chen, H. H., Lee, Y. C., Sun, G. Z., Lee, H. Y., Maxwell, T., and Giles, G. L., High order correlation model for associative memory, AIP Conf. Proc., 151, p. 86 (1986); Psaltis, D., and Park, C. H., Nonlinear discriminant functions and associative memories, AIP Conf. Proc., 151, p. 370 (1986).

53. Moerner, W. E., Length, W., and Bjorklund, G. C., Frequency domain optical

storage and other applications of persistent spectral hole-burning, in (3), p. 251; Szabo, A., Frequency selective optical memory, *U. S. Patent No.3,896,420, 1975; Castro, G., Haarer, D., Macfarlane, R. M., and Trommsdorf, H. M., Frequency selective optical data storage system, U.S. Patent No. 4,101,976, 1978; Rebane, K. K., Rebane, L. A., Gorokhovskii, A. A. and Kikas, J. V., Carrier of information (in Russian), USSR Patent No 943260, April 1, 1982; Kim, M. K., and Kachru, R., Storage and phase conjugation of multiple images using backward-stimulated echoes in Pr^{3+}* : LaF_3, Opt. Lett., 12, 593 (1987); Kröll, S., and Elman, U., Photon-echo-based logical processing, *Opt. Lett.*, *18*, 1834 (1993); Yano, R., Mitsunaga, M., and Uesugi, N., Nonlinear laser spectroscopy of Eu^{3+} : Y_2SiO_5 and its application to time-domain optical memory, *JOSA, B9*, 992 (1992); Samartsev, V. V., Usmanov, R. G., and Khadyev, L. K., *Pis'ma Zh. Eksp. Teor. Fiz.*, *22*, 32 (1975) [JETP Lett., *22*, 14 (1975)]: Zuikov, V. A., Samartsev, V. V., and Usmanov, R. G., *Pis'ma Zh. Eksp. Teor. Fiz.*, *32*, 293 (1980) [Sov. Phys. JETP Lett., *32*, 270 (1980)]; Carlson, N. W., Rothberg, L. J., Yodh, A. G., Babbitt, W. R., and Mossberg, T. W., Storage and time reversal of light pulses using photon echoes, *Opt. Lett.*, *8*, 483 (1983).

54. Rebane, I., Theory of two-step spectral hole burning by pulses, *Phys. Stat. Sol.*, *B145*, 749 (1988); Rebane, I. K., Theory of two-step pulsed photoburning of limiting narrow spectral holes, *J. Phys. B: Atm. Mol. Opt. Phys.*, 22, 2411 (1989); Rebane, I., Theory of two-step pulsed spectral hole burning in four-level system, *Opt. Commun.*, 76, 225 (1990).

55. Mossberg, T. W., Swept-carrier time-domain optical memory, *Optics Lett.*, *17*, 535 (1992).

56. Burge, R. E., Fiddy, M. A., Greenway, A. H., and Ross, G., The phase problem, *Proc. Roy. Soc. London, A350*, 191 (1976).

57. Gabor, D., *Nature*, *217*, 1288 (1968).

58. Ollikainen, O., to be published.

59. Bauer, R., Attenberg, T., Sildos, I., Bogner, U., and Maier, Max, Persistent spectral hole-burning by cw lasers in defect-aggregates induced by particle irradiation, in (7), p. 50; Sildos, I., Persistent spectral hole-burning in neutron-irradiated sapphire and diamond crystals, in (7), p. 142.

60. Holliday, K., Wei, C., Croci, M., and Wild, U. P., Spectral hole-burning measurements of optical dephasing between 2–300K in Sm^{2+} doped substitutionally disordered microcrystals, *J. Lumin.*, *53*, 227 (1992).

61. Jaaniso, R., Hagemann, H., and Bill, H., Inhomogeneous broadening and spectral hole burning in $Sr_yBa_{1-y}FCl_xBr_{1-x}$: Sm^{2+}, in (7), p. 79; Wei, C., Holliday, K., Meixner, A. J., Croci, M., Wild, U. P., *J. Lumin.*, *50*, 89 (1991); Furusawa, A., and Horie, K., High-temperature photochemical hole burning and laser-induced hole filling in dye-doped polymer systems, *J. Chem. Phys.*, *94*, 80 (1991).

14

Holographic Spectral Hole Burning: From Data Storage to Information Processing

Cosimo De Caro, Stefan Bernet, Alois Renn, and Urs P. Wild

Swiss Federal Institute of Technology
Zürich, Switzerland

I. INTRODUCTION

Features of spectral hole burning (SHB) that relate to molecular electronics applications, such as data storage and processing as well as molecular based optical devices, are described in different articles of this book. In our contribution we demonstrate that a combination of these properties with holography supplies us with various new possibilities for practical applications.

In contrast to implementations of "molecular electronics" illustrated in other chapters, where information is stored spatially, on a nm scale, another approach to the molecular level is chosen in SHB. Here, molecular selectivity is not reached by addressing very small spot areas, as is attained with a scanning tunnelling microscope, but is instead reached by addressing the molecules by their extremely pronounced selectivity to the light frequency and externally applied fields. These properties even allow the addressing of single molecules without having spatial resolution in the range of atomic scales, as has been already shown in papers concerning single-molecule spectroscopy (1).

Using holography as an information storage technique of the "one molecule one bit" relation is not intended. The reason is that holography is always an ensemble effect, i.e., each bit of information is spatially delocalized over a considerable sample area. On the other hand, such a hologram usually contains a huge amount of binary information or a whole image. The molecule set corresponding to a single-frequency hologram in an SHB material is still very small as compared to conventional recording materials like silver halide films or magnetic storage devices. Therefore holographic hole burning is a possible approach to the goal of "molecular electronics" where the number of molecules per bit is strongly decreased towards the ideal "one bit per molecule" mapping.

The advantage of holography shows up in various technical applications. As a natural feature of holography one gains the possibility of parallel data manipulation. One hologram usually contains a whole array of binary data or image information that can be processed in a single step. In addition, using the holographic technique a new parameter is obtained, the hologram phase, which can be freely manipulated and turns out to be the key to new possibilities for data storage and processing. In this chapter we describe some of these applications. For instance, using phase-controlled holography it has been possible to store 2000 gray-level holograms in one SHB sample. On the other hand, logical operations, so called "molecular computing," between data arrays are possible by generating electric field induced interferometric superpositions of holograms with a controlled phase relationship. This method has been used to implement a parallel binary full adder, demonstrating a parallel addition of two number arrays, each containing 400 binary coded four-bit numbers. Computing is performed at the molecular level in the sense that it uses the ability of SHB to address individual molecular subsets by an externally applied electric field.

In this contribution the principles of holographic hole burning will be explained. A brief overview of the physical mechanisms of spectral hole burning, and a presentation of the experimental arrangement, can be found in Sec. II, and the relevant results will be illustrated in Sec. III.

II. HOLOGRAPHY IN HOLE BURNING MATERIALS: FUNDAMENTALS

Spectral hole burning allows frequency selective storage at scales much less than the inhomogeneous broadening of doped guest–host systems and has found extensive application in the study of doped crystals and amorphous matrices, such as organic-dye-doped polymer films (2–6). The first experimental SHB results have been independently achieved by two Russian scientific groups in 1974 (4).

At liquid helium temperature (1–4K), the inhomogeneous absorption band of doped solids is composed of a continuum of homogeneous contributions absorbing at slightly different frequencies (2,3). The homogeneous absorption lines are mainly determined by purely electronic transitions without contributions from lattice phonons of the host crystal, and are called zero-phonon lines (ZPL). In the case of organic molecules, the transition from the ground (S_0) to the first excited singlet state (S_1) is called the $S_1 \leftarrow S_0$ transition. The homogeneous line width of the ZPL is composed of contributions arising from the lifetime of the first excited state of the guest molecule and by the dephasing time, which takes into account fluctuations of the optical transition frequency due to interactions with the host. If a hole burning material is illuminated with monochromatic light within the inhomogeneous absorption band, only the subset of molecules centered around the exposure frequency within a spectral range of a homogeneous line width is excited. The excitation leads to photoinduced absorption changes in this frequency range, the so-called spectral holes. Two types of spectral holes can be distinguished (2,3):

1. Permanent hole burning: spectral holes are persistent on a time scale of several hours as long as the samples are kept at liquid helium temperature. Furthermore, we can distinguish between photochemical (photochemical reactions of the guest molecule, e.g., an intramolecular proton shift) and photophysical permanent mechanisms (reorientation of the local environment and/or the guest impurity after laser excitation).

2. Transient hole burning: the spectral holes decay from a metastable state to the ground state after a time varying between μs and a few minutes, e.g., a population decay from a long-lived triplet state in organic molecules such as porphyrins (6).

Spectral holes are very sensitive to small external perturbations such as the application of external electric and magnetic fields, or strain (hydrostatic pressure). This allows the determination of detailed information about guest–host interactions (for a review, see Refs. 2 and 6).

A. Experimental

Chlorin (2,3-dihydroporphin) was used to dope polymer films of polyvinyl butyral (PVB). The chlorin molecule and its related compounds are very well known and have been extensively studied due to their relevance in biology (7). Chlorin was synthesized in our laboratory (8) and used without further purification for the preparation of the doped polymer film. The film was then placed between two glass plates (dimensions 3×3 cm^2) with conducting coatings in order to apply a voltage (Stark cell). The Stark cell was immersed in liquid helium in a bath cryostat.

The mechanism of the photoreaction is demonstrated in Fig. 1. At liquid helium temperature the inhomogeneous absorption band used for hole-burning is centered at 635 nm, with an inhomogeneous line width (FWHM) of 10 THz (14 nm). Monochromatic illumination in this frequency interval leads to a frequency-selective photoexcitation from the S_0 to the S_1 state. After excitation the molecule is able to undergo a permanent but reversible photoreaction, where the two inner protons rotate simultaneously by 90°. The photoproduct absorbs at 578 nm, i.e., outside of the inhomogeneous absorption band used for hole burning. The homogeneous line width of chlorin in PVB is very narrow (approximately 160

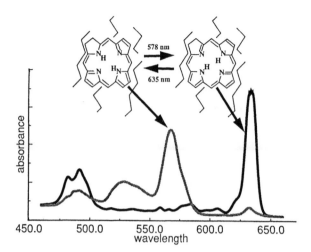

Figure 1   Photoreaction of chlorin. In the graphics, both tautomers involved are drawn. The PVB polymer environment is schematically symbolized by the irregular lines around the molecules. The whole system corresponds to a so-called supermolecule whose absorption frequency strongly depends on the position of the chlorin molecule within its environment due to guest–host interactions. Below, the absorption spectrum of both tautomers at a temperature of 4.2K is drawn. Normally, the equilibrium is shifted to the side of the species absorbing at 635 nm. At liquid helium temperatures the photoproduct, absorbing at 568 nm, is also stable. Exposure with light of the corresponding wavelength leads to reversible photoreaction.

MHz at 1.7K (9) as compared to the inhomogeneous absorption band width, and the system shows a pronounced electric field induced signal splitting owing to the properties of the molecular electronic states of chlorin (10).

An experimental arrangement for holographic recording and detection of spectral holes (11,12) is shown in Fig. 2. A single-mode dye laser pumped by an argon ion laser is used as a tunable light source. The laser bandwidth of 1 MHz is negligible as compared to the homogeneous line width (ca. 160 MHz) of the doped PVB sample. The frequency of the laser can be scanned under computer control over the whole inhomogeneous absorption band (10 THz) centered at 635 nm. The laser beam is enlarged to a diameter of 5 cm by a telescope and then divided into reference and object beams by a cube beam splitter. The two beams are superposed on the sample, forming an interference pattern that is recorded photochemically as a permanent spectral hole. Typical recording times are in the order of 10 s with power densities of 100 $\mu W/cm^2$. Pictures can be recorded by inserting a slide or any other image information carrier (e.g., by a liquid crystal display) into the object beam. Reading of the holograms is performed by illuminating the sample with the attenuated reference beam, in order

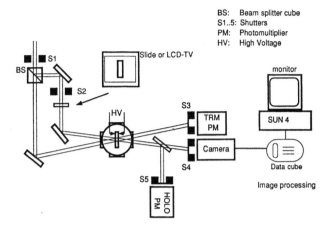

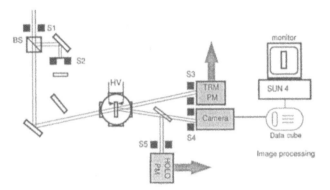

Figure 2 Experimental arrangement for the storage and reconstruction of images recorded as holograms in a hole burning sample. Details are explained in the text.

to avoid further photochemical transformations (attenuation factor 10–100). Readout scans can be performed in either the laser frequency or the applied electric field dimensions. The reference beam is diffracted by the photochemical grating, and the object wave front is reconstructed. The holographic signal is monitored by a photomultiplier or by a video camera (Hamamatsu C 2400–25 VIM CCD camera with image intensifier). The video images can be digitized by an image processing unit and finally stored on a real-time disk. An additional photomultiplier allows for the detection of the light intensity transmitted through the sample. A comparison between the transmitted and the diffracted signals as shown in Fig. 3 clearly illustrates the increased sensitivity of the holographic detection: in contrast to transmission, the holographic signal is background free.

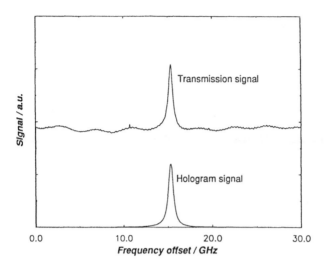

Figure 3 Comparison of the transmission signal with the diffraction signal of a hologram recorded in the center of a spectral region of 30 GHz (8). In contrast to the transmission signal, the hologram is background free.

A special feature of the holographic setup is an active control of the hologram phase. The phase can be shifted by mounting a mirror in the object beam on a piezoelectric transducer. Application of a voltage to the transducer results in a precisely controllable mirror shift. If for example such a mirror shift reduces the object path length by half a wavelength, then the interference grating in the sample plane performs a spatial phase shift of π (the interference maxima shift into their adjacent minima). The phase can be monitored by superposing small parts of the object and the reference beam on a linear photodiode array where they form a copy of the interference grating in the sample plane. The signal from the diode array is monitored by an oscilloscope showing sinusoidal fringes whose phase behaves exactly like the phase of the fringes in the sample plane.

All parameters of the experiment, e.g., the voltage of the piezoelectric transducer, the Stark field, the laser frequency, and the various devices like the shutters, the video camera, and the photomultipliers are controlled by a SUN workstation over GPIB interfaces and D/A-converters.

B. Single-Frequency Holograms

Single-frequency holograms are recorded at selected frequency positions with monochromatic laser light. During the recording of such a hologram, the interference grating of the light intensity leads to an absorption grating and a refractive index grating in the sample (13,14). The spectral extensions of both gratings are related by the Kramers–Kronig relations. During hologram reconstruction, the electric field amplitude of the reference wave is diffracted by both gratings. The spectral shape of the field amplitude diffracted by the absorption grating is a Lorentzian, with a hole width (FWHM) $2\Gamma_h$, where Γ_h is the homogeneous line width of the hole burning material. The field amplitude of the light diffracted by the refractive index grating is a dispersion curve. These features are shown in Fig. 4 in a complex representation. The Lorentzian line in the vertical plane symbolizes the field amplitude of the light diffracted by the absorption grating

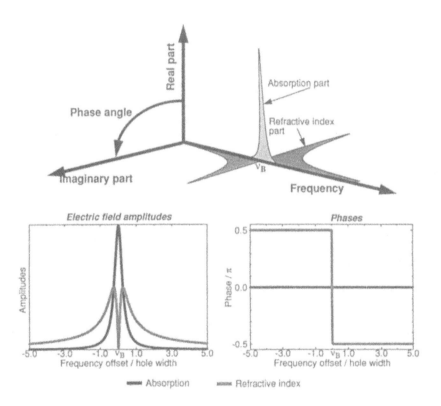

Figure 4 (Top) Complex representation of the electric field amplitudes diffracted by the absorption and the refractive index gratings of a single frequency hologram with center frequency ν_B (17). The spectral shape of the absorption contribution is a Lorentzian, whereas the refractive index contribution is a dispersion-shaped curve. The contributions have a relative phase shift of $\pm \pi/2$ as indicated by the different phase angles in the complex representation. (Bottom) A second representation of the same situation is illustrated, with the amplitudes and phases of the electric field components diffracted at both gratings drawn separately.

as a function of the readout frequency. The "dispersion shaped" curve in the horizontal plane corresponds to the field amplitude diffracted by the refractive index grating. The angle between the two curves corresponds to the phase shift between both contributions. Note that the phase of the absorption part is constant over the whole frequency range, whereas the phase of the refractive index part has a jump of π at the burning frequency, and both contributions are out of phase by $\pm\pi/2$. At the bottom of Fig. 4, the field amplitude diffracted by both gratings is drawn a second time. The left-hand graph shows the amplitudes of the electric field components diffracted at both gratings, while the right-hand graph shows the corresponding phases. For data storing purposes it is interesting to note that the spectral extension of the refractive index contribution is much larger than that from the absorption contribution. To get the total diffracted field amplitude, both contributions have to be added under consideration of their relative phase relation by a complex summation. The total diffraction efficiency is the squared absolute value of the complex result.

C. Frequency and Phase Swept Holograms

Frequency and phase swept (FPS) holograms are a novel type of holography that is only possible in frequency selective materials (15). During hologram exposure, the frequency of narrow band laser light is swept over a spectral range corresponding to a few homogeneous line widths of the spectrally selective recording material. Simultaneously the phase of the hologram is controlled as a function of the frequency—the so called phase sweep function. Depending on the phase sweep function, this hologram type shows very interesting properties such as asymmetric diffraction into conjugated diffraction orders or an increased diffraction efficiency. The diffraction properties are closely related to the causality principle (16). A detailed study of this hologram type is given in (17), and various properties have been investigated in (15,18,19). In this article, the use of this hologram type for high-density data storage applications is described.

For an illustration of FPS holograms, the results of linear sweeps over total phase ranges of -2π, 0, and 2π are schematically drawn in Fig. 5. The spatial population gratings formed in the sample are shifted to the left or to the right side, depending on the sign of the phase sweep. In the same representation as in Fig. 4, a swept hologram can be considered as a series of single-frequency holograms, recorded at closely spaced frequency positions. A phase shift as a function of the frequency corresponds in this picture to a rotation around the frequency axis by a phase angle determined by the phase sweep function. The continuous case is reached when the spacing becomes closer and the phase angle is adapted correspondingly. The total diffraction efficiency of such a structure can be calculated by a complex summation over the discrete lines. This method

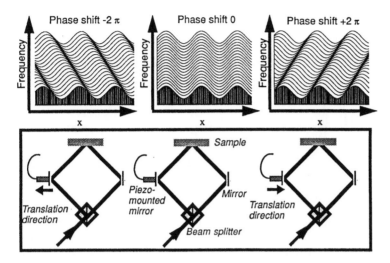

Figure 5 Effect of a frequency and phase sweep on the spectral and spatial molecule density in the sample (18). If no phase shift is applied during the frequency sweep (middle), then the population grating is only spatially modulated (illustrated in the top figures). If the piezoelectric transducer shifts the mirror into the indicated directions (left and right graphics) during a frequency increase, then the spatial population grating shifts to the left or to the right side as a function of its spectral position.

is used in Fig. 6 to determine the amplitudes and phases of a hologram that is linearly swept over a phase range of -2π in a frequency range of 2 hole widths. As compared to the single-frequency hologram in Fig. 4, the diffracted electric field in the center of the burning range is now formed from a contribution of the absorption grating as well as from a contribution of the refractive index grating. Both contributions are exactly in phase in the center and therefore interfere constructively. For high-density data storage applications it is interesting to note that both diffraction contributions fall off rapidly with increasing spectral distance from the burning center. The total diffraction efficiency of such a structure, as determined by a complex addition of both contributions and forming the squared absolute value of the result, will be compared later with a single-frequency hologram and with experimental data.

D. Influence of an External Electric Field: The Linear Stark Effect

According to classical electromagnetism, the energy of a dipole in a static electric field is linearly proportional to the applied field. Therefore the Stark shift of an $S_1 \leftarrow S_0$ transition is proportional to the static electric field and the dipole moment

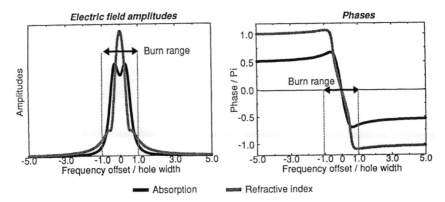

Figure 6 Amplitudes and phases of the electric field components of a light wave diffracted at an FPS hologram (17). The hologram is linearly swept in a frequency range of 2 hole widths with a phase shift of -2π. As compared to the analog representation for a single frequency hologram in Fig. 4, the diffraction efficiency in the burning center is composed of an absorption contribution as well as a refractive index contribution. At the spectral wings, the refractive index contribution falls off much faster than in the case of the single-frequency hologram.

difference between the first excited state S_1 and the ground state S_0. The electric dipole moment is a specific property of each molecule. Its value and direction give information about the charge distribution of the system studied. Not every molecule has a permanent dipole moment: in molecules with inversion symmetry the dipole moment is zero. On the other side, the dipole moment of a molecule embedded in a crystal or an amorphous system can be composed of a permanent contribution and an additional induced term that arises from an electrostatic interaction between the molecule and its microenvironment (local effective field) (20–23). Thus, Linear Stark effect can even be observed on molecules without a permanent dipole moment if these molecules are embedded in a host material (12,21,24). In amorphous systems, the absolute values of the induced contributions are described by a Gaussian distribution (12,23,21,25), and the orientations of the total dipole moments are isotropic. The analysis of the line shape and line width of a spectral hole under the influence of an external electric field requires that all parameters determined by the molecular and experimental arrangement geometries be considered (12,21,26). In particular, the orientation of the applied electric field with respect to laser light polarization plays an important role. Depending on the angle between these two vectors, different line shapes have been observed in organic-dye-doped polymers (20,21,24). Meixner et al. (12,20,21) have analyzed this problem and studied the influence of an electric field on spectral holes in chlorin/PVB samples, a well-known system used in

Stark effect measurements (20,21,27–30). They found that the effect of photoselection by excitation with linear polarized light has to be taken into account. Maximum absorption of the exciting light is achieved only when the transition dipole moment orientation is parallel to the exciting light polarization. On the other hand, the Stark effect of these molecules depends on the direction of the external field with respect to the dipole moment difference. Thus, the Stark effect and the molecule selection during burning are influenced by different geometrical angles, which are connected by the fixed intramolecular angle between the dipole moment difference and the transition dipole moment. A theoretical analysis of the line shapes therefore includes a rather complicated geometric averaging over the absorption and the Stark effect properties of all molecules in the isotropic system.

In Fig. 7, a single spectral hole stored by means of the holographic technique in a chlorin/PVB sample is shown as a function of the laser frequency and the applied electric field (31). The electric field was parallel to the direction of the incoming laser light. The chlorin molecule has a 90° angle between the transition dipole moment and the dipole moment difference (10,21). Typical features of such a system are the splitting, the broadening, and the decrease of the intensity with increasing field strength. The Stark splitting components can be seen clearly. The hologram arises as a narrow peak with zero background in the center of the plotted section of the so called frequency–electric field plane. The narrow peak width of the holographic signal in both dimensions suggests that many holograms can be recorded independently in such a frequency–electric field plane. For data storage purposes it should be noted however, that the electric field does not

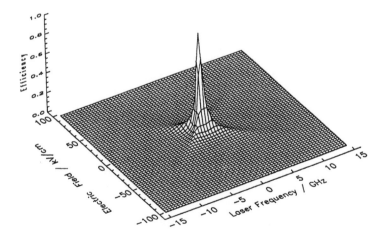

Figure 7 Diffraction efficiency of a hologram recorded in the center of the plotted section of the frequency–electric field plane (31).

provide a completely independent recording dimension. Due to the isotropic orientation of the molecules selected during exposure, the Stark effect leads to different frequency shifts within this subset. Depending on their orientation with respect to the applied electric field, the molecules are tuned out of their resonance frequency in different ways. Simultaneously, other spectrally adjacent molecules are tuned into resonance, resulting in a new distribution of the transition frequencies. The statistical averaging leads to a broadening and filling of the spectral holes. In the case of holography, where the diffracted signal is proportional to the squared number of resonant molecules, the signal disappears rapidly with increasing distance from the recording position. Under these conditions a second hologram can be recorded at an adjacent electric field position. When the electric field is turned off, the original conditions are restored and the first hologram can be detected again. These applications will be investigated in the next section.

III. APPLICATIONS

A. Data and Image Storage

In 1975 and 1978 two patents were published describing the optical recording of a series of bits in the inhomogeneously broadened absorption bands of doped guest–host systems (32,33). The optical frequency provides a new storage dimension in addition to the spatial dimensions defined by the laser spot size on the sample. Due to the large ratio between the inhomogeneous and the homogeneous line widths of doped amorphous systems at low temperature it is conceivable to store up to 10^6 bits as spectral holes within the inhomogeneous band. Binary information may be stored by associating the presence (absence) of a spectral hole at a selected optical frequency with a digital "1" ("0"). Figure 8 illustrates the principle of frequency multiplexed data storage (31,35): spectral holes were recorded with a constant wavelength separation of about 1 nm within the whole $S_1 \leftarrow S_0$ absorption band of a chlorin-doped polyvinyl butyral film at 1.7K between 622 and 638 nm. The inset of this figure shows a wavelength scan magnified by a factor of 200. Using this wavelength resolution, up to 10 holes have been stored in a spectral range of only 1 cm^{-1} (0.04 nm, 30 GHz). The storage density can even be increased by using the Stark effect. Independent spectral holes can be recorded at a constant optical frequency using different electric fields (31,36,37). Thus the whole accessible frequency–electric field plane can be used for two-dimensional data storage. Examples for this recording scheme are presented in the next section.

Image storage by SHB techniques is achieved with the same principle (38), using whole images (like slides or images on a liquid crystal display) as an input in the object beam. Like binary information, the holograms of whole images can be recorded at different coordinates in the frequency–electric field plane

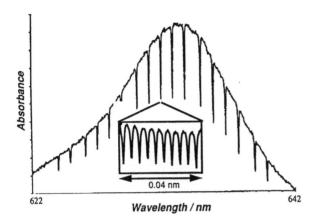

Figure 8 Spectral holes in the inhomogeneous absorption band of a chlorin/PVB sample at a temperature of 4.2K. The inset shows a high resolution scan, magnified by a factor of 200 with respect to the wavelength axis, where 10 absorption holes are resolved within a spectral range of only 1 cm^{-1} (0.04 nm).

(31,34,38). Readout of the images is performed with a video camera. The left column of Fig. 9 shows some pictures used as input for holographic recording in a chlorin/PVB polymer film (8). The corresponding holographically read-out images are presented in the right column. The comparison illustrates the quality and resolution of the image storing system. The information content that can be stored by a single image may be estimated by determining the number of pixels necessary to achieve the same spatial resolution, for example on a computer monitor. In the actual case, the image resolution is diffraction limited by the sample aperture, and an effective sample area of 250 μm^2 per bit (but spatially delocalized) may be estimated. Even more information may be included if gray-level information is also considered. In this case the dynamic range of the distinguishable intensities of different image pixels is regarded as an extra information source. In the next section several examples of holographic data storage by spectral hole burning will be presented, and different aspects, in particular with regard to storage capacity, will be discussed.

1. Multiple Hologram Storage Using Single-Frequency Holograms

In order to exploit the advantageous frequency multiplexing ratio provided by hole burning materials, the holograms have to be recorded as densely as possible in both storage dimensions, frequency and electric field (8,39). When the spectral separation between two holograms is less than 10 line widths, disturbing short-range interactions (cross talk) due to interference are observed (8,40). The interference is due to a superposition of the spectral wings of the holographic gratings. An increase or decrease of the background signal can be observed in both

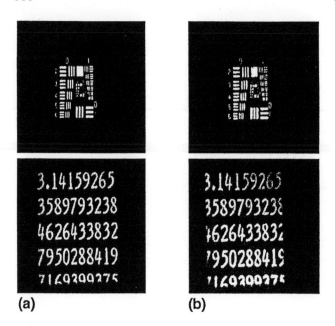

Figure 9 (Left) Two slides as directly read with the CCD camera through the optical setup for the holographic experiments. The slides were used as input for holographic image storage experiments. (Right) The corresponding holographically reconstructed images show that almost no quality loss occurs as a result of the storage and readout process (8).

recording dimensions depending on the phase difference chosen during recording (41,42). In fact, two spectrally adjacent holograms stored at different frequencies but at the same electric field position show constructive interference in the center of their recording positions for a phase difference of π, while destructive interference appears for no phase difference (8,12,41). A similar behavior is observed when the hologram pair is stored at a single frequency and adjacent electric field values. In this case, constructive interference is obtained by applying no phase difference, whereas a phase difference of π leads to destructive interference. A detailed study of the diffraction properties of spectrally adjacent holograms is presented by Renn et al. (41,42) based on Kogelnik's coupled wave theory (43). They found that the interference effects can be explained in terms of diffraction contributions from both the absorption and the refractive index gratings, formed during interferometric recording (12,13). The signs of these contributions determine whether constructive or destructive interference is dominant. When more than two holograms are stored in both recording dimensions, the coherent background produced by each single hologram—especially from the refractive index contributions, as indicated in Fig. 4—may sum up leading to an intolerable disturbance of the hologram signals (8). In the case of a constant phase of a whole hologram group, a characteristic increase of the background is clearly observed (39,40). The experiments suggest the use of a selected phase difference in high-density storage applications (40). Figure 10 shows 25 holograms that were stored in a chlorin-doped PVB polymer film at 1.7K (39). The

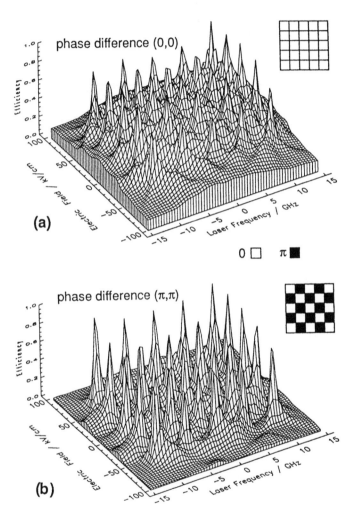

Figure 10   Twenty-five single-frequency holograms recorded within a section of the frequency–electric field plane (39). The spectral separation of the holograms in the frequency–electric field plane is 5 GHz and 33 kV/cm respectively. (a) All holograms were recorded with the same phase symbolized by a single-colored "chessboard" in the upper right corner. (b) The holograms were stored with a "chessboard-like" phase pattern by alternating the phases between 0 and π in a way symbolized by the white–black (0–π) "chessboard" pattern.

spectral separation between the holograms was about 5 GHz (0.17 cm^{-1}), and a 33 kV/cm separation was used in the electric field domain. The hologram arrays were recorded in two different ways. In Fig. 10a, all hologram phases were kept constant during recording, resulting in a strong increase of the background signal. In Fig. 10b, a "chessboard phase pattern," alternating between the phase values 0 and π for adjacent holograms, has been applied during recording. As compared to figure 10a, a large suppression of the cross talk and the background in the whole storage domain is observed. These results indicate that the "chessboard" phase pattern is a suitable solution for the reduction of undesired background and cross talk in high-density storage applications. Using this method it was possible to store 100 holograms in a spectral range of only 30 GHz (1 cm^{-1}) in a chlorin/PVB film (40). The results are shown in Fig. 11. Groups of 10 holograms were successively stored at a fixed frequency and equally spaced electric field values with a phase difference of π between adjacent holograms. After recording each group of 10 holograms, the laser frequency was stepped to its next value, and the described procedure was repeated until 100 holograms were stored. In this experiment, the separations between adjacent spectral holes was reduced to 20 kV/cm (electric field domain) and 3 GHz (frequency dimension). The effective storage density of 100 holograms in a spectral range of 30 GHz corresponds to 1 hologram per 300 MHz, a value that cannot be obtained in the same hole burning material by using exclusively the frequency dimension, since the typical spectral hole width is in the order of 660 MHz.

These results indicate that the hologram phase is a relevant parameter in order

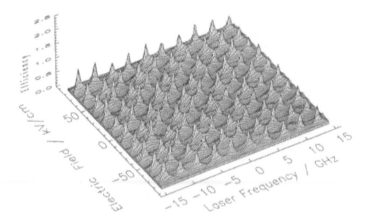

Figure 11   Diffraction efficiency of 100 single-frequency holograms recorded within a frequency interval of only 1 cm^{-1} (30 GHz) using the "chessboard" phase pattern as explained in figure 10b (40).

to optimize the storage density achievable by holographic hole burning. For optimal results, the appropriate values for the hologram phase have to be chosen in both recording dimensions. As will be shown later, the hologram phase also represents the key element for information processing experiments based on controlled interferometric hologram superpositions in hole burning materials.

2. 2000 Gray Level Images Using FPS Holography

In Fig. 12 the spectral shapes of single-frequency holograms and FPS holograms are compared. The simulation on the left side is made for a single-frequency hologram (dashed curve) and for a FPS hologram linearly swept in a frequency range of 1 GHz (2 hole widths) with a total phase shift of 2π. On the right side, the corresponding experimental results are compared, and a good agreement with the simulations is observed. The calculations are made as already described in Sec. II.C. An important feature of the FPS hologram is the fast decay of the diffraction efficiency with increasing distance from the burning center as compared to the single-frequency hologram. As illustrated in Fig. 6, this is due to the suppressed refractive index contribution at the spectral wings of the FPS hologram. A theoretical analysis (17) has shown that the diffraction efficiency of a single-frequency hologram falls off as $1/(\Delta\nu)^2$, whereas the diffraction efficiency of a linear 2π swept hologram falls off as $1/(\Delta\nu)^4$, where $\Delta\nu$ is the spectral distance from the burning center. This suggests a reduced cross talk with respect to a discrete recording scheme when storing several holograms at spectrally adjacent positions.

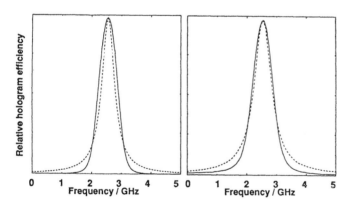

Figure 12 Comparison of the line shapes of a single-frequency hologram (dotted) and an FPS hologram in theory (left side) and experiment (right side). Both holograms are exposed with the same energy. The "normal" hologram is recorded at a single frequency position in the center of the plotted frequency range, while the FPS hologram is swept continuously over a 1 GHz range with a linear phase shift of -2π during recording (19).

This hologram type has been used to store 2000 gray-level cartoon images, with reduced cross talk, using both the frequency and the electric field dimensions of a chlorin/PVB sample (44). The input images were digitized frames from an animated cartoon that could be displayed on the LCD TV placed in the object beam. Electric field multiplexed groups of eight holograms were stored at intervals of 5 GHz along the frequency axis using the FPS method as described in Sec. II.C. An electric field range of -150 kV/cm to $+150$ kV/cm and a frequency range of 1.25 THz (42 cm^{-1}, 1.7 nm) were used. Figure 13 shows a frequency scan over 200 holograms at one discrete electric field value. On this scale the discrete lines are not resolved. A magnified scan over a subgroup of six holograms in a 30 GHz (1 cm^{-1}) range is shown below. Each hologram contains another gray-level image that is read by the CCD camera and looks similar to the example on the right side. For demonstration purposes an animated video movie corresponding to a sequence of 80 s duration was realized from the

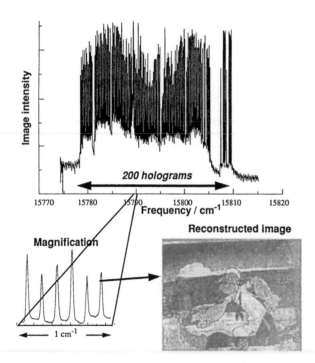

Figure 13 (Top) Diffraction efficiency of 200 FPS holograms stored with a spectral separation of 5 GHz at one electric field position. (Below) A magnified scan over a frequency range of 30 GHz showing the well-resolved holograms. Each hologram contains gray-level image information, as shown nearby for one of the reconstructed holograms (44).

holographically reconstructed images. It may be noted that in this experiment only a spectral range of 1.4 nm of the accessible inhomogeneous absorption line (14 nm) has been used for recording. Therefore a total storage capacity of more than 10,000 images may be estimated. Recently, the recording of 6,000 holograms containing binary data arrays has been realized using a similar FPS method (45).

B. Data Processing

A property of holography is the parallel access to a data array during recording and retrieval of information. This great advantage has excited a strong interest in computer technology and represents a promising way for future developments in computer systems. The parallel access provided by the technique opens new possibilities for optical computing applications such as associative memories (46). The great interest in this field is well documented by the large number of related publications.

In the previous section, a new technique for data storage based on the imaging properties of holography, optical spectroscopy, and the Stark effect has been presented. To achieve high-density data storage, the spectral separation between adjacent holograms has to be decreased, leading to undesired interferometric hologram interactions. On the other hand, this feature can be used advantageously for data processing applications. In contrast to the last section, where the phase control has been used to suppress undesired hologram cross talk, now the phase control allows for achieving logical operations between holographically stored data arrays. These applications are known as "molecular computing" (47,48). The concept is based on spectral hole burning, the interaction of the molecular energy levels with an externally applied electric field, and the interferometric properties of holography. In particular, the interference behavior of two holograms can be controlled by setting a specific phase difference between them. The different superpositions of the two holograms correspond to the fundamental logical operations AND, OR, and XOR.

1. From Hologram Superpositions to Logical Operations

The data processing is separated into two steps. In the first step the data arrays to be superposed are stored as spectrally adjacent holograms with an appropriate phase difference, depending on the desired logical operation. The next step is to apply an external electric field that induces the superposition of the Stark split components of both holograms (see Sec. II.D). Readout at the center frequency between the two holograms results in a coherently superimposed hologram and shows constructive or destructive interference—i.e., addition or subtraction—depending on the phase difference chosen during recording. In this sense, hole burning materials are a new generation of recording materials in which both data storage and information processing are performed. The molecular system itself represents a model of a parallel processor: different data fields stored as two-dimensional pixel arrays can be logically connected in parallel without using any external processor. The principle of hologram superposition is illustrated in Fig. 14 and has been tested as follows: two slides of a horizontal and a vertical bar were stored at the same frequency but at two different electric field values, E_1 and E_2, in a chlorin-doped polymer film (8,47). The typical Stark splitting behavior of the holographic signal has already been described in Sec. II.D. The separation of the recording coordinates has been chosen so that the hole contours overlap within the accessible frequency and electric field range. The two images can be reconstructed individually by adjusting exactly the experimental parameters as applied during recording. The coherent superposition of the images is obtained in the frequency—electric field region, where the Stark "wings" overlap—at the central electric field and a shifted frequency position. Here a superposition of both input images, the vertical and horizontal bar, can be observed, i.e., a cross. If both holograms are recorded with zero phase difference, then the two images are "added" constructively, which leads to an increase of the light intensity in the center of the cross. On the other hand, if the holograms are stored with a phase difference of π, then destructive interference occurs and

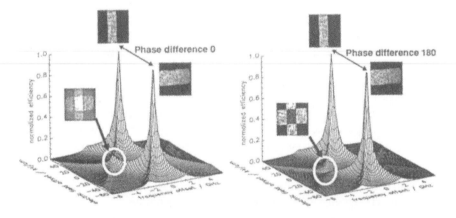

Figure 14   Electric field induced interference between two different holograms (47). The two input patterns, i.e., a horizontal and a vertical bar, are stored at the same frequency but with a Stark-field difference of 12 kV/cm. Application of an external electric field splits the holograms and shifts them symmetrically to lower and higher frequencies, producing Stark components—the "wings"—of the holograms. The white circles indicate two regions in the frequency–electric field plane where the wings of adjacent holograms are overlapping. At these positions, constructive or destructive interference is obtained, depending on the phase relation between the recorded holograms. In the figure this is illustrated for phase differences of 0 and π.

the two images are "subtracted." Parts of the resulting superposed images are shown in Fig. 15 on top of the two tables. In the case of constructive interference, an intensity value of 4 appears in the center of the cross, because the electric field amplitudes rather than the intensities are added interferometrically. In order to get logical truth tables a suitable light level discrimination is employed in the next step. By choosing a discrimination level of 0.5 the truth table corresponding to the logical operation XOR can be derived. From the constructively superposed images the logical operations OR and AND can be derived using two different discrimination levels 0.5 and 2.5 respectively. It is straightforward to see that

Phase difference 0   Phase difference 180°

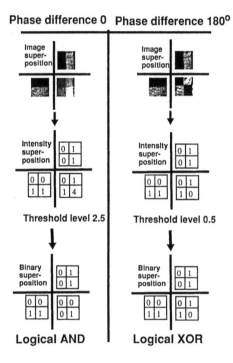

Figure 15 Implementation of logical operations using hologram interferences. The upper table show the results of interferences between input holograms with phase differences of 0 and π. The normalized intensities of the input patterns are listed at the borders of the tables. The resulting intensities of the superposed holograms are shown in the tables. Note that the intensity 4 at the position where two input intensities 1 are superposed (phase difference 0) is the result of the coherent processing. In the case of phase difference 0, a threshold at the intensity level 2.5 leads to the Boolean logic operation AND (shown below). In the case of a phase difference π, the logical XOR truth table is obtained with the threshold level 0.5.

this kind of information processing works in parallel for data arrays of any size. Experiments using these operations have been performed to demonstrate the parallel addition of octal numbers (34,49).

2. Parallel Binary Four-bit Full Adder

In the previous section, the way from hologram superpositions to the logical operations AND, OR, and XOR has been shown. In this section we use these logical operations to perform a parallel binary addition of two sets of 400 binary coded four-bit numbers (49).

The experimental setup is the same as already shown in Fig. 2, with the exception that now the object beam passes through a liquid crystal television (LCD TV) and a polarizer in front of the cryostat before illuminating the sample. The LCD TV is used to display the input data sent by the SUN workstation. In the object beam path a mirror mounted on a piezoelectric transducer allows fine adjustments of the phase by changing the driver voltage under computer control. The phase of the interfering waves is controlled by the oscilloscope as illustrated in Sec. II.A. Reading of the holograms is performed with the CCD camera. The reconstructed images are then digitized and discriminated by the computer. The result is sent back to the LCD TV in the object beam and used as an input pattern for the next logical operation.

The Coding of Four-Bit Number Arrays. For demonstration purposes the following figures and explanations use (3×3) matrices instead of the experimentally used (20×20) matrices. In the experiment all matrices are coded as spatial images, where dark fields correspond to a logical 0 and bright fields to a logical 1. In Fig. 16 the binary coding of two sets (each set consisting of 9 numbers) of four-bit numbers that shall be added in parallel is shown. Each set is represented by four matrices. Each of the numbers consists of the four bits that are at corresponding locations in the four matrices. Numbers on identical places in both sets will be added.

The Binary Addition. The calculation is performed as shown in the flow diagram of Fig. 17, according to the algorithm of a ripple carry full adder (50). The principle is a two-step addition where in the first step the numbers are added without a carry, while in the second step the carry is added to the result of the first step, using iteratively the same procedure again until no carry shows up any more (in an n-bit addition, this condition is fulfilled automatically after $n + 1$ iterative steps). Using binary coded numbers, the first step (addition without carry) is performed by a logical XOR operation between the corresponding bits of the two numbers to be added. The carry for the second step is obtained by a logical AND operation between the same bits. The great advantage of the algorithm is that the whole procedure works in a parallel way, and consequently the time consumption does not depend on the number of additions performed.

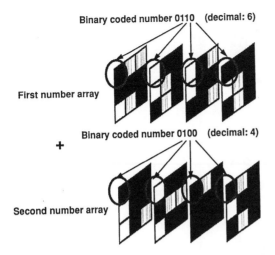

Figure 16   Coding of four-bit number arrays consisting of 3 × 3 numbers. Each array is composed of four matrices where bits belonging to the same number are on the same spatial locations. The four-bit numbers at equal positions in both sets are added in parallel. In the experiment all matrices are stored as spatial patterns, where dark fields correspond to a logical 0 and bright fields to a logical 1.

The Experiment.   In the flow diagram of Fig. 17 the recording positions in the frequency–electric field plane are indicated. In contrast to the diagram, all holo-gram pairs that are processed with XOR operations as well as with AND opera-tions had to be recorded twice during the experiment, with phase differences of 0 and 180° for the AND and XOR operations respectively. The superposition of the holograms is achieved by recording them at the same frequency but at different electric fields (separated by 12 kV/cm) and with a relative phase shift of 0 for the AND and 180° for the XOR operation. At the central field and a frequency shift of 1.5 GHz, the result of the interference between the input patterns is read with the camera, then digitized, thresholded, sent back to the input LCD TV as new input pattern, and recorded again at the indicated position in the frequency–electric field plane for further processing.

In order to demonstrate the experimental quality of the images, the recon-structed holograms are shown in Fig. 18 for one of the calculation steps. In the upper row, the two input patterns sent to the LCD TV and recorded as holograms are shown. Those patterns can be read undisturbed at their exact storage positions (second row). The holograms can also be reconstructed in their processed forms at the appropriate interference positions in the frequency–electric field plane (indicated in Fig. 14) after recording with the corresponding phase relations (third row). Finally, after thresholding of these patterns by the computer at the

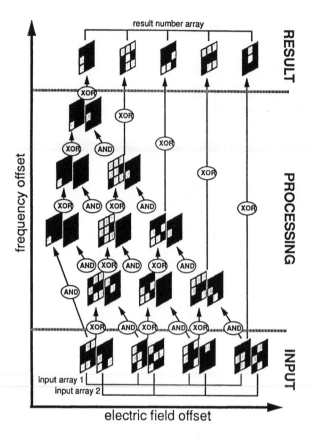

Figure 17 Data flow diagram of addition demonstrating the principle of the binary ripple carry full adder (50). The diagram indicates the experimentally used recording positions in the frequency–electric field plane. The input patterns are stored at equal frequencies and adjacent electric fields. A separation of 80 kV/cm between different hologram pairs in the electric field dimension is sufficient to exclude interactions between them. In the lowest row, the two input number sets are stored, each matrix being connected with the corresponding matrix of the other array. The arrows marked by AND and XOR show where the results of the logical operations are recorded for the next processing steps. During the experiment, each logical operation (AND and XOR) needs a separate recording step with the correct phase difference. A total of 10 AND and 13 XOR operations is necessary for the whole calculation (49).

threshold levels 0.5 and 2.5 (Fig. 15), the AND and XOR superposed matrices in the fourth row are obtained. The images are already in input format and can be sent back to the LCD for the next recording stage according to Fig. 17.

Input patterns

Holographically
read-out input
patterns

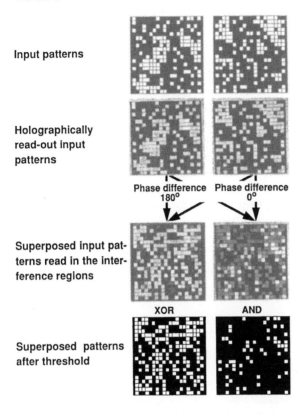

Phase difference   **Phase difference**
     180°   **0°**

Superposed input pat-
terns read in the inter-
ference regions

       XOR   **AND**

Superposed patterns
after threshold

Figure 18 Example for one calculation step of binary addition. The two input patterns of the upper row are sent to the LCD TV and recorded as holograms. They can be read without cross talk at their exact storage positions in the frequency–electric field plane as shown in the second row. Hologram interferences (third row) are obtained at the overlapping points of the Stark-split holograms. After a threshold by the computer, new patterns corresponding to the logical operations AND and XOR between the input patterns are obtained (fourth row). They are recorded again for the next processing step (49).

For the whole calculation a total of 13 XOR and 10 AND connections have been used, corresponding to 46 recording and 23 reconstruction steps, and 9200 bitwise logical operations. In the first experiment, 396 out of 400 additions have been performed correctly. The remaining errors are due to a slightly inhomogeneous light distribution in the expanded laser beam, leading to problems with the threshold operations.

In principle, all of the operations performed by the computer, especially the time-consuming threshold operations, could be done by ''nonintelligent'' optical devices such as electrooptic light valves. In contrast to interference methods where holograms are stored directly on top of each other (51), no additional external storage device is necessary, because all intermediate results are recorded in the sample directly after processing and can be accessed undisturbed. Thus the sample is both a processor and a memory. The number of parallel processed bits is limited by the electrooptic devices (LCD and CCD camera). By using the full spectral and spatial resolution of the SHB material, more than 10^4 logical operations between pattern pairs, each consisting of more than 10^6 pixels, might be possible before erasure of the sample content would be necessary. Because of the large storage capacity of hole burning samples, this method of computing might become especially useful in calculations where large amounts of intermediate results are produced and used frequently, as for instance in recursive calculations. Nevertheless, for practical applications the record time must be decreased drastically, and fast electrooptic devices for data input and thresholding have to be employed.

IV. CONCLUSION

The experiments illustrated in this contribution clearly demonstrate the potential of holographic hole burning, not only with respect to optical data storage but also in parallel information processing. A possible approach to improving the storage capacity by means of this technique has been presented: frequency multiplexing data storage. The optical frequency and the electric field represent two additional parameters for optical data recording in a single piece of hole burning material, thus increasing the storage density by orders of magnitude. An interesting technical application of spectral hole burning was suggested by a new method of parallel information processing. It has been shown that the combination of holography with molecular properties (the Stark splitting of absorption holes in an externally applied electric field) allows nontrivial processing of data. In this case a four-bit full adder has been realized, and two arrays of 400 octal numbers have been added in parallel.

Both data storage and processing of the recorded information can be performed exclusively in the same component, the hole burning material, without further use of additional devices. Moving parts need not be involved, and the information processing is based only on the interactions of the molecular energy levels with an external electric field. The system can be considered as a molecular processor able to perform parallel data processing: a molecular computer. Even though the basic feasibility of frequency multiplexed data storage and information processing by spectral hole burning were demonstrated, some problems still must be solved before any practical applications can be attempted. First, high-density data storage is only achieved at extremely low temperatures that are not easily reached. One has the possibility of compromising with a lower storage density at a somewhat higher temperature. In this sense, the search for adequate recording systems represents a relevant task for fundamental investigations and for the material sciences. Second, the readout of the recorded information occurs at the same wavelength used for recording, which leads to an erasure of the information. Using photon-gated processes, i.e., two different frequencies for recording and only one for the readout, is a very promising solution to this problem. The development of suitable photochemical systems is a challenge to researchers. Finally, the recording time currently required, in the order of 5 s, is too long as compared to electronic recording devices. The writing time has to be considerably decreased in order to become competitive with respect to conventional computing techniques: in this sense, materials with fast photoinduced changes are required. To summarize, although severe problems are present and limit the practical realization of an optical computer based on holographic hole burning, the principles of this technique have been successfully demonstrated. Whereas electronic information processing is based on the properties of electrons in an electric field, the concept of a molecular processor relies on spectroscopic properties of molecules, i.e., the behavior of molecular energy levels in an electric field. This may be considered as a new branch of optical computing.

ACKNOWLEDGMENTS

We should like to thank Eric Maniloff, Alfred J. Meixner, Bern Kohler, Markus Traber, Peter Nyffeler, and Jürg Keller for their help and support in this project. This work was financially supported by the Swiss National Science Foundation, the KWF (Kommission zur Förderung der wissenschaftlichen Forschung), and the SPP Optik (Schwerpunktprogramm Optische Wissenschaften).

REFERENCES

1. Moerner, W. E., and Basché, T., Optical spectroscopy of single impurity molecules in solids, *Angew. Chem. Int. Ed. Engl.*, 32; 475 (1993).
2. Moerner, W. E., ed., *Persistent Spectral Hole-Burning: Science and Applications*, Topics in Current Physics 44, Springer-Verlag, Berlin, New York, 1988.
3. Holliday K., and Wild, U. P., *Spectral Hole-Burning, Molecular Luminescence Spectroscopy* (St. G. Schulman, ed.), Chemical Analysis Series 77, John Wiley, New York, 1993, p. 149.
4. Gorokhovskii, A. A., Kaarli, R. K., and Rebane, L. A., Hole burning in the contour of a pure electronic line in a Shpol'skii system, *JETP Lett.*, *20*, 216 (1974), Kharlamov, B. M., Personov, R. I., and Bykovskaya, L. A., Stable gap in absorption spectra of solid solutions of organic molecules by laser irradiation, *Optics Commun.*, *12*, 191 (1974).
5. Friedrich, J., and Haarer, D., Photochemisches Lochbrennen und optische Relax-

ationsspektroskopie in Polymeren und Gläsern, *Angew. Chem.*, *96*, 96 (1984) (*Angew. Chem. Int. Ed. Engl.*, *23*; 113 (1984)).

6. Völker, S., Spectral hole-burning in crystalline and amorphous organic solids. Optical relaxation processes at low temperatures, *Relaxation Processes in Molecular Excited States* (J. Fünfschilling, ed.), Kluwer, Dordrecht, 1989, p. 113.

7. Burkhalter, F. A., Photophysikalische Untersuchungen an synthetischen metallfreien Isobacteriochlorinen und an Chlorin, Ph.D. thesis, ETH, Zürich, Nr. 7492, 1984.

8. De Caro, C. A., Von der Bildspeicherung zu den Bildkorrelations experimenten: Beitrag zum holographischen spektralen Lochbrennen, Ph.D. thesis, ETH, Zürich, Nr. 9526, 1991.

9. Locher, R., Renn, A., and Wild, U. P., Hole-burning spectroscopy on organic molecules in amorphous silica, *Chem. Phys. Lett.*, *138*(5), 405 (1987).

10. Petke, J., Maggiora, G., Shipman, L. L., and Christoffersen, R. E., Stereoelectronic properties of photosynthetic and related systems, *J. Mol. Spectrosc.*, *73*, 311 (1978).

11. Renn, A., Meixner, A. J., Wild, U. P., and Burkhalter, F. A., Holographic detection of photochemical holes, *Chem. Phys.*, *93*, 157 (1985).

12. Meixner, A. J. R., Spektrales Lochbrennen: Entwicklung der holographischen Detektionsmethode, Stark-Effekt Experimente an spektralen Löchern, Ph.D. thesis, ETH, Zürich, Nr. 8726, 1988.

13. Meixner, A. J., Renn, A., and Wild, U. P., Spectral hole burning and holography l. Transmission and holographic detection of spectral holes, *J. Chem. Phys.*, *91*(11), 6728 (1989).

14. Bräuchle, Chr., and Burland, D. M., Holographische Methoden zur Untersuchung photochemischer und photophysikalischer Eigenschaften von Molekülen, *Angew. Chem.*, *95*, 612 (1983) (*Angew. Chem. Int. Ed. Engl.*, *22*, 582 (1983)).

15. Bernet, S., Altner, S. B., Graf, F. R., Maniloff, E. S., Renn, A., and Wild, U. P., Frequency and phase swept holograms in spectral hole-burning materials, *Appl. Opt.*, *34*, 4674 (1995).

16. Rebane, A., Bernet, S., Renn, A., and Wild, U. P., Holography in frequency selective media: hologram phase and causality, *Optics Com.*, *86*, 7 (1991).

17. Bernet, S., Phasenkontrollierte Holographie in frequenzselektiven Materialien, Ph.D. thesis, ETH, Zürich, Nr. 10292, 1993.

18. Bernet, S., Kohler, B., Rebane, A., Renn, A., and Wild, U. P., Spectral hole-burning and holography V. Asymmetric diffraction from thin holograms, *J. Opt. Soc. Am.*, *B9*(6), 987 (1992).

19. Bernet, S., Kohler, B., Rebane, A., Renn, A., and Wild, U. P., Holography in frequency selective media ll. Controlling the diffraction efficiency, *J. Lumin.*, *53*, 215 (1992).

20. Meixner, A. J., Renn, A., Bucher, S. E., and Wild, U. P., Electric field dependent photochemical hole-burning, *Proceedings of The Xlth Molecular Crystal Symposium*, Lugano, Switzerland, 1985, pp. 198–201.

21. Meixner, A. J., Renn, A., Bucher S. E., and Wild, U. P., Spectral hole-burning in glasses and polymer films: The Stark effect, *J. Phys. Chem.*, *90*, 6777 (1986).

22. Marchetti, A. P., and Scozzafava, M., Crystal field induced dipole moments: the Stark spectrum of naphthacene in benzophenone, *Mol. Cryst. Liq. Cryst.*, *31*, 115 (1975).

23. Bogner, U., Schätz, P., Seel, R., and Maier, Max, Electric-field-induced level shifts of perylene in amorphous solids determined by persistent hole-burning spectroscopy, *Chem. Phys. Lett.*, *102* (2,3), 267 (1983).

24. Renn, A., Bucher, S. E., Meixner, A. J., Meister, E., and Wild, U. P., Spectral hole-burning: electric field effect on resorufin, oxazine − 4, and cresylviolet in polyvinylbutyral, *J. Luminescence*, *39*, 181 (1988).

25. Kador, L., Investigation of electric-field effects on hole-burning spectra in doped polymers, Ph.D. thesis, University of Bayreuth (D), 1988.

26. Schätz, P., and Maier, Max, Calculations of electric field effects on persistent spectral holes in amorphous host–guest systems, *J. Chem. Phys.*, *87*(2), 809 (1987).

27. Samoilenko, V. D., Razumova, N. V., and Personov, R. I., Stark effect in narrow gaps in the absorption bands of complex molecules, *Opt. Spectrosc.* (USSR), *52*(4), 346 (1982).

28. Kador, L., Haarer, D., and Personov, R. I., Stark effect on polar and unpolar dye molecules in amorphous hosts, studied via persistent spectral hole-burning, *J. Chem. Phys.*, *86*(10), 5300 (1986).

29. Burkhalter, F. A., Suter, G. W., Wild, U. P., Samoilenko, V. D., Razumova, N. V., and Personov, R. I., Hole burning in the absorption spectrum of chlorin in polymer films: Stark effect and temperature dependence, *Chem. Phys. Lett.*, *94*(5), 483 (1983).

30. Dicker, A. J., Johnson, L. W., Noort, M., and van der Waals, J. H., Stark effect on the $S_1 \leftarrow S_0$ transition of the two tautomeric forms of chlorin studied by photochemical hole-burning in *n*-hexane and *n*-octane single crystals at 1.2K, *Chem. Phys. Lett.*, *94*(1), 14 (1983).

31. De Caro, C., Renn, A., and Wild, U. P., Spectral hole-burning: applications to optical image storage, *Ber. Bunsenges. Phys. Chem.*, *93*, 1395 (1989).

32. Szabo, A., U.S. Patent 3'896'420, 1975.

33. Castro, G., Haarer, D., Macfarlane, R. M., and Trommsdorff, H. P., U.S. Patent 4'101'976, 1978.

34. Wild, U. P., Bernet, S., Kohler, B., and Renn, A., From supramolecular photochemistry to the molecular computer, *Pure Appl. Chem.*, *64*(9), 1335 (1992).

35. Bucher, S. E., Spektrales Lochbrennen und optische Datenspeicherung, Ph.D. thesis, ETH, Zürich, Nr. 8541, 1988.

36. Wild, U. P., Bucher, S. E., and Burkhalter, F. A., Hole burning, Stark effect, and data storage, *Appl. Opt.*, *24*(10), 1526 (1985).

37. Bogner, U., Beck, K., and Maier, Max, Electric field selective optical data storage using persistent spectral hole burning, *Appl. Phys. Lett.*, *46*(6), 534 (1985).

38. Renn, A., and Wild, U. P., Spectral hole-burning and hologram storage, *Appl. Opt.*, *26*(19), 4040 (1987).

39. Wild, U. P., Renn, A., De Caro, C., and Meixner, A. J., Spectral hole burning and holographic image storage in polymer films, *Proceedings of PME'89*, Tokyo, 1989, pp. 508–518.

40. De Caro, C., Renn, A., and Wild, U. P., Hole burning, Stark effect, and data storage 2. Holographic recording and detection of spectral holes, *Appl. Opt.*, *30*(20), 2890 (1991).

41. Renn, A., Meixner, A. J., and Wild, U. P., Spectral hole burning and holography II. Diffraction properties of two spectrally adjacent holograms, *J. Chem. Phys.*, *92*(5), 2748 (1990).

42. Renn, A., Meixner, A. J., and Wild, U. P., Spectral hole burning and holography III. Electric field induced interference of holograms, *J. Chem. Phys.*, *93*(4), 2299 (1990).

43. Kogelnik, H., Coupled wave theory for thick hologram gratings, *Bell System Tech. J.*, *48*(9), 2909 (1969).

44. Kohler, B., Bernet, S., Renn, A., and Wild, U. P., Storage of 2000 holograms in a photochemical hole-burning system, *Opt. Lett.*, *18*, 2144 (1993).

45. Maniloff, E. S., Altner, S. B., Bernet, S., Graf, F. R., Renn, A., and Wild, U. P., Recording of 6000 holograms by use of spectral hole burning, *Appl. Opt.*, *34*, 4140 (1995).

46. Becker, P.-J., Bolle, H., Munser, R., and Wagner, U., Grundlagen und physikalische Grenzen integrierbarer holographischer Speicher mit und ohne assoziativen Zugriff, Forschungsbericht DV 83–002, Bundesministerium für Forschung und Technologie, Bundesrepublik Deutschland, 1983.

47. Wild, U. P., Renn, A., De Caro, C., and Bernet, S., Spectral hole burning and molecular computing, *Appl. Opt.*, *29*(29), 4329 (1990).

48. Wild, U. P., De Caro, C., Bernet, S., Traber, M., and Renn, A., Molecular computing, *J. Luminescence*, *48&49*, 335 (1991).

49. Bernet, S., A., Renn, Kohler, B., and Wild, U. P., Molecular computing: parallel binary additions, *Technical Digest 1992*, Optical Society of America, Washington, D.C., 1992, vol. 22, p. 218.

50. See e.g. Fink, Donald G., *Electronic Engineers' Handbook*, McGraw Hill, New York.

51. Huignard, J. P., Herriau, J. P., and Micheron, F., *Appl. Phys. Lett.*, *26*, 256 (1975).

15

Biomolecular Electronics and Optical Computing

Robert R. Birge, Richard B. Gross, and Albert F. Lawrence

Syracuse University
Syracuse, New York

I. INTRODUCTION

This chapter gives an overview of the use of biological molecules in optically coupled molecular electronic devices. Our emphasis is on the use of the protein bacteriorhodopsin in optical memories and optical computing devices. We describe the use of this protein in selected applications that include holography, spatial light modulators, neural network optical computing, and volumetric and associative optical memories. There are significant advantages inherent in the use of biological molecules, either in their native form or modified via chemical or mutagenic methods, as active components in optoelectronic devices (1–29). These advantages derive in large part from the natural selection process, since nature has solved through trial and error problems of a similar nature to those encountered in harnessing organic molecules to carry out logic, switching, or data manipulative functions (10). Individual biological chromophores have been linked synthetically to produce optically coupled gates and switches with unique properties (30,31). Light transducing proteins such as visual rhodopsin (3), bacteriorhodopsin (1,2), chloroplasts (32,33), and photosynthetic reaction centers (34) are salient examples of protein based systems that have been investigated for optoelectronic applications.

This review will concentrate primarily on the current and proposed applications of bacteriorhodopsin in optical computing and optical memories. This protein has received more attention with respect to optoelectronics than any other biomolecule studied to date (1–23). The significance of bacteriorhodopsin stems from its biological function as a photosynthetic proton pump in the bacterium *Halobacterium salinarium* (also called *Halobacterium halobium*). A combination of serendipity and natural selection has yielded a native protein with characteristics near optimum for many linear and nonlinear optical applications. The development of genetic engineering combined with chemical modification and chromophore substitution methods also provides an additional flexibility in the adaptation of this material for individual applications. We shall explore a number of optoelectronic applications of bacteriorhodopsin as well as the methods of modifying the protein in this chapter. In order to reduce overlap with the chapter of Christoph Bräuchle, we shall limit our discussion of the properties of the protein to those aspects directly relevant to our applications.

II. PHOTOPHYSICS OF BACTERIORHODOPSIN

Optically coupled devices based on bacteriorhodopsin are driven by linear (one-photon) or nonlinear (two-photon) excitation. A majority of these devices operate by switching between two of the intermediates that populate the complex photocycle of the protein. In order to understand their operation, it is necessary to understand the nature of the bacteriorhodopsin photocycle. The following paragraphs are devoted to a brief overview of the photochemistry of this protein. For additional details, the chapter by C. Bräuchle can be consulted.

Bacteriorhodopsin (MW $\cong$ 26,000) is the light harvesting protein in the purple membrane of a microorganism called *Halobacterium salinarium* (*Halobacterium halobium*) (35,36). This bacterium thrives in salt marshes where the concentration of salt is roughly six times higher than that of sea water. The purple membrane, which constitutes a specific functional site in the plasma membrane of the bacterial cell, houses semicrystalline protein trimers in a phospholipid matrix (3:1 protein to lipid). The bacterium synthesizes the purple membrane when the concentration of dissolved oxygen in its surroundings becomes too low to sustain ATP production through aerobic respiration. The light absorbing chromophore of bacteriorhodopsin is all-trans retinal (Vitamin A aldehyde (Fig. 1). It is bound to the protein through a protonated Schiff base linkage to a lysine residue attached to one of the seven α-helices that make up the protein's secondary structure. The absorption of light energy by the chromophore initiates a complex photochemical cycle characterized by a series of spectrally distinct thermal intermediates and a total cycle time of approximately 10 milliseconds (see Fig. 2). As a result of this process, the protein expels a proton from the intracellular to the extracellular side of the membrane. This light induced proton pumping generates an electrochemical gradient that the bacterium uses to synthesize ATP. Accordingly, *Halobacterium salinarium* can switch from aerobic respiration to photosynthesis in response to changing environmental conditions. The functioning of the

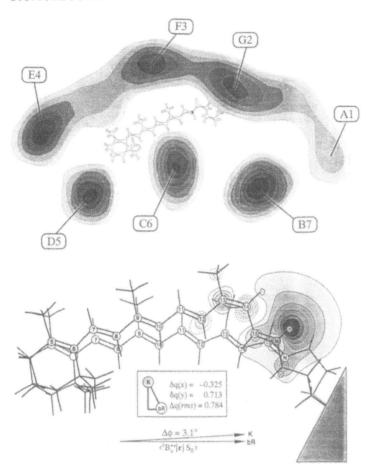

Figure 1 The chromophore binding site and the primary photochemical event in bacteriorhodopsin. The upper diagram shows electron density profiles (data from Ref. 80) of bacteriorhodopsin viewed from the cytoplasmic side showing the seven transmembrane spanning segments and the presumed location of the chromophore in relation to the helices based on the available experimental data (3). FTIR studies indicate that the polyene chain of the chromophore in bacteriorhodopsin lies roughly perpendicular to the membrane plane (81). The retinyl chromophore is rotated artifically into the membrane plane to show more clearly the polyene chain (the imine nitrogen is indicated with a solid black circle) and the β-ionylidene ring. The bottom diagram shows a model of the primary photochemical event [**bR** (gray; underneath) → **K** (black; above)] and the shift in charge that is associated with the motion of the positively charged chromophore following 13-trans → 3-cis photoisomerization. It is believed that the initial photoelectric signal is due primarily to the motion of the chromophore. The conformations of the lysine residue in the **bR** and **K** states are tentative (82).

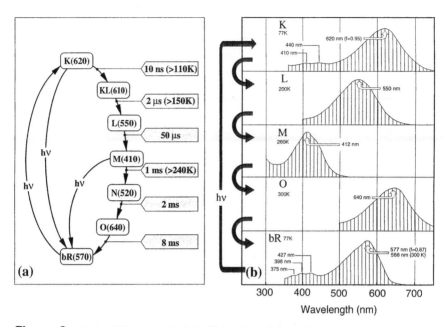

Figure 2 A simplified model of the light adapted bacteriorhodopsin photocycle (a) and the electronic (one-photon) absorption spectra of selected intermediates in the photocycle (b). The height of the symbols in (a) is representative of the relative free energy of the intermediates, and the key photochemical transformations relevant to device applications are shown. Note that not all of the intermediates are shown, and that there are in fact two species of M (M_{fast} and M_{slow}), but only one is shown for convenience. (M_{fast} and M_{slow} have virtually identical absorption spectra.) Band maxima are indicated in nanometers. Oscillator strengths (f) determined by log-normal fits of selected λ_{max} bands are indicated in parentheses.

purple membrane in the harsh environment of a salt marsh requires a robust light transducing protein resistant to both thermal and photochemical damage. The cyclicity of the protein (i.e., the expected number of times the protein can be photochemically cycled between intermediates before denaturing) exceeds 10^6, a value considerably higher than values observed in known synthetic photochromic materials. The excellent value for cyclicity is due to the protective features of the integral membrane protein that serves to isolate the chromophore from potentially reactive oxygen, singlet oxygen, and free radicals. Thus the common misperception that biological materials are too fragile to be used in technological devices outside the laboratory does not apply to bacteriorhodopsin. The optoelectronic characteristics of bacteriorhodopsin have been reviewed in detail (2,3,10).

Although the bacteriorhodopsin photocycle is comprised of at least five thermal intermediates (Fig. 2a), only three of the intermediates (**bR**, **K**, and **M**) have recognized potential in device applications. The absorption spectra of the key intermediates are shown in Fig. 2b. The unique absorption spectra exhibited by each intermediate is associated with the changing electronic environment of the chromophore binding site during the course of the photocycle. The internally bound chromophore carries a net positive charge and interacts electrostatically with neighboring charged amino acids in the binding site. These interactions in large part determine the protein's photochemical and spectral properties during the photocycle and are therefore targeted areas for genetic engineering and/or chemical modification. When the chromophore absorbs a photon of light energy, an instantaneous ($<10^{-15}$ s) shift of electron density occurs with negative charge moving along the polyene chain towards the nitrogen atom. The "shifted" electrons interact with nearby negatively charged residues and activate a rotation around the $C_{13} = C_{14}$ double bond, thereby generating a 13-cis chromophore geometry (see Fig. 1). The result of this photoisomerization process, which occurs in less than one picosecond (37–40), is the formation of the **K** spectral intermediate. The reason for the unusually high isomerization speed is a barrierless excited state potential surface (3). In this regard, bacteriorhodopsin is the biological analog of high electron mobility transistor (HEMT) devices (41). The isomerization of the protonated chromophore induces a shift in positive charge perpendicular to the membrane sheet containing the protein and generates a measurable and potentially useful electrical signal. The rise time of this signal is less than 5 picoseconds and correlates with the formation time of **K** (6,42). Another feature in addition to charge transfer of the chromophore trans–cis isomerization process is its photoreversibility. Thus irradiation of the protein with a wavelength within the absorption band of **K** results in the reformation of the ground state. Many of the early proposed optical memories based on bacteriorhodopsin utilized the photochemical switching between **bR** and **K**. These devices suffered from the requirement that liquid nitrogen temperatures were needed to arrest the photocycle at the **K** intermediate (e.g., Ref. 43). While these devices were potentially efficient and very fast (the **bR**↔**K** interconversions take place in a few picoseconds), the use of cryogenic temperatures and the small change in absorption maxima associated with the **bR** to **K** transition mandate expensive operating hardware and preclude general use.

The most significant photochemical intermediate both from the physiological and the engineering points of view is the blue light absorbing **M** intermediate. The formation of **M** ensues after a series of protein conformational changes occurring ~50μs after the absorption of a photon of light by **bR**. In this stage of the photocycle, the Schiff base proton on the chromophore is transferred to an amino acid of the protein. In doing so, the electrostatic nature of the chromophore and the electric potential of the chromophore binding site is dramatically changed, as reflected in this intermediate's highly blue shifted absorption spectrum. Under normal conditions, **M** thermally reverts to the ground state with a time constant of about 10 ms. Very significantly, for the applications **bR** can also be photochemically regenerated from **M** by the absorption of blue light. This property of a material, where a ground state photoinitiated reaction results in a relatively long-lived thermal intermediate that can also be photochemically driven back to the ground state, is called photochromism. The photochromic properties of bacteriorhodopsin are summarized by

$$\textbf{bR} \ (\lambda_{max} \cong 570 \text{ nm}) \ (\textit{State 0}) \ \underset{\Phi_2 \sim 0.65}{\overset{\Phi_1 \sim 0.65}{\underset{\longleftarrow}{\longrightarrow}}} \ \textbf{M} \ (\lambda_{max} \cong 410 \text{ nm}) \ (\textit{State 1})$$

where the quantum yields of the forward reaction (**bR** to **M**) and reverse reaction (**M** to **bR**) are indicated by Φ_1 and $\Phi_2$hr, respectively. One inherent advantage of bacteriorhodopsin as an optical recording medium is the high quantum efficiency with which it converts light into a state change. Complementing this property is the relative ease of prolonging the thermal decay of **M**. The **M → bR** thermal transition is highly susceptible to temperature, chemical environment, genetic modification, and chromophore substitution. This property is exploited in many optical devices based on bacteriorhodopsin.

III. SPATIAL LIGHT MODULATORS AND HOLOGRAPHIC OPTICAL MEMORIES

Thin films of bacteriorhodopsin fabricated by incorporating the protein into optically transparent polymers and polymer blends have shown good holographic performance and the capability of real time optical processing (1–3,5,9,14,18,22,23,44–48). The mechanism of diffracting light from a volume hologram produced in a photochromic material is important to understand, as it is fundamental to many optical applications. We therefore provide a brief overview of this process.

The optical protocol used to record a plane wave hologram in a thin film of bacteriorhodopsin is schematically shown in Fig. 3. Two laser beams derived from the same laser and of a wavelength (λ_w) absorbed by **bR** are overlapped at the plane of the film. Both beams are polarized perpendicular to the plane of incidence and make an angle ϕ_w with respect to the film normal. Due to the coherence properties of laser light, a three-dimensional interference pattern is imposed on the film. The resulting periodic spatial light intensity distribution is schematically shown in Fig. 4 and can be mathematically described in one dimension by using Eq. (1):

$$I(x) = (I_1 + I_2) \left[1 + V\cos \frac{2\pi x}{P} \right] \tag{1}$$

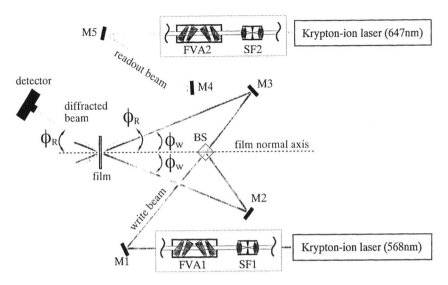

Figure 3 A schematic representation of the experimental apparatus used in writing and reading a hologram. A bacteriorhodopsin hologram is written by overlapping two 568 nm beams derived from a krypton-ion laser. The hologram is nondestructively read at the Bragg angle by using a probe beam not strongly absorbed by the protein (see text).

where I_1 and I_2 represent the intensities of the individual beams, V is the contrast ratio of the interference pattern, $V = 2 (I_1 I_2)^{1/2} / (I_1 + I_2)$, and P is the fringe spacing of the grating (Fig. 4) given by $\lambda_w / (2 \sin \phi_w)$. Thus in places of constructive interference **bR** is driven to **M**, and in regions of destructive interference no photochemistry is initiated. The film records both the amplitude and the phase information contained in the two incident beams as a periodic spatial concentration distribution of **bR** and **M** (since neither beam contains an object, the hologram is "structureless" as compared to most holograms). The spatial concentration distribution of **bR** and **M** can be more conveniently viewed as a spatial modulation of the material's absorption coefficient and index of refraction. These material properties, which are fundamental to the diffraction or reconstruction process in holography, are not independent and are related through the Kramers–Kronig transform (49,50):

$$\Delta n(\lambda) = \frac{2.3026}{2\pi^2 t} P.V. \int_0^\infty \frac{A_M(\lambda') - A_{bR}(\lambda') \, d\lambda'}{1 - \frac{\lambda'^2}{\lambda^2}} \qquad (2)$$

where $P.V.$ represents the principal value of the Cauchy integral, $\Delta n(\lambda)$ is the change in refractive index at the readout wavelength, A_M and A_{bR} represent the

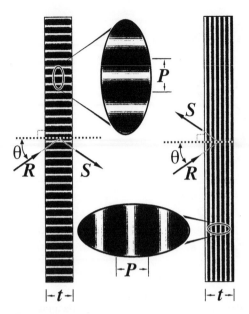

Figure 4   Volume transmission (left) and reflection (right) gratings and the key variables that define the properties of the gratings. The dark regions represent regions of high refractive index (or increased absorptivity), and the light regions represent regions of low refractive index (or decreased absorptivity); the magnitude of the difference in refractive index (or absorbtivity) determines in part the diffraction efficiency of the phase (or absorption) hologram. The incident light vector is indicated by **R** and the scattered light vector is indicated by **S**.

absorbances of the ground state and the thermal intermediate, respectively, t is the thickness of the hologram, and λ is the wavelength of the readout beam. The absorbance $A(\lambda')$ is related to the absorption coefficient of a material $\alpha(\lambda')$ (units of reciprocal length) by noting that

$$\alpha(\lambda') = 2.3026 \frac{A(\lambda')}{t} \tag{3}$$

In general, for a photochromic material to exhibit useful holographic properties it should possess a high quantum efficiency and a large shift in absorption maxima between photochromic states. The latter property generally results in a large photochemically induced change in refractive index. As we have seen in the previous section, the large blue shift in absorption maxima generated by deprotonation of the chromophore during the **bR** to **M** phototransformation makes bacteriorhodopsin an ideal optical recording material. Figure 5 shows a simulation of the refractive and diffractive properties of a thin film of bacteriorhodopsin based on application of the Kramers–Kronig transform and coupled wave theory (see below). It should be clear from the figure that the largest change in the

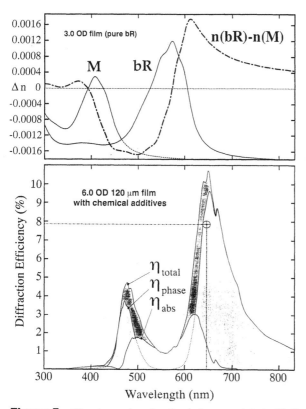

Figure 5 The change in refractive index associated with the **bR** → **M** photoisomerization for a 30 μm film of bacteriorhodopsin with an optical density (OD) of ~3 is shown as a function of wavelength in the upper panel. The refractive index change is expressed as the value for pure **bR** minus the value for pure **M** and is calculated by using the Kramers–Kronig transformation. The absorption spectra of **bR** and **M** are shown for reference. The diffraction efficiency associated with a 6 O.D. film for **bR** (100%) → **bR** (50%) + **M** (50%) photoconversion is shown in the lower panel and is calculated based on the observed absorption spectra by using the Kramers–Kronig relationship and Kogelnik approximation (51). The dot at ~640 nm and ~8% diffraction efficiency represents a recent experimental result from our laboratory using the holographic spatial light modulator described in Fig. 7.

refractive index is expected when the hologram is produced with a write wavelength that efficiently drives the **bR** to **M** photoconversion, and when readout wavelengths are used that yield nondestructive readout (not strongly absorbed by **bR** or **M**).

The photodiffractive process can be analyzed using the coupled wave theory developed by Kogelnik (51). The spatial modulations of the absorption coefficient and the index of refraction are described by the truncated Fourier expansions

$$\alpha(x) \cong \alpha_{avg} + \alpha_1 \cos \frac{2\pi x}{P} \tag{4}$$

$$n(x) \cong n_{avg} + n_1 \cos \frac{2\pi x}{P} \tag{5}$$

where P has been defined previously as the fringe spacing of the grating (Fig. 4), $\alpha(x)$ and $n(x)$ are the spatial dependent values of the absorption constant and index of refraction, respectively, α_{avg} and n_{avg} are the average values of the absorption coefficient and the refractive index, respectively, and α_1 and n_1 represent the modulation amplitudes of the absorption coefficient (amplitude) and index of refraction. The latter parameters contribute to the total diffraction, the absorptive part through absorptive modulation of the light electric field amplitude and the refractive component through phase or optical path modulation of the light electric field amplitude. These parameters can be estimated through the use of Eq. (2) and taking into account the electric field description of the absorption coefficient described in Eq. (4).

The diffraction efficiency of a hologram is defined as the ratio of the diffracted light intensity I_D to the intensity of the reading beam I_0. As before, the diffraction process can have both an absorption and a phase component, and in the case of bacteriorhodopsin both contribute (52):

$$\eta_{total} = \frac{I_D}{I_O} = \eta_{abs} + \eta_{phase} \tag{6}$$

$$\eta_{abs} = \sinh^2 \left\{ \frac{\alpha_1 \lambda_R) \, t}{2 \cos \theta_R} \right\} D \tag{7}$$

$$\eta_{phase} = \sin^2 \left\{ \frac{\pi \, n_1(\lambda_R) \, t}{\lambda_R \, 2 \cos \theta_R} \right\} D \tag{8}$$

$$D = \exp \left\{ \frac{-\alpha_{ave}(\lambda_R) \, t}{\cos \theta_R} \right\} \tag{9}$$

where η_{total} is the total diffraction efficiency ($1 = 100\%$), η_{abs} is the diffraction efficiency due to absorption, η_{phase} is the diffraction efficiency due to refraction,

t is the thickness of the hologram (Fig. 4), λ_R is the wavelength of the read laser, ϕ_R is the angle of incidence of the read laser (Fig. 3), $\alpha_1(\lambda_R)$ is the modulation amplitude of the absorption coefficient at the read wavelength, $n_1(\lambda_R)$ is the modulation amplitude of the refractive index at the read wavelength, and $\alpha_{ave}(\lambda_R)$ is the average absorption coefficient of the hologram. Although the read angle ϕ_R is an experimentally adjustable variable, maximum efficiency is achieved by satisfying the Bragg condition:

$$\phi_R = \sin^{-1}\left\{\frac{\lambda_R \sin(\phi_W)}{\lambda_W}\right\} \tag{10}$$

where ϕ_W is the angle of the write beam relative to the hologram film normal (Fig. 3) and λ_W is the wavelength of the write beam. The D term defined in Eq. (9), and that appears in Eqs. (7) and (8), places a constraint on the maximum value of the absorptive component of the diffraction efficiency because the absorption modulation change $\alpha_1(\lambda_R)$ that is required to generate diffraction also contributes to the average absorption $\alpha_{ave}(\lambda_R)$. The contribution of D limits the η_{abs} value to 0.037 (3.7%) or less. In contrast, η_{phase} is determined entirely by the change in refractive index, and values approaching unity (100%) are possible. Thus for applications requiring diffraction efficiencies exceeding 3.5%, phase holograms or mixed absorptive and phase holograms are preferred. This situation is found in the more traditional type of irreversible recording materials such as silver halide photographic films and dichromated gelatin. Figure 5 shows the results of the theoretically predicted diffraction efficiency of a 6 OD film of chemically enhanced bacteriorhodopsin as a function of varying readout wavelength. An experimental measurement is also shown indicating that the excellent diffraction efficiency that is predicted can, in fact, be experimentally realized. Holograms can be recorded in pure phase, pure absorption, or mixed modes with recording wavelengths of 400–700 nm and readout of 400–850 nm. The recording sensitivity at ambient temperature is in the range 1–80 mJ/cm². An additional advantage of using this protein as an optical recording medium is its small size (~50 nm diameter) relative to the wavelength of light. This results in diffraction-limited performance (>5000 lines/mm for thin films). Diffraction efficiencies can also be improved by using genetically modified proteins (18,44), chromophore analogs, and chemical enhancement of the native protein (for reviews see (1–3).

It should be emphasized that maximizing holographic efficiency requires careful adjustment of the laser write and read beam intensities. In contrast to the ideal plane wave situation described above, two interferring Gaussian laser beams do not produce the simple sinusoidal concentration grating depicted in Fig. 4 but one that has an overall profile determined by the Gaussian intensity distribution within the interacting beams as depicted in Fig. 6. If the intensity is too

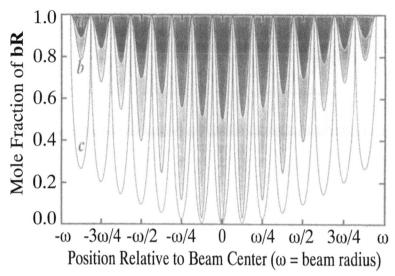

Figure 6 A schematic representation of the grating pattern induced in a thin film of bacteriorhodopsin by using two interfering green laser beams. In regions of constructive interference, the forward (**bR** → **M**) photochemical reaction is driven decreasing the mole fraction of **bR** relative to **M** (vertical axis). The horizontal axis represents the position relative to the center of the overlapping (interfering) Gaussian laser beams where ω represents the beam radius. Three situations are depicted. The top graph (a) shows a situation where the intensity of the laser light is sufficiently low to generate only partial photochemical conversion even in regions of maximal constructive interference. The middle graph (b) shows a grating generated with a laser intensity that is at the higher value possible without ''overdriving'' the photochemistry. This situation generates the optimal diffraction efficiency without introducing any nonlinearity. The lower graph (c) shows a situation where the intensity of the laser light is sufficiently high to ''overdrive'' the photochemistry and generate nonlinear performance. In general, a small amount of nonlinearity will improve diffraction efficiency without diminishing the quality of the hologram. However, the situation depicted in graph (c) represents a serious level of nonlinearity that will diminish both diffraction efficiency and holographic image quality.

low, the extent of conversion may be adequate at the beam center but inadequate off-center (graph (a) in Fig. 6). If the intensity is too high, photochemistry is overdriven in the higher intensity regions (graph (c) in Fig. 6). The optimal linear photochemical transformation is shown in graph (b) of Fig. 6. In some cases, however, overdriving the photochemistry to generate a nonlinear grating as shown in graph (c) of Fig. 6 can produce enhanced diffraction efficiencies. Unfortunately, this is often accompanied by a loss of resolution that can have a deleterious impact on the image quality or data density when the higher laser intensities are used in optical memory applications or in pattern recognition systems. Simply stated, optimizing the laser intensity to generate a maximum diffraction efficiency does not always generate the optimal optical excitation levels for optoelectronic applications requiring high resolution.

There are a number of polymer matrices that can be used to solubilize bacteriorhodopsin, including polyvinyl alcohol, bovine skin gelatin, methylcellulose, and polyacrylamide. Polymeric films containing bacteriorhodopsin are usually sealed from the outside environment to prevent humidity and pH changes, which dramatically influence the photochromic properties exhibited by the film. The design shown in Fig. 7 yields bacteriorhodopsin films with excellent long-term stability and high diffraction efficiencies (see Fig. 5).

Spatial Light Modulators. Research in optical engineering during the past decade has demonstrated the unique capability of two-dimensional optical processing systems to perform complex mathematical processing functions such as pattern recognition, image processing, solution of partial differential and integral equations, linear algebra, and nonlinear arithmetic (22,46,52–62). Interest in exploring optical processing architectures is prompted by the inherent speed and massive parallel processing and interconnection capabilities of optical systems.

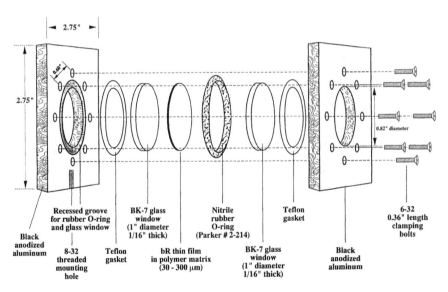

Figure 7 Schematic design of a reversible holographic spatial light modulator based on a thin film of bacteriorhodopsin. A key feature of this design is the use of a compressed nitrile rubber O-ring to seal the protein thin film in order to prevent dehydration of the polymer matrix. The long-term optical and shelf stability of the holographic media is excellent.

Spatial light modulators (SLMs) are integral components in the majority of one-dimensional and two-dimensional optical processing systems. These devices modify the amplitude, intensity, phase, or polarization of a spatial light distribution as a function of an external electrical signal or the intensity of a secondary light distribution. The observation that a thin film of bacteriorhodopsin can act as a photochromic bistable optical device (either $bR \leftrightarrow K$ or $bR \leftrightarrow M$ photoreactions) or as a voltage-controlled bistable optical device ($bR \leftrightarrow M$ photoreaction) suggests that it has significant potential as the active medium in SLMs (1,2,22,44,46,52,60,62). Soviet scientists were the first to exploit this potential and deserve much of the credit for bringing bacteriorhodopsin to the attention of researchers working in optical engineering (45,47,48,63–65). The most successful bacteriorhodopsin based SLM device has been recently demonstrated by German researchers. Their work exploits the $bR \leftrightarrow M$ photoreaction of a mutant protein film in a Fourier optical architectural scheme that implements edge enhancement (spatial frequency filtering) on an input image (2,9,18,62,66). Additional discussion of this area may be found in the chapter by C. Bräuchle.

Holographic Associative Memories. Associative memories operate in a fashion quite different from the serial memories that dominate current computer architectures (1,9,46,52,67,68). These memories take an input data block (or image) and, independently of the central processor, "scan" the entire memory for the data block that matches the input. In some implementations, the memory will find the closest match if it cannot find a perfect match. Finally, the memory will return the data block that satisfies the matching criteria. Because the human brain operates in an associative mode, many computer scientists believe that the implementation of large capacity associative memories will be required if we are fully to achieve artificial intelligence. Optical associative memories using Fourier transform holograms have significant potential for applications in optical computer architectures, optically coupled neural network computers, robotic vision hardware, and generic pattern recognition systems. The ability to change rapidly the holographic reference patterns via a single optical input while maintaining both feedback and thresholding increases the utility of the associative memory, and in conjunction with solid state hardware it opens up new possibilities for high-speed pattern recognition architectures.

One application currently under investigation is the use of bacteriorhodopsin thin films as the holographic storage components in a real time optical associative memory (1,3,46). Our current design is shown in Fig. 8 (52). The optical design, which employs both feedback and thresholding, is based on the closed-loop autoassociative design of Paek and Psaltis (59). During the write operation, reference images stored in an electronically addressable spatial light modulator (ESLM) are optically fed into the loop by plane wave illumination ($\lambda_W = 568$ nm) from a krypton-ion laser. The reference images are stored as Fourier transform holograms on thin polymer films containing bacteriorhodopsin [**H1** and **H2**] (Fig.

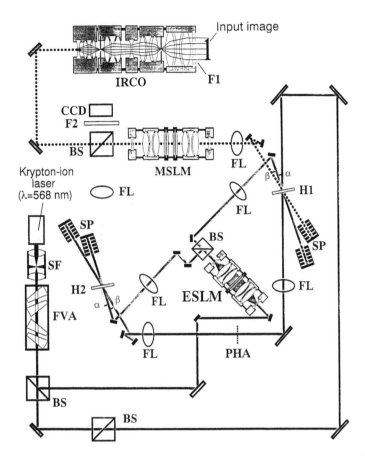

Figure 8   Schematic diagram of a Fourier transform holographic (FTH) associative memory with read/write FTH reference planes using thin polymer films of bacteriorhodopsin to provide real time storage of the holograms. The optical design is a modification of one proposed by John Izgi (26). The following symbols are used: BS (beam splitter), CCD (charge coupled device two-dimensional array), CL (condensing lens), ESLM (electronically addressable spatial light modulator), FL (Fourier lens), FVA (Fresnel variable attenuator), F1 (broadband filter for image), F2 (interference filter with transmission maximum at laser wavelength; different from λ_{max} of F1), H1 and H2 (holographic spatial light modulator, Fig. 7), IRCO (image reduction and condensing optics), MSLM (multichannel plate spatial light modulator), PHA (pinhole array), SF (spatial filter to select TEM_{00}), SP (beam stop).

7). For this real time application, no chemical additives are used to enhance the **M** state lifetime. Accordingly, the hologram stores the reference image for approximately 10 ms before reverting to the ground state. During the readout operation, the input image (from transparencies or another ESLM) is read into the loop by using the optical imaging system shown in Fig. 8. The best results are obtained by illuminating the object by using plane wave illumination from a second krypton-ion laser operating at a wavelength of 676.5 nm. The input image beam is passed through a microchannel plate spatial light modulator (MSLM) operating in thresholding mode. Thereafter, the Fourier transformed product of the image-reference is formed and retransformed at the plane of a pinhole array (PHA). The resulting correlation patterns are sampled by the pinholes (diameter ~500 μm), which are precisely aligned with the optical axis of the reference images. Light from the pinhole plane is retransformed and superimposed with the reference image stored on the second bacteriorhodopsin hologram (**H2**). The resulting cross correlation pattern represents the superposition of all images stored on the multiplexed holograms and is fed back through the microchannel plate spatial light modulator for another iteration. Thus each image is weighted by the inner product between the pattern recorded on the MSLM from the previous iteration and itself. The output locks on to that image stored in the holograms that produces the largest correlation flux through its aligned pinhole.

The real time capability of the associative loop is made possible by using bacteriorhodopsin films as the transient holographic medium. The high speed of phototransformation during the write operation (<50 μs) coupled with the quick relaxation time of the **M** state (~ 10 ms) allow for framing rates up to 100 frame/sec. Slower or faster framing rates can be attained by simply altering the **M** lifetime with chemical additives and/or by intermittently erasing the hologram with an external blue light source. The write and read wavelengths of the krypton-ion lasers as well as the respective angle of incidence are chosen to optimize the diffraction efficiency of the bacteriorhodopsin holograms. During the process of sampling the correlation patterns, it is interesting to note that the inclusion of the pinholes destroys the shift invariance of the optical system. If the input pattern is shifted from its nominal position, the correlation peak shifts as well, and the correlation light flux will miss the pinhole. If the pinholes were removed, however, ghost holography would seriously impair image quality. The two apertures within the image reduction and collimation optics (*IRCO*) serve to provide correct registration, but the input image must still be properly centered to generate proper correlation. The problem of shift invariance represents one of the fundamental design issues that will have to be resolved before optical associative memories will reach their full potential. While there are a number of optical "tricks" that can be used to counteract poor registration, the most easily implemented approach is to use the controller of the *ESLM* to scale and translate the reference images to maximize the correlation light flux as measured by the intensity of the image falling on the CCD output detector.

V. TWO-PHOTON VOLUMETRIC MEMORIES

Two-photon three-dimensional optical addressing architectures offer significant promise for the development of a new generation of ultra-high-density random access memories (1,21,69–72). Two-dimensional optical memories have a storage capacity that is limited to $\sim 1/\lambda^2$, where λ is the wavelength, which yields approximately 10^8 bit/cm^2. In contrast, three-dimensional memories can approach storage densities of $1/\lambda^3$, which yields storages in the range 10^{11} to 10^{13} bit/cm^3. These memories read and write information by using two orthogonal laser beams to address an irradiated volume (1–50 μm^3) within a much larger volume of a nonlinear photochromic material. Because the probability to a two-photon absorption process scales as the square of the intensity, to a first approximation photochemical activation is limited to regions within the irradiated volume. (Methods to correct for photochemistry outside the irradiated volume are described below.) The three-dimensional addressing capability derives from moving the media or the location of the beam crossing. The volumetric memory described below is designed to store 18 Gbytes (1 Gbyte = 10^9 bytes) within a data storage cuvette with dimensions of 1.6 $\times$ 1.6 $\times$ 2 cm. Our current storage capacity is well below the maximum theoretical limit of $\sim$512 Gbytes for the same $\sim$5 cm^3 volume.

Bacteriorhodopsin has four characteristics that contribute to its advantage as a two-photon volumetric medium (1,21). First, it has a large two-photon absorptivity due to the highly polar environment of the protein binding site and the large change in dipole moment that accompanies excitation (73). Second, bacteriorhodopsin exhibits large quantum efficiencies in both the forward and the reverse direction. Third, the protein gives off a fast electrical signal that indicates its state when light activated (1). Fourth, the protein can be oriented in optically clear polymer matrices permitting photoelectric state interrogation (21). The two-photon induced photochromic behavior is summarized in the scheme.

$$\textbf{bR} \; (\lambda_{max} \cong 570 \text{ nm}) \; (\textit{State 0}) \quad \underset{h\omega^2; \; \Phi_2 \sim 0.65}{\overset{h\omega^2; \; \Phi_1 \sim 0.65}{\underset{\longleftarrow}{\longrightarrow}}} \quad \textbf{M} \; (\lambda_{max} \cong 410 \text{ nm}) \; (\textit{State 1})$$

We arbitrarily assign **bR** to binary state 0 and **M** to binary state 1. The chromophore in **bR** has an unusually large two-photon absorptivity that permits the use of much-lower-intensity laser excitation to induce the forward photochemistry. The above wavelengths are correct to only $\leq$40 nm, because the two-photon absorption maxima shift as a function of temperature and polymer matrix water content.

The optical design of the two-photon three-dimensional optical memory is shown in Fig. 9 (1,20,21). The bacteriorhodopsin is contained in a cuvette and is oriented by using electric fields prior to polymerizing the polyacrylamide gel

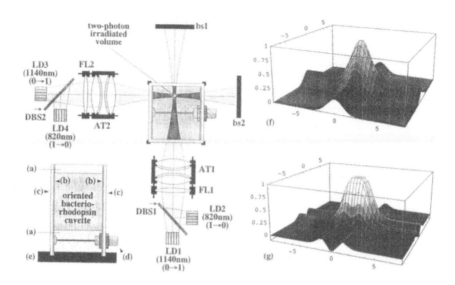

Figure 9 Schematic diagram of the principal optical components of a two-photon three-dimensional optical memory based on bacteriorhodopsin. The write operation involves the simultaneous activation of LD_1 and LD_3 ($0 \rightarrow 1$) or LD_2 and LD_4 ($1 \rightarrow 0$) to induce two-photon absorption within the irradiated volume and partially convert either **bR** to **M** ($0 \rightarrow 0$) or **M** to **bR** ($1 \rightarrow 0$). The write operation uses a 20 ns pulse and a pulse simultaneity of 1 ns. The protein is oriented within the cuvette by using an electric field prior to polymerization of the polyacrylamide gel. A polymer sealant is then used to maintain the correct polymer humidity. The *SMA* connector is attached to the indium–tin-oxide conducting surfaces on opposing sides of the cuvette and is used to transfer the photoelectric signal to the external amplifiers and box-car integrators. Symbols and letter codes are as follows: (a) sealing polymer, (b) indium–tin-oxide conductive coating, (c) *BK7* optical glass; (d) *SMA* or *OS50* connector; (e) Peltier temperature-controlled base plate (0–20°C); AT (achromatic focusing triplet); bs (beam stop); DBS (dichroic beam splitter); LD (laser diode); FL (adjustable focusing lens). Computer simulations of the probability of two-photon induced photochemistry (vertical axis) as a function of location relative to the center of the irradiated volume (ΔX_{focus} and ΔY_{focus}) in microns are shown in (f) and (g). The upper right contour plot (f) shows the probability after two 1140 nm laser beams have been simultaneously directed along orthogonal axes crossing at the center of the irradiated volume. The lower right contour plot (g) shows the probability after two 820 nm "cleaning pulses" have been independently directed along the same axes. The maximum conversion probability at $x = 0$, $y = 0$ is normalized to unity for both contour plots.

matrix. This orientation is required in order to observe and use the photoelectric signal to monitor the state of the proteins occupying the irradiated volume. A write operation is carried out by firing simultaneously the two 1140 nm lasers (to write a 1) or the two 820 nm lasers (to write a 0). To eliminate unwanted photochemistry along the laser axes, the lasers not used in the original write operation are fired nonsimultaneously immediately following the write operation. The position of the cube is controlled in three dimensions by using a series of actuators that independently drive the cube in the x, y, or z direction. For slower-speed maximum-density applications, electrostrictive micrometers can be used. For higher-speed lower-density applications, voice-coil actuators can be used. Parallel addressing of large data blocks can also be accomplished by using holographic lenses or other optical architectures (20,70).

However, there are a number of technical problems that remain to be solved before parallel addressing is reliable. A key requirement of the two-photon memory is to generate an irradiated volume that is reproducible in terms of xyz location over lengths as large as 2 cm. In the present case, our cubes are typically ~ 1.6 cm in the x and y dimensions and ~ 2 cm in the z direction (see Fig. 9). These dimensions are variable up to 2 cm on all sides, and can be as small as 1 cm on a side depending upon the desired storage capacity of the device. By using a set of fixed lasers and lenses, and moving the cube by using orthogonal translation stages, excellent reproducibility can be achieved (± 1 μm for electrostrictive micropositioners, ± 3 μm by using voice-coil actuators). Refractive inhomogeneities that develop within the protein–polymer cube as a function of write cycles adversely affect the ability to position the irradiated volume with reproducibility. This problem is due to the change in refractive index associated with the photochemical transformation (see Fig. 3). The problem is minimized by operating with a relatively large irradiated volume (30 μm^3) and by limiting the photochemical transformation to 60:40 versus 40:60 in terms of relative **bR:M** percentages. The techniques of reading and writing data without corrupting data outside of the irradiated volume are complicated, and the interested reader is referred to (1,20,21) for the details. Ultimately, the full potential of two-photon volumetric memories will only be reached if parallel addressing can be implemented or phased arrays can be used to provide ultra-high-speed binary read/write capability (20,69,70,72).

VI. RELIABILITY OF MOLECULAR ELECTRONIC DEVICES

The important issue of the reliability of molecular electronic devices has been studied in detail in the literature with mixed conclusions (16,71,74–79). Adverse opinions have been used repeatedly by semiconductor scientists and engineers as reasons to view molecular electronics as impractical. Indeed, an article by Birge and coworkers (76) has been referenced by others to argue (incorrectly) that the need to use ensemble averaging in optically coupled molecular gates and switches demonstrates the inherent unreliability of molecular electronic devices. This point of view is comparable to suggesting that transistors are inherently unreliable because more than one charge carrier must be used to provide satisfactory performance. The majority of ambient temperature molecular and bulk semiconductor devices use more than one molecule or charge carrier to represent a bit for two reasons: (1) ensemble averaging improves reliability and (2) ensemble averaging permits higher speeds (79). The implicit use of ensemble averaging does not, however, rule out reliable monomolecular or monoelectronic devices. We explore the issue of reliability in ensemble averaged and monomolecular devices briefly.

The probability of correctly assigning the state of a single molecule p_1 is never exactly unity. This less-than-perfect assignment capability is due to quantum effects as well as inherent limitations in the state assignment process. The probability of an error in state assignment P_{error} is a function of p_1 and the number of molecules n within the ensemble used to represent a single bit of information. P_{error} can be approximated by (77).

$$P_{error}(n, p_1) \cong -\text{erf} \left[\frac{(2 p_1 + 1) \sqrt{n}}{4 \sqrt{2 p_1(1 - p_1)}}; \frac{(2 p_1 - 1) \sqrt{n}}{4 \sqrt{2 p_1(1 - p_1)}} \right] \tag{11}$$

where erf $[Z_0; Z_1]$ is the differential error function defined by

$$\text{erf} [Z_0; Z_1] = \text{erf} [Z_1] - \text{erf} [Z_0] \tag{12}$$

where

$$\text{erf} [Z] = \frac{2}{(\pi)^{1/2}} \int_0^z \exp(-t^2) \, dt \tag{13}$$

Equation (11) is approximate and neglects error associated with the probability that the number of molecules in the correct conformation can stray from their expectation values based on statistical considerations. Nevertheless it is sufficient to demonstrate the issue of reliability and ensemble size. First, we define a logarithmic reliability parameter ξ that is related to the probability of error in the measurement of the state of the ensemble (device) by the function $P_{error} = 10 - \xi$. A value of $\zeta = 10$ is considered a minimal requirement for reliability in non-error-correcting digital architectures.

If we assume that the state of a single molecule can be assigned correctly with a probability of 90% ($p_1 = 0.9$), then Eq. (1) indicates that 95 molecules must collectively represent a single bit to yield $\xi > 10$ ($P_{error}(95, 0.9) \simeq 8 \times 10^{-11}$). We must recognize that a value of $p_1 = 0.9$ is larger than is normally observed, and some examples of reliability analyses for specific molecular based devices is provided in (77). In general, ensembles larger than 10^3 are required for reliability unless fault-tolerant or fault-correcting architectures can be implemented.

The question then arises whether or not we can design a reliable computer or memory that uses a single molecule to represent a bit of information. The answer is yes provided one of two conditions apply. The first condition is architectural. It is possible to design fault-tolerant architectures that either recover from digital errors or simply operate reliably with occasional error due to analog or analog-type environments. An example of digital error correction is the use of additional bits beyond the number required to represent a number. This approach is common in semiconductor memories, and under most implementations these additional bits provide for single-bit error correction and multiple bit error detection. Such architectures lower the required value of ξ to values less than 4. An example of analog error tolerance is embodied in many optical computer designs that use holographic or Fourier architectures to carry out complex functions. The second condition is more subtle. It is possible to design molecular architectures that can undergo a state reading process that does not disturb the state of the molecule. For example, an electrostatic switch could be designed that could be "read" without changing the state of the switch. Alternatively, an optically coupled device can be read by using a wavelength that is absorbed or diffracted but that does not initiate state conversion. Under these conditions, the variable n that appears in Eq. (1) can be defined as the number of read "operations" rather than the ensemble size. Thus our previous example indicating that 95 molecules must be included in the ensemble to achieve reliability can be restated as follows: a single molecule can be used provided we can carry out 95 nondestructive measurements to define the state. Multiple state measurements are equivalent to integrated measurements and should not be interpreted as a start-read-stop cycle repeated n times. A continuous read with digital or analog averaging can achieve the same level of reliability.

VII. SUMMARY AND CONCLUSIONS

The current and potential use of bacteriorhodopsin in optical computing and memory devices has been reviewed. This protein has significant potential for use in these applications due to unique intrinsic photophysical properties and the range of chemical and genetic methods available for optimizing performance for specific application environments. Although further research and development is required before protein based optical devices and memories will be competitive with current technologies, such efforts are underway in many countries. The future impact of biomolecular optoelectronics on computer architecture remains to be fully revealed, but many in the scientific and business communities have concluded that it may represent one of the key emerging technologies of the next decade.

ACKNOWLEDGMENTS

The authors thank Zhongping Chen, Deshan Govender, and John Izgi for interesting and helpful discussions. The research from the authors' laboratory was sponsored in part by grants from the W. M. Keck Foundation, the U.S. Air Force Rome Laboratory, the National Institutes of Health, and the corporate Affiliates Program of the W. M. Keck Center for Molecular Electronics.

REFERENCES

1. Birge, R. R., *IEEE Computer*, 25, 56–67 (1992).
2. Oesterhelt, D., Bräuchle, C., and Hampp, N., *Quart. Rev. Biophys.*, 24, 425–478 (1991).
3. Birge, R. R., *Annu. Rev. Phys. Chem.*, 41, 683–733 (1990).
4. Chen, Z., and Birge, R. R., *Trends Biotech.* 11, 292–300 (1993).
5. Kumar, G. R., Wategaonkar, S. J., and Roy, M., *Opt. Comm.*, 98, 127–131 (1993).
6. Rayfield, G., *Adv. Chem.*, 240 (in press).
7. Lanyi, J. K., *Adv. Chem.*, 240 (in press).
8. Hong, F. T., *Adv. Chem.*, 240 (in press).
9. Hampp, N., Thoma, R., Zeisel, D., and Bräuchle, C., *Adv. Chem.*, 240 (in press).
10. Birge, R. R., edi., *Molecular and Biomolecular Electronics*, Advances in Chemistry Series 240, American Chemical Society, Washington, D.C. (in press). This reference is abbreviated as *Adv. Chem.* 240 in this reference set.
11. Werner, O., Fisher, B., and Lewis, A., *Opt. Lett.*, 17, 241–243 (1992).
12. Werner, O., Daisy, R., Fisher, B., and Lewis, A., *Opt. Comm.*, 92, 108–110 (1992).
13. Zaitsev, S. Y., Kozhevikov, N. M., Barmenkov, Y. O., and Lipovskaya, M. Y., *Photochem. Photbiol.*, 55, 851–856 (1992).
14. Thoma, R., and Hampp, N., *Opt. Lett.*, 17, 1158–1160 (1992).
15. Renner, T., and Hampp, N., *Opt. Comm.*, 96, 142–149 (1992).
16. Mirkin, C. A., and Ratner, M. A., *Ann. Rev. Phys. Chem.*, 43, 719–754 (1992).
17. Haronian, D., and Lewis, A., *Appl. Phys. Lett.*, 61, 2237–2239 (1992).
18. Hampp, N., Popp, A., Bräuchle, C., and Oesterhelt, D., *J. Phys. Chem.*, 96, 4679–4685 (1992).
19. El-Sayed, M. A., *Accts. Chem. Res.*, 25, 279–286 (1992).
20. Lawrence, A. F., and Birge, R. R., *Proc. SPIE*, 1773, 401–412 (1992).
21. Birge, R. R., Gross, R. B., Masthay, M. B., Stuart, J. A., Tallent, J. R., and Zhang, C. F., *Mol. Cryst. Liq. Cryst. Sci. Technol. Sec. B. Nonlinear Optics*, 3, 133–147 (1992).
22. Song, Q. W., Zhang, C., Blumer, R., Gross, R. B., Chen, Z., and Birge, R. R., *Opt. Lett.*, 18 (in press).
23. Song, Q. W., Zhang, C., Gross, R., and Birge, R. R., *Opt. Lett.*, 18, 775–777 (1993).

24. Albrecht, O., Sakai, K., Takomoto, K., Matsuda, H., Eguchi, K., and Nakagiri, T., *Adv. Chem.*, 240 (in press).
25. Marx, K. A., Samuelson, L. A., Kamath, M., Sengupta, S., Kaplan, D., Kumar, J., and Tripathy, S., *Adv. Chem.*, 240 (in press).
26. Shashidar, R., and Schnur, J. M., *Adv. Chem.*, 240 (in press).
27. Stayton, P. S., Olinger, J. M., Wollman, S. T., Bohn, P. W., and Sligar, S. G., *Adv. Chem.*, 240 (in press).
28. Wada, T., Hosoda, M., and Sasabe, H., *Adv. Chem.*, 240 (in press).
29. Zhang, J., and Sponsler, M. B., *Adv. Chem.*, 240 (in press).
30. O'Neal, M. P., Niemczyk, M. P., Svec, W. A., Gosztola, D., Gaines, G. L., and Wasielewski, M. R., *Science*, 257, 63–65 (1992).
31. Birge R. R., in *Nanotechnology: Research and Perspective* (B. C. Crandall and J. Lewis, eds.), MIT Press, Cambridge, Mass., 1992, pp. 149–170.
32. Greenbaum, E., *J. Phys. Chem.*, 96, 514–516 (1992).
33. Greenbaum, E., *J. Phys. Chem.*, 94, 6151–6153 (1990).
34. Boxer, S. G., Stocker, J., Franzen, S., and Salafsky J., in *Molecular Electronics—Science and Technology* (A. Aviram, eds.), vol. 262, American Institute of Physics, New York, 1992, pp. 226–241.
35. Oesterhelt, D., and Stoeckenius, W., *Nature* (London), *New Biol.*, 233, 149–152 (1971).
36. Birge, R. R., *Biochim. Biophys. Acta*, 1016, 293–327 (1990).
37. Polland, H. J., Franz, M. A., Zinth, W., Kaiser, W., Kolling, E., and Oesterhelt, D., *Biophys. J.*, 49, 651–662 (1986).
38. Dobler, J., Zinth, W., Kaiser, W., and Oesterhelt, D., *Chem. Phys. Lett.*, 144, 215–220 (1988).
39. Mathies, R. A., Lugtenburg, J., and Shank, C. V., in *Biomolecular spectroscopy* (R. R. Birge and H. H. Mantsch, eds.), vol. 1057, International Society for Optical Engineering, Bellingham, Washington, 1989, pp. 138–145.
40. Mathies, R. A., Brito Cruz, C. H., Pollard, W. T., and Shank, C. V., *Science*, 240, 777 (1988).
41. Reed, M., and Seabaugh, A. C., *Adv. Chem.*, 240 (in press).
42. Simmeth, R., and Rayfield, G. W., *Biophys. J.*, 57, 1099–1101 (1990).
43. Birge, R. R., Zhang, C. F., and Lawrence, A. F., in *Molecular Electronics* (F. Hong, ed.), Plenum press, New York, 1989, pp. 369–379.
44. Hampp, N., Bräuchle, C., and Oesterhelt, D., *Biophys. J.*, 58, 83–93 (1990).
45. Vsevolodov, N. N., and Poltoratskii, V. A., *Sov. Phys. Tech. Phys.*, 30, 1235 (1985).
46. Birge, R. R., Fleitz, P. A., Gross, R. B., Izgi, J. C., Lawrence, A. F., Stuart, J. A., and Tallent, J. R., *Proc. IEEE EMBS*, 12, 1788–1789 (1990).
47. Bazhenov, V. Y., Soskin, M. S., and Taranenko, V. B., *Sov. Tech. Phys. Lett.*, 13, 382–384 (1987).
48. Bazhenov, V. Y., Soskin., M. S., and Taranenko, V. B., and Vasnetsov, M. V., in *Optical Processing and Computing* (H. H. Arsenault, T. Szoplik and B. Macukow, eds.), Academic Press, New York, 1989, pp. 103–144.
49. Loudon, R., *The Quantum Theory of Light*, Clarendon, Press, Oxford, 1973.

50. Landau, L. D., and Lifshitz, E. M., *Electrodynamics of Continuous Media*, Pergamon Press, New York, 1960.
51. Kogelnik, H., *Bell Syst. Tech. J.*, 48, 2909–2947 (1969).
52. Gross, R. B., Izgi, K. C., and Birge, R. R., *Proc. SPIE*, 1662, 186–196 (1992).
53. Tanguay, A. R., Jr., *Optics News*, Feb., 23–26 (1988).
54. Tanguay, A. R., Jr., *Opt. Eng.*, 24, 2–18 (1985).
55. Casasent, D., *Proc. IEEE*, 65, 143–157 (1977).
56. Fisher, A. D., and Lee, J. N., *Proc. SPIE*, 634, 352–371 (1986).
57. Neff, J. A., Athale, R. A., and Lee, S. H., *Proc. IEEE*, 78, 826–855 (1990).
58. Javidi, B., Tang, Q., Gregory, D. A., and Hudson, T. D., *Appl. Opt.*, 30, 1772–1775 (1991).
59. Paek, E. G., and Psaltis, D., *Opt. Eng.*, 26, 428–433 (1987).
60. Song, Q. W., and Yu, F. T. S., *Opt. Eng.*, 28, 533–535 (1989).
61. Kirkby, C. J. G., and Bennion, I., *IEE Proc.*, 133, 98–104 (1986).
62. Thoma, R., Hampp, N., Brauchle, C., and Oesterhelt, D., *Opt. Lett.*, 16, 651–653 (1991).
63. Druzhko, A. B., and Zharmukhamedov, S. K., in *Photosensitive Biological Complexes and Optical Recording of Information* (G. R. Ivanitskiy and N. N. Vsevolodov, eds.), USSR Academy of Sciences, Biological Research Center, Institute of Biological Physics, Pushchino, 1985, pp. 119–125.
64. Ivanitskiy, G. R., and Vsevolodov, N. N., *Photosensitive Biological Complexes and Optical Recording of Information*, USSR Academy of Sciences, Biological Research Center, Institute of Biological Physics, Pushchino, 1985, pp. 1–209.
65. Savranskiy, V. V., Tkachenko, N. V., and Chukharev, V. I., in *Photosensitive Biological Complexes and Optical Recording of Information* (G. R. Ivanitskiy and N. N. Vsevolodov, eds.), USSR Academy of Sciences, Biological Research Center, Institute of Biological Physics, Pushchino, 1985, pp. 97–100.
66. Bräuchle, C., Hampp, N., and Oesterhelt, D., *Adv. Mater.*, 3, 420–428 (1991).
67. Lu, T., Choi, K., Wu, S., Xu, X., and Yu, F. T., *Appl. Opt.*, 28, 4722–4724 (1989).
68. Casasent, D., *Opt. Eng.*, 24, 026–032 (1985).
69. Parthenopoulos, D. A., and Rentzepis, P. M., *Science*, 245, 843–845 (1989).
70. Hunter, S., Kiamilev, F., Esener, S., Parthenopoulos, D. A., and Rentzepis, P. M., *Appl. Opt.*, 29, 2058–2066 (1990).
71. Lawrence, A. F., and Birge, R. R., *Adv. Chem.*, 240 (in press).
72. Dvornikov, A. S., and Rentzepis, P. M., *Adv. Chem.*, 240 (in press).
73. Birge, R. R., and Zhang, C. F., *J. Chem. Phys.*, 92, 7178–7195 (1990).
74. Haddon, R. C., and Lamola, A., *Proc. Natl. Acad. Sci. USA*, 82, 1874–1878 (1985).
75. Crandall, B. C., and Lewis, J., *Nanotechnology: Research and Perspective*, MIT Press, Cambridge, Mass., 1992, pp. 1–381.
76. Birge, R. R., Ware, B. R., Dowben, P. A., and Lawrence, A. F., in *Molecular Electronics—Science and Technology* (A. Aviram, ed.), Engineering Foundation, New York, 1989, pp. 275–284.

77. Birge, R. R., Lawrence, A. F., and Tallent, J. A., *Nanotechnology*, 2, 73–87 (1991).

78. Launay, J. P., in *Molecular Electronics—Science and Technology* (A. Aviram, ed.), Engineering Foundation, New York, 1989, pp. 237–246.

79. Keyes, R. W., in *Molecular Electronics—Science and Technology* (A. Aviram, ed.), Engineering Foundation, New York, 1989, pp. 197–204.

80. Hayward, B. S., Grano, D. A., Glaeser, R. M., and Fisher, K. A., *Proc. Natl. Acad. Sci. USA.*, 75, 4320–4324 (1978).

81. Earnest, T. N., Roepe, P., Braiman, M. S., Gillespie, J., and Rothschild, K. J., *Biochemistry*, 25, 7793–7798 (1986).

82. Birge, R. R., Findsen, L. A., and Pierce, B. M., *J. Am. Chem. Soc.*, 109, 5041–5043 (1987).

16

Bacteriorhodopsin—Optical Processor Molecules from Nature?

C. Bräuchle

University of Munich
Munich, Germany

N. Hampp

University of Marburg
Marburg, Germany

D. Oesterhelt

Max-Planck-Institute of Biochemistry
Martinsried, Germany

I. INTRODUCTION

The field of molecular electronics, which includes molecular photonics, is characterized by an attempt to engineer and use devices on a molecular level. Supramolecular chemistry is one way to produce such molecular devices and is a subject of this book too. However, long before supramolecular chemists tried to synthesize highly organized artificial systems often employing principles of living nature, nature itself has produced and optimized biological systems on a molecular device level during a long period of evolution. The retinal protein bacteriorhodopsin (BR) (1), which is related to the human visual pigment, represents such a molecular device. It is the key protein of the photosynthetic center of the *Halobacterium salinarium* and acts as a light driven proton pump (for reviews see e.g. 2–4). By absorption of light BR pumps protons across the membrane, generating a chemical potential that is a source of energy for the living bacterium. As a molecular device BR offers three interesting light driven functions: (1) as a proton pump, (2) as a charge separator, and (3) as a photochromic material. Due to these interesting functions and other attractive properties such as its high stability, different technical applications of BR have been proposed (for references see 5,6), among them desalination of sea water and conversion of sunlight into electricity. So far, however, none of these has reached a technical standard. For the field of molecular electronics and photonics two functions of BR are of high interest, its light driven charge separation and its photochromism. In this article we shall focus mainly on photochromism.

In recent years it has been shown that not only the naturally occuring BR, the so-called wild-type (BR_{WT}), has interesting properties for e.g. optical processing but also and mainly its genetically modified mutants (for reviews see 6–8). Since the mutants could be produced by genetic engineering, this opened up a unique and highly exciting way to generate desirable properties. A detailed knowledge of the structure–function relation can help to do this in a relatively predicted way.

A further attractive point of this type of molecular engineering is that nature itself synthesizes new molecules of high complexity according to the changed gene code, which would not be obtainable with conventional chemical techniques. This high complexity with its precise stereochemical arrangements is often the key to the high efficiency by which naturally occuring systems are characterized. The well-defined cage for the retinal chromophore in BR made up by the protein moiety is one of many examples.

In this article we give a brief description of (1) the structure and function of BR and (2) the generation of mutants with new properties. This will be followed by applications of BR and its mutants as optical processor molecules in (3) spatial light modulation, (4) logical operations, (5) holographic recording, (6) real time pattern recognition, and (7) E-DRAW storage. In conclusion we summarize the wide range of properties that we have obtained with BR and its mutants, which span from molecules for fast processing in the ps time regime to ones for permanent storage.

II. STRUCTURE AND FUNCTION OF BACTERIORHODOPSIN

Bacteriorhodopsin (BR) is contained in the purple membrane (PM) that is part of the cell membrane of the *Halobacterium salinarium* (1–3). PM consists of a two-dimensional hexagonal lattice of trimers of BR in a lipid bilayer. BR has a sequence of 248 amino acid residues arranged in seven trans-membrane α-helices and has a molecular weight of 26,000 daltons. The retinal chromophore is bound to the protein moiety through a protonated Schiff base linkage at the ε-amino group of Lys_{216} (lysine residue in position 216 in the amino acid sequence) and is located inside the pore formed by the seven α-helices. This is schematically shown in Fig. 1a. The basic biochemical and photophysical properties of BR are summarized in Table 1.

In the living cell BR acts as a light driven proton pump (Fig. 1b) and thereby converts light energy into chemical energy. The proton pumping activity is closely related to the photocycle (Fig. 1c) and can be understood as follows (9–11): In the dark BR shows an equilibration of the B- and the D-state with

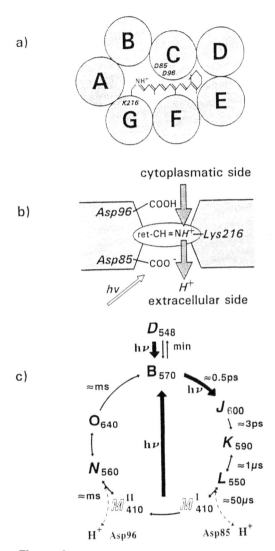

Figure 1 Schematic views of (a) topography of the seven amino acid helices, (b) cross-section of the proton pore, and (c) the photocycle of bacteriorhodopsin. Photochemical conversions (thick arrows), thermal relaxations (thin arrows), relaxation times, and absorption maxima are given for the different states (letters). (From Ref. 50.)

an all-trans and 13-cis configuration respectively of the retinylidene group. By illumination the system turns into the so-called light adapted B-state. From there by absorption of a photon in the yellow to red spectral range the J-state is formed by a very fast 13-cis photoisomerization within less than one ps. The high

Table 1 Biochemical and Photophysical Properties of Bacteriorhodopsin

Biochemical properties		Photophysical properties	
Molecular weight	26,000 daltons	Initial absorption	$\epsilon_{570} = 63,000$ $\text{lmol}^{-1} \text{cm}^{--1}$
Structure		Quantum	$\Phi = 64\%$
Primary	248 amino acids	efficiency	
Secondary	7 α-helical domains		
Tertiary	Cage with proton pore	Angle of retinal	20°
Quarternary	Trimers, 2-D crystalline	to membrane	
Lipid bilayer	Transmembrane, 10 lipid	Photoactive	At least 4
	molecules per BR	intermediates	
Chromophoric	Retinal	Photochromism	cis–trans Isomerisation
group			Protonation change
Biol. function	Light driven proton pump	Refractory period	None
		after relaxation	
Stability	Constant illumination	Instabilities	UV-light
	Oxygen + light		Organic solvents
	Temperatures < 80°C		
	pH 3–10		
	Most proteases		

Source: Ref. 8.

quantum yield ($\Phi = 0.64$) and the high absorption ($\epsilon_{570} = 63,000 \text{ l mol}^{-1}$ cm^{-1}) of the B-state guarantees an efficient stimulation of the photocycle. As a result of continued protein conformation changes, BR thermally relaxes from the J-state through the K-state to the L-state. The Schiff base releases its proton to the Asp_{85} during the transition from L→M^I, causing a strong hypsochromic shift of 160 nm compared to the initial B-state. The M^I→M^{II} transition is considered to be the dominating irreversible step in the photocycle with a strong change in the pK value of the Schiff base. The reprotonation of the Schiff base from M^{II}→N is catalyzed by the internal proton donor Asp_{96}, which in turn is reprotonated from the outer medium. Finally the system returns from the O-state to the initial B-state with a typical overall cycle time of 10 ms for BR_{WT} in suspension at $T = 20°C$ and pH $= 7$.

Figure 2 summarizes the absorption range, the configuration of the $C_{13} = C_{14}$ double bond, and the state of protonation of the Schiff base, the Asp_{85} and the Asp_{96}, for the different states of the photocycle. From the above it is obvious that by absorption of a photon BR translocates a proton from the cytoplasmatic side to the extracellular side using Asp_{96} and Asp_{85} as proton donor and acceptor, respectively, and the photoisomerization of the retinylidene group as the light driven step of the proton movement between proton donor and acceptor (Fig. 1b).

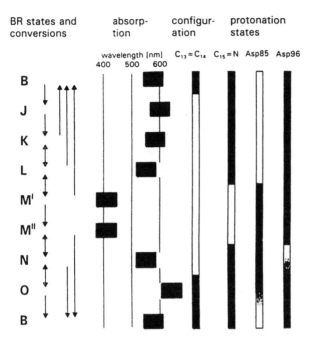

Figure 2 Scheme of the temporal course of the changes of BR during the photocycle and the related proton transport. The spectral absorption and the width of absorption bands of the different BR states are given in the middle. The filled/open bars in the right part symbolize the temporal relation of the configurational (filled = all trans, open = 13-cis) and protonation changes (filled = protonated, open = deprotonated) to the absorption shifts. (From Ref. 50.)

In Fig. 2 it is indicated that all intermediate states without O of the photocycle can be converted back to the initial B-state by photoexcitation within their absorption bands. This is of interest with respect to two points. First, it gives an explanation for the photostability of the system, because it shows that all photoexcitations (for $\lambda \geq 400$ nm) will bring the system back to its initial state and no permanent photoproducts will be formed. Second, it allows us to use BR as a light driven molecular switch. This is indicated in Fig. 1c using the B- and the M-state as two photochromic states that can be interconverted into one another using yellow to red light for the B→M and blue light for the M→B transition. In addition there is a thermal pathway that sets the system back by itself to the initial B-state. Since the M-state shows the strongest spectral shift and the longest lifetime, i.e., the highest population of all intermediates under steady-state light conditions, the photochromism between the B- and the M-state is most pronounced. Therefore we shall concentrate in most of our examples on the photochromism of these two states. However, the fast rising K-state and the special photochemistry of the O-state will offer further exciting possibilities. Besides the photochromism the translocation of charge during the photocycle is of interest for electrooptical applications of BR (12–16), but this will not be outlined in this article.

III. PROPERTIES OF BR AND GENERATION OF MUTANTS

The most attractive properties of BR for optical applications and engineering (2–8) are (1) its efficient photochemistry with high absorption cross-sections and quantum yields (e.g., for B-M or B-K in both directions), which result in good photosensitivities, (2) its excellent reversibility of the photochromism of e.g. B-M allowing more than 10^6 write and erase cycles without any degradation, (3) its high photochemical, chemical, and thermal stability, which allow permanent illumination over years, a pH range from 3–10, and temperatures up to 80°C, (4) the capability of forming films in polymers, gels, etc., with high optical quality and of sufficiently large areas, and (5) the possibilities of modifying its properties by physical, chemical, and genetic engineering methods. Physical methods include e.g. lowering of temperature and thereby stabilizing intermediates as M (at 230K) (17) K or (at 80K) (18). By chemical methods one can modify both the kinetics of the photocycle and the spectral range of absorption. The kinetics can be influenced by changing the humidity (dry films) (5,6,19) or solvent (less protic) (5,6,20), this slowing down the proton uptake of the Schiff base and increasing the lifetime of the M-state. Replacing the retinal chromophore by synthetic analogs (5,6,21) is a way to accomplish a shift in the absorption spectrum. In the past, these methods have been studied intensively and used in a variety of applications by several groups (5–8,22–26), but they are limited. So for example the photophysics of bacteriorhodopsin reconstituted with a retinal analog cannot be predicted and often changes in an unfavorable way, loosing e.g. the photocycle or the reversibility. Today it seems that the key to a real success in the technical use of BR lies in the possibilities of genetically engineering its properties (6–8). This can be done because the structure–function relation of BR is known in great detail, so it can be used as a basis for a controlled modification. In order not to perturb the structure of BR, only a few—but functionally relevant—amino acid residues are changed. In this way the system can be adapted to different applications such as long-term data storage at room temperature on the one hand and fast optical processing in real time pattern recognition on the other.

The modification of the amino acid sequence can be done by site-directed mutagenesis (27–31). With the expression in *H. salinarium* the stable PM form of BR variants is obtained. Once a mutated halobacterial strain producing the desired BR variant is generated, the preparation of virtually unlimited amounts of material can be achieved by conventional biotechnological methods. Two mutants, BR_{D96N} and $BR_{D85,96N}$, which are the most interesting ones so far, will be briefly described.

In BR_{D96N} asparagine (Asn) replaces Asp_{96} (exchange of carbonyl group by carboxyamino group) and can no longer serve as a proton donor (Fig. 1b). In BR_{WT} the Asp_{96} controls the availability and the uptake of the protons by the Schiff base. Therefore in BR_{WT} the lifetime τ_M of the M-state is independent of the pH of the suspending medium. However, in BR_{D96N} the lack of this local regulation by a proton donor allows the lifetime τ_M to be changed over several orders of magnitude by the pH of the sample (32). Using buffered samples with pH from 4 to 9, τ_M ranges from 100 ms to over 100 s. The increase of the lifetime of the M-state leads for a given light intensity to a higher population of the M-state and thus increases the photobleaching of the initial B-state. This is of interest for many transient holographic and spatial light modulating applications (6–8).

For permanent optical storage applications a mutant is needed in which an intermediate state occurs with a lifetime of years or more. Since in the BR-system the 13-cis to all-trans photoisomerization of the retinylidene group is catalyzed, this step should be avoided. The O-photochemistry of BR_{WT} (33), and the photochemistry of the blue membrane (34) that is formed at pH values below 3.5, give the correct hint. Without going into details it will only be mentioned that in both cases the Asp_{85} residue is protonated and cannot act as a proton acceptor. Photoexcitation of the all-trans confirmation leads in both cases to a state with 9-cis configuration which is thermally stable and is no longer catalyzed to transform to the all-trans configuration of the initial state. Therefore the 9-cis configuration has to be achieved for permanent storage (35). However, blue membrane itself is chemically not stable, and the O-photochemistry of BR_{WT} cannot be induced directly and efficiently from the B-state. This leads to the double mutant $BR_{D85,96N}$, in which both the proton acceptor Asp_{85} and the proton donor Asp_{96} are replaced by Asn residues. In addition to the examples above, in this mutant neither deprotonation nor reprotonation of the Schiff base can be supported by proton acceptor or donor. Thus, in a sense, photoexcitation leads to a short circuit in the photocycle, and the effective formation of a thermally stable 9-cis retinal configuration is observed at pH 5–7 and glycerol as a less protic solvent. The photoproduct state was named P-state. It absorbs at 490 nm and can be photochemically reconverted to the initial state at 610 nm (8,35).

IV. SPATIAL LIGHT MODULATORS WITH BR AS OPTICAL PROCESSOR

In optical computing, a spatial light modulator (SLM) can be used as a transducer, a processor, or a memory or any combination of these functions. Combined with optical Fourier transformation (FT) an SLM is capable of performing useful processing operations including image addition and subtraction, logic operations, edge enhancement, and thresholding (36). How these processing functions can be carried out with BR (6–8,37) in a useful way will be demonstrated in this chapter. To make the approach clear to a reader who is not an expert in optical processing, we shall describe e.g. image subtraction as one example in more detail.

In optics it is well known that a lens can produce an FT of an object in its back focal plane if the object is placed in its front focal plane. This is demonstrated in a schematic way in Fig. 3a for the letter K. The FT consists of a two-dimensional distribution of light spots where the distance from the center gives the spatial frequency (1/m) and the brightness the weight of the Fourier component represented by the respective spot. It is obvious that by using a second FT lens a retransformation can be made and the letter K is retrieved as an image. The most attractive aspect of this simple arrangement is that in the Fourier plane selected Fourier components can be e.g. blocked by spatial filtering, which results in a manipulated image at the output. Thus by using an SLM in the Fourier plane for spatial filtering, image processing can be achieved. In Fig. 3b a setup is shown in which a BR film is used advantageously as an SLM in the Fourier plane (37). With this setup an operation can be demonstrated that shows e.g. the subtraction of the lower bar from the three bars of the letter K. Following the blue beam (λ = 413 nm) after expansion it illuminates the letter K as a

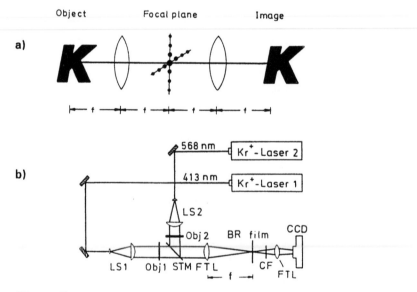

Figure 3 (a) Principle of Fourier transformation with a lens. (b) Setup for dynamic spatial filtering in the Fourier plane with BR films. LS1 and LS2 = lenses, STM = semitransparent mirror, FTL = Fourier transform lens, CF = color filter.

transparency forming the object 1. This is Fourier transformed by the FT lens (FTL) onto the BR film. Since the BR film is transparent at this wavelength, no filtering takes place, and the FT is retransformed by the second FTL and can be seen with the CCD camera. However, when the second beam of yellow light (λ = 568 nm) enters the common light path carrying the lower bar of the letter K (object 2) as information, its FT will overlap with the FT of the full letter K on the BR film. Since the yellow light will initiate the BM phototransformation, all Fourier components belonging to the lower bar of letter K will be populated by the M-state, and thus these components will be effectively suppressed in the blue beam. As a result, the CCD-camera sees a letter K without the lower bar; thus a subtraction was executed.

This is shown in Fig. 4a. In addition to the subtraction described above, we have in Fig. 4b edge-enhancement of the letter E, and in Fig. 4c, that of a more detailed picture, a house (37). Edge-enhancement is achieved in principally the same way, by suppression of the low-frequency part of the FT. From these

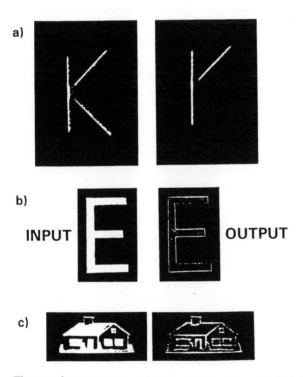

Figure 4 (a) Selective filtering (subtraction operation) for the letter K, (b) edge-enhancement for the letter E, and (c) the same for a house, with BR film as processor medium.

examples it is obvious that by using an optical setup with two inputs of different color the photocontrolled switching of BR between its B- and M-states allows optical processing. It should be emphasized with respect to the title of this book, *Molecular Electronics*, that the central part of the system, the BR film in the Fourier plane, represents a molecular processor unit.

From the function of the SLM as described above it is clear that such BR-mutants with long M-lifetimes as BR_{D96N} are most useful because there the B-state can be bleached easily and the spatial filtering is most effective. In Table 2 a comparison is given with state-of-the-art liquid crystal SLMs, which were put on the market by Hamamatsu (38) recently. These SLMs have a sandwich structure consisting of a photoactive α-Si:H layer, a dielectric mirror, and a (ferroelectric) liquid crystal layer between two transparent electrode-coated glass plates. An input optical image is incident on the side of the α-Si:H layer producing an electrooptical response that in turn switches the alignment of the liquid crystal. With ferroelectric liquid crystals (FLC-SLM), the device can also act as a memory (bistable state of the FLC), whereas for nematic liquid crystals (LC-SLM) this is not possible. Table 2 clearly demonstrates the advantages and disadvantages of the different systems. For fast response time and high spatial resolution a BR-SLM would be the system of choice, i.e., for fast processing with a high parallel throughput. However, because of the lower contrast ratio, the data may have to be cleaned earlier by thresholding, i.e., after fewer sequential processing steps as compared to LC- and FLC-SLMs.

The high input intensity in mW/cm^2 for BR_{D96N} in Table 2 can be misleading. Since BR_{D96N} has a faster response time, i.e., a higher switching rate, the switching energy per switching event is already much less. Further, by taking into account the much higher spatial resolution, i.e., the smallest resolvable area, as a processor or memory unit—called a pixel—it can be calculated that the switching energy per pixel for BR_{D96N} is of the order of 0.2 pJ and is by far

Table 2 Comparison of Properties for Spatial Light Modulators Made of BR_{D96N} Film, Nematic Liquid Crystals (LC-SLM) and Ferroelectric Liquid Crystals (FLC-SLM)

Property	BR_{D96N}	LC-SLM	FLC-SLM
Spatial resolution (lines/mm)	>5000	80	130
Contrast ratio	1:10	1:40	1:30
Response time (μs)	50	4×10^4	100
Input intensity (mW/cm^2)	100	0.1	0.2
Switching energy (mJ/cm^2)	5×10^{-3}	4×10^{-3}	2×10^{-5}
Switching energy per pixel (pJ/pixel)	0.2	600	1.3
Electric power supply	No	Yes	Yes

Source: Hamamatsu Corp.

the lowest in Table 2. Since we have taken as a pixel for the BR film the highest possible resolution of 5000 line/mm, the switching energy of 0.2 pJ/pixel is an upper limit. However,it is interesting to note that this value lies already in the range (0.05–0.5 pJ/pixel) of the very best so-called ''smart pixels'' as quantum-well-devices that the semiconductor industry can produce in its research laboratories to date (39).

V. PHOTOINDUCED ANISOTROPY OF BR FOR LOGICAL OPERATIONS

Besides the photochromic switching of BR between its intermediate states with light of different color, photoinduced anisotropy is another way for optical processing with BR using differently polarized light. Several groups have used this effect (6, 23). Here we want to present an example from our laboratory (40). Since in BR films the BR molecules are fixed in their positions, and in the B-state the retinylidene residues exhibit a transition dipole moment along their long axis, an angle dependent interaction with polarized light occurs. Thus the BR molecules with their retinylidene residue parallel to the electric field vector of the light are preferentially bleached. This results in a photoinduced anisotropy of the absorption and of the index of refraction, i.e., in photoinduced dichroism and birefringence. Both effects can be used for optical processing. In Fig. 5 the realization of an exlusive or (XOR) as logical operation with the photoinduced anisotropy of BR is shown. Actinic light in the yellow to red spectral range is

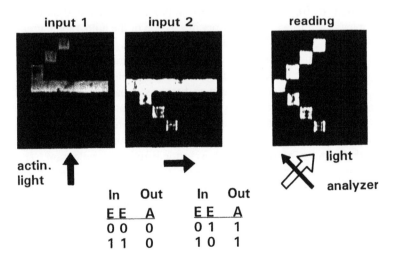

Figure 5   Exclusive OR as an example for logical operations with BR film as processor medium.

used in two perpendicular polarizations carrying the input information 1 and 2 in Fig. 5, respectively. The input patterns are overlapped on a BR film, and photoinduced anisotropic bleaching occurs. However, since the horizontal bar of input 1 is common to both inputs, the polarization bleaching in this part of the pattern is balanced and no net anisotropy results. The reading for the photochemically impressed pattern is accomplished with a light beam outside of the absorption region and with a polarization of 45° with respect to both input beams. In front of the detector, this reading beam is blocked with an analysator in case there is no BR film present or the BR film shows isotropic behavior. Since the state of polarization of the reading beam is changed by the birefringent effect in the anisotropic regions of the BR film, the reading pattern shown in Fig. 5 is observed. Identifying the bright squares with the digital 1 and the dark parts with the digital 0, the truth table in Fig. 5 clearly indicates that an XOR logical operation has been achieved. Other logical operations (AND . . .) can be done as well and form the basis for digital optical computing. Without going into further details it should be emphasized that also in this case the high parallelism of the optical approach helps to improve the overall switching parameters of such a device. Again, the switching itself is based on molecules and their phototransitions, i.e., on molecular processing in the sense that an ensemble of molecules and their states are responsible for the operation.

VI. HOLOGRAPHIC PROPERTIES OF BACTERIORHODOPSIN FILMS

Several applications of BR in optical information processing are based on holographic principles. Therefore an introductory discussion of the holographic properties of BR (5–8,22,23,25,41–45) will be given.

A typical setup for the recording and reading of holograms in BR films is given in Fig. 6. The beam of the recording laser 1 (wavelength λ_1) is split into a reference and an object beam. The object beam is expanded and spatially modulated in phase and amplitude by the object. After refocusing it is overlapped with the reference beam, and their interference pattern is recorded by the BR photochemistry into the BR containing film. The modulation of the absorption coefficient and the index of refraction thus formed gives the hologram. The hologram can be reconstructed by another beam from laser 2 (wavelength λ_2) at an angle corresponding to Bragg's law. The reconstructed wave contains the holographic image, i.e., the full information about the object. This setup can be used either with two cw lasers or with laser 1 in a pulsed version and laser 2 as a cw or delayed pulsed laser.

So far with cw lasers, mainly two types of holograms have been used, the so-called B-type and M-type holograms. In the B-type hologram the phototransformation B→M is used to record the hologram ($\lambda_1 = 570$ nm), and nondestruc-

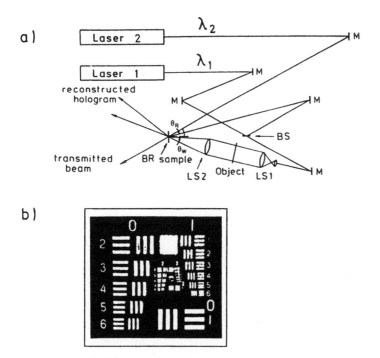

Figure 6 (a) Setup for recording and reading of holograms in BR film. (b) USAF test pattern simultaneously recorded and reconstructed with an M-type hologram. (From Ref. 7.)

tive readout can be accomplished with wavelengths λ_2 from 670–800 nm. The decay of the B-type hologram is determined by the thermal relaxation of the M-state, but it can also be induced with a flash of blue light. For the M-type hologram, a uniform population of the M-state has to be installed first by optical pumping with yellow to red light (B→M) before the information can be recorded with blue light ($\lambda_1 = 413$ nm) in an M→B transition. Since the beam for optical pumping can be used simultaneously as reading beam, it is constructive instead of being destructive with respect to the hologram formation. For dynamic applications with fast response of the rise and decay of holograms, M-type holograms in films containing BR_{D96N} with a long M-lifetime are advantageous because they consist of a purely photocontrolled switching between the B and M states and are not determined by slow thermal relaxation processes. The detailed properties of B- and M-type holograms are summarized in Table 3. A comparison with other holographic recording materials relevant to the applications here will be made in Sec. VII.

For holograms with very fast rise times in the ps time regime, the B→K

Table 3 Holographic Properties of BR Films

Spectral range	
Recording	400–700 nm
Readout	400–800 nm
Resolution	$\geq$5000 line/mm
Recording sensitivity (at 22°C)	
B type	1–80 mJ/cm^2
M type	30 mJ/cm^2
Diffraction efficiency	0.1–7%
Reversibility	$>10^6$
	Write/erase-cycles
Polarization recording	Possible
Relaxation to the initial B state	
Thermal	10 ms–100 s
Photochemical	-50 μs
Thickness	10 μm–400 μm

Source: Ref. 50.

transition can be used. Since the decay time of the K-state is much longer (ns) than its rise time (ps), a short laser pulse of laser 1 (Fig. 6) can produce a strong population grating. Together with the change of Δa and Δn, this determines the holographic grating efficiency. So far, hologram efficiencies of up to 1% have been achieved with the K-state (45), which is comparable to B- and M-type holograms and sufficient for most of the applications discussed here. If the decay of these holograms is too slow, they can be set back with a second laser pulse using the K→B phototransformation, which is also very efficient. Thus holograms with the K-state using pulsed lasers open up an additional time regime of very fast transient optical recording and switching in the ps to ns time regime.

VII. REAL TIME HOLOGRAPHIC PATTERN RECOGNITION

Real time pattern recognition is a rapidly growing field of interest (46) for many practical applications where the identification and localization of a special pattern out of a complex picture is necessary, e.g., in robot vision or automatic inspection of large data bases like those of satellites or medical imagery. Whereas the Vander Lugt filter (47) as the first correlator for shift-invariant pattern recognition was restricted in its use because of its inflexibility, new developments in the design of correlators and electrooptical devices as input modulators make the technical application of optical correlators nowaday realistic. Especially, their extremely high speed as well as their easy optical implementation is unchallenged by digital electronic computers, which are relatively slow because of the large computational capacities required for pattern recognition. BR films turned out to be competitive materials in real time holographic pattern recognition (6–8,48–50).

For real time holographic pattern recognition with BR films, a modified dual-axis joint Fourier transform (DAJFT) correlator was used as shown in Fig. 7. Laser 1 supplies the two axes via a polarizing beam splitter (BS) with coherent light ($\lambda_1 = 413$ nm). For dynamic real time applications, two liquid crystal TV (LCTV) screens are used to transfer the scenes 1 and 2 as seen by the two TV cameras (TVC) into the two optical axes. Scene 1 and optical axes 1 may be identified with the full picture or master pattern, whereas scene 2 and its corresponding optical axes contain the search criterion or filter pattern. Both patterns are post-lens Fourier transformed by the Fourier transform lens (FTL) into the same Fourier plane and recorded into the BR film. Since the light of both axes is mutually coherent, overlapping parts of the Fourier components will form a hologram (M-type). These overlapping parts have the same spatial frequencies and thus identify common patterns of both objects. The hologram is read out with the beam of laser 2 ($\lambda_2 = 530$ nm), which in turn is also the pumping beam for the M-type hologram. Finally, a third Fourier transform lens (FTL) leads to light spots on the CCD camera indicating the location and degree of correlation between common patterns in scenes 1 and 2.

Figure 8 shows a typical experimental result. A DNA sequence in seven lines represents the master pattern, and the correlation with a much shorter part (15 base pairs, mirrored for technical reasons) is made. In Fig. 8b the autocorrelation

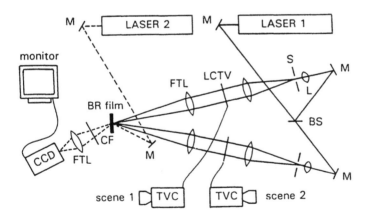

Figure 7 Dual-axis joint Fourier transform correlator for real time pattern recognition with BR films. M = mirror, BS = beam splitter, L = lens, S = slit, LCTV = liquid crystal television screen, FTL = Fourier transform lens, CF = color filter, TVC = TV camera. (From Ref. 49.)

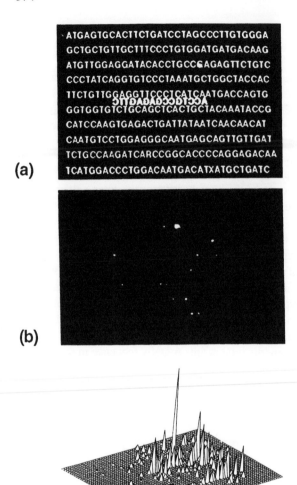

(a)

(b)

(c)

Figure 8   Real time holographic pattern recognition with BR films. (a) Nine lines of DNA sequence as master pattern and a small sequence of 15 bases (mirrored for technical reasons) as search criterion. (b) Autocorrelation (identity) and cross correlation (similarity) as observed on the screen of the monitor and (c) in the pseudo-3D representation.

($\rightarrow$identity) is given as the brightest spot, whereas smaller cross correlations ($\rightarrow$similarities) are indicated by less intensive spots. This is demonstrated more quantitatively in the pseudo $-$3-D plot of Fig. 8c. The interesting point of this example is that not only the identity but also the degree of similarity defined according to the shift-invariant Fourier transformation is given by this procedure. Without going into further details it should be mentioned that this gives a direct link to the possibility of associative recognition, which is the domain of the human brain, and in principle can be installed optically as described above. In Fig. 9 a further example shows the dynamic real time properties of pattern recognition as it can be obtained especially with BR films as processing media. A sail boat (upper left corner of Fig. 9) was tracked during its passage through the viewing range of one of the two TV cameras, while the other TV camera saw the fixed sail boat as filter pattern. The figure shows a long-term exposure of a video film where the thickness of the line indicates the variable speed and even stops of the movement of the sail boat. It should be emphasized that so far the experimental data throughout, i.e., the experimentally achievable frame rate of the pattern analysis, is not limited by the processing speed of the BR film (calculated frame rate ~4 kHz) but by the available electrooptical spatial light modulators (LCTVs), which have a frame rate of 25 pattern/s. For further informations about the performance of the pattern recognition system and its relation to the molecular properties of BR, Table 4 gives a detailed analysis.

So far, photorefractive crystals such as $LiNbO_3$, BSO, and others have been mainly used for holographic pattern recognition and many other holographic applications including holographic optical storage (51). Table 5 shows a comparison of some relevant parameters. It should be pointed out that for BR especially the easy, inexpensive, and reproducible film preparation is one major advantage, where large areas for large optical apertures of thin or thick holographic media

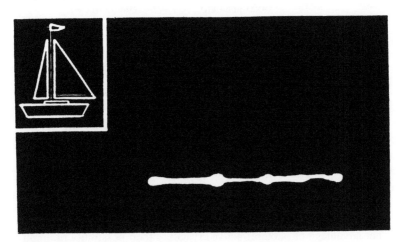

Figure 9 A sailboat, shown in the upper left-hand corner, was tracked during its passage through the viewing range of a TV camera. (From Ref. 50.)

Table 4   Relation Between the Molecular Properties of BR, the Holographic Properties of BR Films, and the Performance of a Real Time Pattern Recognition System Based on BR Films

Molecular properties of BR	Properties of BR films	Performance of the pattern recognition system
Mol. absorption & quantum efficiency (0 = 64%) (ϵ = 63.000 lmol^{-1} cm^{-1})	Light sensitivity (M type: 30 mJ/cm^2)	Laser power required (writing = 25 mW/cm^2) (reading = 60 mW/cm^2)
M lifetime (10 ms–100 sec)		
Transition times B→M & M→B (= 50 μs)	Hologram rise and decay times (<40 ms)	Data throughout (25 frame/s)
Anisotropic chromophore (linear)	Polarization recording	Signal-to-noise ratio (45 dB)
2-dimensional crystal lattice of BR and BR variants (purple membrane)	Reversibility (>10^6) Thermal stability	Lifetime of BR films (years)
Absorption range of the B (570 ± 60 nm) and M (410 ± 50 nm) state	Spectral range (400–700 nm)	Suitable laser light sources (Krypton gas laser, frequency doubled NdYag, HeNe, laser diodes 650–670 nm)
Packing density of BR molecules (crystalline)	Thickness of BR films (10 μm–40 μm)	Plane or volume holograms
Size of BR (5 nm)	Spatial resolution (5000 line/mm)	Influence of Bragg condition Space-bandwidth product

Source: Ref. 50.

in good optical quality can be produced. In contrast, the growth of photorefractive crystals is still a very difficult task, the size is limited, and each crystal has different properties that makes its use for an industrial product difficult. Another advantage of BR films is their much broader range of response times (see Sec. IX) including especially faster response. Permanent storage can be done with both media (see Sec. VIII), but photorefractive crystals need special heating as a fixing procedure and an even higher heating for erasure. Photorefractive crystals, however, have the advantage of showing higher diffraction efficiencies and nearly pure phase holograms, which is not possible for a photochromic medium like BR. A further advantage of photorefractive crystals is their possibility of amplifying crossed optical beams. This is a characteristic property and cannot be obtained with BR films.

Table 5 Comparison of BR Films and Photorefractive Crystals for Holographic Applications

Parameter	BR films	LiNbO$_3$	Bi$_{12}$SiO$_{20}$
Spectral range (nm)	400–700	350–500	350–500
Resolution line/mm	>5000	5000	5000
Exposure μJ/cm^2	10^2–10^5	10^5	10^3
Diffraction Efficiency %	1–7	20	25
Response time	ps-μs-min	ms	ms
Permanent storage	Yes	Heating	
Hologram	Amplitude/phase	Phase	Phase
Amplification	No	Yes	Yes

VIII. REVERSIBLE OPTICAL DATA STORAGE AT ROOM TEMPERATURE

In optical data storage the three fundamental operations "write," "read," and "erase" are demanded. Media that allow these operations are called E-DRAW (Erase Direct Read After Write) media (52). For these materials the storage time should exceed 10 years, at least 10^6 reversible write and erase cycles should be possible, and the reading process should be nondestructive, i.e., frequent reading should not change the stored information. Especially these three requirements are critical for photochrome systems.

In Fig. 10 a demonstration of these three fundamental operations for optical storage with a BR$_{D85,96N}$ film (35) is shown. A pattern of 10×10 dots was written into a BR$_{D85,96N}$ film with 650 nm using the O→P photochemistry. Then some of the dots have been photochemically erased with 490 nm using the reconversion of the photoproduct state P. Finally, the resulting pattern was nondestructively read out at 750 nm using the photoinduced birefringence of the BR film already described in Sec. IV. In this way, polarization effects and the photoinduced anisotropy of the photochemistry of BR films can be used helpfully to circumvent the problem of destructive readout by absorption of the reading beam.

As discussed in Sec. III, the photoproduct state P is thermally stable at room temperature because of its 9-cis retinal conformation, which is not catalyzed by the protein moiety of BR to thermally reconvert to the all-trans conformation. At pH 5–7 and in a glycerol containing sample, the measured reconversion of the P-state of BR$_{D96,85N}$ was less than 20% after eight months. From the kinetics observed so far, it can be concluded that after 10 years still more than 50% of the photoproduct will be present, which would be enough for technical applications. Other important parameters like reversibility and photosensitivity for write

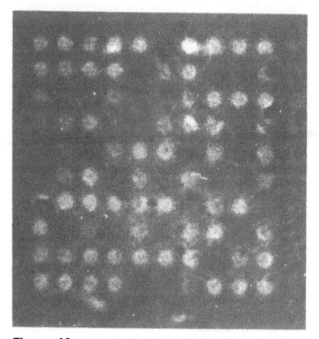

Figure 10 Pattern of 10×10 dots written into a $BR_{D85,96N}$ film. Some of the dots have been photochemically erased. The whole pattern was nondestructively read out by using the photoinduced birefringence of the BR film.

and erase cycles have not been studied in great detail so far, but preliminary results show that at least 10^4 write and erase cycles are possible without reaching an upper limit and that the quantum yield for the photoproduct state is in the range of about 1%. Especially the latter values indicate that further research is necessary to improve some of the data.

In summary, however, it can be clearly shown that by genetic engineering a mutant could be generated in a predicted way, which opens up the field for optical storage with BR at room temperature, although the naturally occurring system BR_{WT} cannot be used for this application. In principle, different techniques of optical storage can use this mutant; some of them are compact disk techniques (10^8 bit/cm^2) (52), holographic techniques (10^{13} bit/cm^3) (53), or optical near field techniques (10^{10} bit/cm^2) (54), with upper limits of storage capacities given in parentheses.

IX. CONCLUSIONS

The applications described here differ widely in the specifications of the required properties. With the naturally occurring system BR_{WT} this wide range cannot be

Table 6 Bacteriorhodopsin: From Fast Processing to Permanent Storage

Time regime	ps	µs–min	∞
Photoreaction	B ↔ K	M ↔ B	O ↔ P
	all-trans	13-cis	all-trans
Retinal	↕	↕	↕
	13-cis	all-trans	9-cis
Spectral range	(570–610) nm	(410–570) nm	(490–640) nm
Spectral shift	Δ = 40 nm	Δ = 160 nm	Δ = 150 nm
Quantum yield	64%	64%	2%
Absorption (ϵ)	63,000 lmol^{-1}cm^{-1}	45,000 lmol^{-1}cm^{-1}	67,000 lmol^{-1}cm^{-1}

covered. Using different BR mutants, specific sample preparation techniques, and various photointermediate states, a large variety of properties could be generated. Table 6 summarizes some of the properties with emphasis on the time regime. It shows that BR systems can cover the wide time range from ps for very fast optical processing up to permanent storage. This wide time range was made possible by molecular engineering including genetic engineering as a very powerful method. The application of genetic engineering to generate new molecular systems with valuable properties not available by conventional synthesis makes the biological approach very useful. It starts with naturally found molecular devices, developed by nature under evolutionary conditions, and can broaden and further optimize the properties of such systems in a remarkable way.

ACKNOWLEDGMENT

We want to acknowledge gratefully the contributions of our coworkers cited in the references. This work was supported by the Bundesministerium für Forschung und Technologie.

REFERENCES

1. Oesterhelt, D., and Stoeckenius, W., *Nature* (London), *New Biol.*, *233*, 1149 (1971).
2. Kouyama, T., Kinosita, K., and Ikegami, A., *Adv. Biophys.*, *24*, 123 (1988).
3. Birge, R. R., *Biochim. Biophys. Acta*, *1016*, 293 (1990).
4. Mathies, R. A., Lin, S. W., Ames, J. B., and Pollard, W. T., *Annu. Rev. Biophys. Biophys. Chem.*, *20*, 491 (1991).
5. Birge, R. R., *Annu. Rev. Phys. Chem.*, *41*, 683 (1990).
6. Oesterhelt, D., Bräuchle, C., and Hampp, N., *Quart. Rev. Biophys.*, *24*, 425 (1991).
7. Bräuchle, C., Hampp, N., and Oesterhelt, D., *Adv. Mater.*, *3*, 420 (1991).

8. Hampp, N., Thoma, R., Bräuchle, C., Kreuzer, F. H., Maurer, R., and Oesterhelt, D., *Molecular Electronics—Science and Technology* (A. Aviram, ed.), American Institute of Physics, New York, 1992, pp., 181–190.

9. Harbison, G. S., Smith, S. O., Winkel, J. A. P. C., Lugtenberg, J., Herzfeld, J. J., Mathies, R., and Griffin, R. G., *Proc. Natl. Acad. Sci. USA, 81*, 1706 (1984).

10. Varo, G., Duschl, A., and Lanyi, J. K., *Biochemistry, 29*, 3798 (1990).

11. Varo, G., and Lanyi, J. K., *Biochemistry, 30*, 5008 (1991).

12. Trissl, H. W., *Photochem. Photobiol., 51*, 793 (1990).

13. Takei, H., Lewis, A., Chen, Z., and Nebenzahl, I., *Appl. Optics, 30*, 597 (1991).

14. Haronian, D., and Lewis, A., *Appl. Optics, 30*, 597 (1991).

15. Haronian, D., and Lewis, A., *Appl. Phys. Lett., 61*; 2237 (1991).

16. Miyasaka, T., Koyama, K., and Itoh, I., *Science, 225*; 342 (1992).

17. Becher, B., Tokunaga, F., and Ebrey, T. G., *Biochemistry, 17*, 2293 (1978).

18. Birge, R. R., Cooper, T. M. Lawrence, A. F., Masthay, M. B., and Vasilakis, C., *J. Am. Chem. Soc., 111*, 4063 (1989).

19. Korenstein, R., and Hess, B., *Nature, 270*, 184 (1977).

20. Oesterhelt, D., and Hess, B., *J. Eur. Biochem., 37*, 316 (1973).

21. Druzhko, A. B., and Zharmukhamedov, S. K., *Photosensitive Biological Complexes and Optical Recording of Information*, USSR Acad. Sci., Biol. Res. Cent., Inst. Biol. Phys., Pushkino, 1985, p. 119. 1985.

22. Vsevolodov, N. N., Iranitskii, G. R., Soskin, M. S., and Taranenko, V. B., *Avtometrya, 2*, 41 (1986).

23. Bazhenov, V. Y., Soskin, M. S., Taranenko, V. B., and Vasnetsov, M. V., *Optical Processing and Computing* (H. H. Arsenault, T. Szoplik, and B. Makucow, eds.), Academic Press, New York, 1989, p. 103.

24. Chen, Z., Lewis, A., Takei, H., and Nebenzahl, I., *Appl. Optics, 30*, 5188 (1991).

25. Hampp, N., Bräuchle, C., and Oesterhelt, D., *Biophys. J., 58*, 83 (1990).

26. Birge, R. R., Zhang, C. F., and Lawrence, A. F., *Molecular Electronics* (F. Hong, ed.), Plenum Press, New York, 1989, p. 369.

27. Dunn, R. J., Hackett, N. R., McCoy, J. M., Chao, B. H., Kimura, K., and Khorana, H. G., *J. Biol. Chem., 262*, 9246 (1987).

28. Nassal, M., Mogi, T., Karnik, S. S., and Khorana, H. G., *J. Biol. Chem., 262*, 9264 (1987).

29. Soppa, J., Otomo, J., Straub, J., Tittor, J., Meeβen, S., and Oesterhelt, D., *J. Biol. Chem., 264*, 13049 (1989).

30. Soppa, J., and Oesterhelt, D., *J. Biol. Chem., 264*, 13043 (1989).

31. Ni, B. F., Chang, M., Duschl, A., Lanyi J., and Needleman, R., *Gene, 90*, 169 (1990).

32. Miller, A., and Oesterhelt, D., *Biochim. Biophys. Acta, 1020*, 57 (1990).

33. Popp, A., Wolperdinger, M., Oesterhelt, D., Bräuchle, C., and Hampp, N., *Biophys. J.*, accepted for publication.

34. Fischer, U. C., Towner, P., and Oesterhelt, D., *Photochem. Photobiol., 33*, 529 (1981).

35. Hampp, N., Bräuchle, C., and Oesterhelt, D., Fourth International Symposium on Bioelectronic and Molecular Devices, R & D Association for Future Electron Devices, Miyazaki, Japan, Dec. 1992.

36. Casent, D., ed., *Optical Data Processing*, Top. Appl. Phys. 23, Springer-Verlag, Berlin, 1978.
37. Thoma, R., Hampp, N., Bräuchle, C., and Oesterhelt, D., *Opt. Lett.*, *16*, 651 (1991).
38. Technical data sheet on FLC-SLM and LC-SLM, X4061, Hamamatsu, Japan (1993).
39. Miller, D. A., AT&T Bell Laboratories, in *Spatial Light Modulators and Applications Technical Digest 1993*, Optical Society of America, Washington, D.C., vol. 6, 1993, p. 150.
40. Hampp, N., Thoma, R., Bräuchle, C., and Oesterhelt, D., to be published.
41. Burykin, N. M., Korchemskaya, E. Y., Soskin, M. S., Taranenko, V. B., Dukova, T. V., and Vsevoldov, N. N., *Opt. Commun.*, *54*, 68 (1985).
42. Hampp, N., Bräuchle, C., and Oesterhelt, D., *SPIE Thin Films, 1125*, 2 (1989).
43. Hampp, N., Popp, A., Bräuchle, C., and Oesterhelt, D., *J. Phys. Chem.*, *96*, 4679 (1992).
44. Zeisel, D., and Hampp, N., *J. Phys. Chem.*, *96*, 7788 (1992).
45. Wu, S., Bräuchle, C., and El-Sayed, M. A., *Adv. Materials*, in press.
46. Arsenault, H. H., *Optical Processing and Computing* (H. H. Arsenault, T. Szoplik, and B. Macukow, eds.), Academic Press, New York, 1989, p. 315.
47. Vander Lugt, A., *IEEE Trans.*, *IT10*, 139 (1964).
48. Hampp, N., Thoma, R., Bräuchle, C., and Oesterhelt, D., *Appl. Optics, 31*, 1834 (1992).
49. Thoma, R., and Hampp, N., *Opt. Lett.*, *17*, 1158 (1992).
50. Hampp, N., Bräuchle, C., and Oesterhelt, D., *Material Research Society Bulletin*, November 1992, p. 56.
51. Günter, P., *Application of Photorefractive Crystals*, Topics in Applied Physics, vol. 61 (P. Günter and J. P. Huignard, eds.), Springer-Verlag, Berlin, 1988, p. 1.
52. Emmelius, M., Pawlowski, G., and Vollmann, H. W., *Angew. Chem. Int. Ed.*, *101*, 1475 (1989).
53. Hariharan, P., *Optical Holography*, Academic Press, Cambridge, 1984.
54. Betzig, E., Trautmann, J. K., Wolfe, R., Gyorgy, E. M., and Finn, P. L., *Appl. Phys. Lett.*, *61*, 142 (1992).

17

Prospects of Molecular Electronics

Günter Mahler

University of Stuttgart, Stuttgart, Germany

Volkhard May

Humboldt University of Berlin, Berlin, Germany

Michael Schreiber

Technical University, Chemnitz, Germany

Molecular electronics is interpreted as part of the rapidly growing scientific activities concerned with electron and energy transfer phenomena on a molecular scale. This emerging field will not only address quantitative aspects of information processing; from a practical point of view, molecular networks can hardly be competitive if they merely try to copy present-day electronics. Optical interconnects are likely to replace other means of communication in future "conventional" electronics, and even more so in molecular electronics. Qualitatively new operational modes like massive parallelism should thus come into reach; others like quantum computation may, in fact, be realizable only on a molecular level.

I. INTRODUCTION

Information processing is concerned with the physical representation of algorithms, which, in turn, has to exploit our algorithmic description of nature (1). Molecular electronics proper is thus the attempt to use a molecular level of description. The respective rules could, of course, be simulated by any universal machine; genetic algorithms e.g. are routinely implemented in electronic computers. Molecular electronics is therefore implicitly based on the additional proposition that using "real" molecular hardware may indeed have substantial advantages.

Research on molecular electronics is still dominated by groups at universities traditionally involved in fundamental research. This volume also gives evidence for this observation. Even though the field of molecular electronics will eventually present an engineering-type challenge motivated by explicit applications, at present industrial laboratories mostly follow those activities in the role of sceptical observers.

This situation, we think, underlines a characteristic trend in physics at the turn of the century (2): fundamental research in the strict sense as exemplified by elementary particle physics and the quest for a ''theory of everything'' is losing some of its previous fascination for other scientists. It is fairly obvious to a solid-state physicist, say, that the codification of those results would have little if any impact on his own work. At the same time, he finds himself engaged in work on artificial semiconductor nanostructures, the details and almost unlimited variations of which would no longer be appreciated as part of basic research, if there were no unifying motives for guidance. Specific functions like nonlinear or bistable properties could be such goals (others could involve the mere demonstration of fundamental quantum features). A substantial part of fundamental research, we believe, is thus becoming concerned with phenomena in recently emergent fields (3). Molecular electronics is just one pertinent example; the increasing popularity of research on complex systems (4,5) is another.

On the other hand, such studies in these emergent fields, despite their appeal for *possible* applications, are not necessarily directed towards *practical* applications that would easily sell on the market. To understand how a given molecular structure might be able to support a specific function is considered a meaningful project, whether or not this system would, as such, be *competitive* with existing technical realizations of that very function e.g. on a traditional semiconductor structure. Unfortunately, such a divergent motivational frame can easily lead to distorted expectations and mutual misunderstandings; it certainly accounts for the reservation of those who are under immediate economic pressure to turn their research objects into new products. Are they right, or might molecular electronics also have something to offer for them, at least in the long run?

Before we try to answer this question we first summarize the present state of the art as it was highlighted in the preceding papers and supplement this with some potential future trends.

II. STATE OF THE ART

Structure has been discussed in this book as a means to define and constrain dynamical paths. This is seen as an important precondition for any system used as a device, which is usually designed to single out one of a few alternative paths. We have restricted ourselves to an overall solid structure that can be decomposed into molecular scale subsystems. A key–lock logic, which could, in principle, be implemented into a *solution* of appropriate ensembles of molecules, and which has been proposed as a type of molecular computer (6), appears to be much too slow and inflexible. Working examples from biology would be hard to adopt for technical purposes.

Mesoscopic or even macroscopic systems with order of atomic precision do not exist. Controlled crystal growth provides some of the best samples though with unavoidable defects, but apparently not yet with the hierarchical complexity desired. Other examples are LB films (7) and quantum dot arrays (8) or custom-made STM structures on a substrate (e.g., the "quantum corral" (9)). In any case, synthetic structures result from a combination of self-assembly and direct external interference. They typically represent structures frozen in a nonequilibrium state. Lowering of the structural entropy is connected with rapidly increasing costs.

An attractive alternative is the use of natural building blocks (bacteriorhodopsin, e.g., also covered in this book (10,11)). Their physical properties can even be optimized genetically. Up to now, however, it has not been possible to let these subsystems interact in a well-defined fashion, a prerequisite for computing by molecules.

Dynamical properties are part of the physical characterization. Examples are the transport measurements of antidot arrays; specific ensemble properties result as dynamical conductivity in dot arrays or the optical properties of spectral hole burning material. The nonlinear optical response can be tailored by selecting appropriate molecular subunits. Studies on single defects in a matrix address the uncontrolled stochastic dynamics of the local neighborhood as monitored by the optical transitions in the respective defect (12). These studies indicate that even low-temperature crystal matrices need not be "quiet."

The only dynamical features that have made their way into *applications* related to molecular information processing are connected with spectral hole burning. It has therefore been covered at some length in this book. It is an advantage that this technique is completely compatible with disorder. Device applications include data storage (frequency multiplexing) as well as restricted information processing by exploiting the dependence of the local molecular states on the external electric field as a parameter. On the other hand, spectral hole burning uses the molecular ensemble only as a passive filter for the optical field. The burning of the holes as "one-shot dynamics" is not part of information processing proper. It is not accidental that optical fields play a dominating role, though: there is hardly any other way left for contact. On the other hand, this approach renders parallel processing possible in a straightforward way. Holographic information storage and retrieval in terms of pictures appears to be especially attractive.

III. FUTURE TRENDS

The *structural complexity* of the system under consideration has to be significantly increased before interesting new properties can be expected to emerge. Combinations and improvements of existing structuring technologies may suffice to do the job. An example of a promising new variant is atomic lithography based on refractive atom optics exploiting wave properties (13): here the effective lens for the atom beams is created by laser fields, thus underlining the importance of light–matter interactions also for the purpose of structuring.

Novel *dynamical* features might include effects of quantum stochastics (14) and quantum coherence, the latter including incomplete decomposability (15). These phenomena require a delicate balance between isolated subsystems (which would have infinite coherence times) and couplings between the subsystems, to the outside world, and to any uncontrolled internal degrees of freedom. There exist design principles (16) to reach that goal; preliminary experimental verifications have been possible even in semiconductor quantum film structures (17). Examples for coherent optics abound in atomic spectroscopy.

Single-particle coherence is fundamental also to what may be called "wave electronics" (18), stressing the analogy of matter waves and electromagnetic waves in terms of interference. The Mach–Zehnder ring wave guide in optics thus has an electron counterpart, for which the phase may be controlled by the Aharonov–Bohm effect. However, contrary to optics, this coherence can be sustained only on microscopic scales, i.e., tiny patches of coherence. The resulting system properties appear to be hard to control.

Novel *tasks* are likely to include analogs to biological information processing, smart sensors, quantum communication, and quantum computation. Quantum communication (19) is a new field, in which one tries to map specific communication scenarios onto quantum-mechanical rules. It is an "ultimate" technology in the sense that the implementation refers to a sequence of elementary events, for which manipulations cannot go unnoticed. Of course, this does not apply to the classical periphery, where those advantages could easily be spoiled.

The basic concept for quantum computation (20) is the superposition principle underlying nonclassical states. Such states are supposed to allow for a novel kind of parallel processing impossible in the classical domain. Some caution is in place, however: a single quantum object is not measured in a superposition state; rather is it found in either one of the possible alternatives defined by the measurement apparatus. To verify the whole set of alternatives, large homogeneous ensembles are required (21), which would interact and disturb each other if sent simultaneously to avoid classical serial testing. However, one can demonstrate that the state space of a *quantum* system, even if embedded in a classical environment, can be considerably larger than in its classical limit. This extended space is said to carry "nonlocal" information (22).

IV. SYSTEM ASPECTS AND CONCLUSIONS

What are the advantages of conventional electronics, besides being well-established? It allows for networks with node interactions in terms of well-defined and robust local rules usually specified on a hydrodynamic level of description (23). Limitations and disadvantages are connected with the way those interconnects are built in: the electric wiring consumes much space and is topologically constrained (24); also the limited velocity has to be taken into account.

Optics has the advantage of highly parallel information transfer which travels at the maximum speed possible. There is no cross talk as the photons do not interact. At the same time this latter property is a severe disadvantage as information processing *requires* specific interactions. Optical computing cannot do without matter even though the information is still carried by the light field. Prototypes of optical computing systems have been successfully demonstrated (25), but it is difficult to see how they might become competitive with modern electronic supercomputers, except for very special purposes.

The combination of optics and electronics may offer the advantages of both without the disadvantages of either field. This is why optoelectronics is believed to have a strong impact in the near future (26). It is a substantial part also of the Real World Computation (RWC) project initiated in Japan in 1993 (27).

Devices based on high T_c superconductivity are being discussed (28). Possible applications are Josephson elements and superconductive wiring. However, due to the rather complicated and until-now-ill-controlled materials, the possible advantages are not yet convincing.

What then is the situation for molecular electronics? It also lacks clearly specified and well-characterized material classes to build on. But it does offer a valuable promise: it should allow for massively parallel *interactions*, novel means for optical interconnects (spectral hole burning being a special example), and, eventually, implementation of quantum effects. These prospects justify, in our opinion, enhanced activity in this field and the hope for real competitive technological applications.

But what kind of applications? We may again look at the RWC project. It aims at realizing human-like, "soft," flexible information processing for the 21st century. It should process the diverse types of information used in the real world, often represented in terms of "patterns." Working with pictures, e.g., is certainly more appropriate for human beings than working with large amounts of numbers; the latter are merely good for transfer to other computers.

Physics is not scale-invariant; this means that there are fundamental limits for any device type. Molecular electronics would come, by principle, much closer to biological information processing than any other technology, simply because physics would work here on the same scale. Molecular electronics should thus be a realistic option to reach those goals of "soft" information processing.

The success of the present information technology rests upon the use of a few primitive operations supported by a digital computer, by which any computable function can be approximated. This tremendous flexibility has its price: it is a "low-level language," even though the user will usually manipulate the system on a much higher (macro) level.

Analog systems have experienced some modest revivals in the form of special purpose systems like the so-called silicon retina (29): the reduced flexibility (parameters cannot be set at will, interactions are defined by the actual system) may be compensated for by increased efficiency (30).

Efficiency is an issue: part of the motivation of the RWC derives from the observation that present-day computing systems are strikingly inefficient for pattern recognition, etc. Efficiency is also an important ingredient of complexity theory (1), in which problem classes are defined. Quantum-mechanical computers appear to reduce the complexity of specific problems (31). If true (and applicable), the impact of molecular electronics as their carrier could hardly be overestimated.

Quantum computation is still far beyond the reach of any established technology. However, as far as the conceptional foundations of information theory are concerned, it may well represent one of the most challenging subjects for the time to come. We are accustomed to the fact that the algorithms by which we represent the laws of nature are computable in the sense that physical systems can easily represent those fundamental operations. But what happens if other operations become trivially computable (as adding is now), operations that we presently call strange, nonclassical? Would this not change also the way in which we perceive and model nature?

We have not included a more detailed exposition of quantum computation in this book (for details consult, e.g., (32)); its relation to material science as well as to concrete applications has not yet been worked out in any detail. Even if the practical applications remained limited, experiencing a different scheme of computation should greatly enhance our understanding of what computation is all about. Before one can appreciate one's mother tongue, one has to get familiar with one or another foreign language.

ACKNOWLEDGMENT

One of us (G.M.) thanks K. Rebane for valuable discussions.

REFERENCES

1. Mühlenbein, H., this volume.
2. Schweber, S. S., *Physics Today*, Nov. 1993, p. 34.
3. Horgan, J., *Sci. Am.*, Dec. 1992, p. 16.

4. Anderson, P. W., *Physics Today*, July 1991, p. 9.
5. M. M. Waldrop, *Complexity: The Emerging Science at the Edge of Order and Chaos*, Simon and Schuster, New York, 1992.
6. Conrad, M., *Commun. ACM*, *28*, 464 (1985).
7. Peterson, I. R., this volume.
8. Ensslin, K., Hanson, W., and Kotthaus, J., this volume.
9. Crommie, M. F., Lutz, C. P., and Eigler, D. M., *Science*, *262*, 218 (1993).
10. Birge, R. R., this volume.
11. Bräuchle, C., this volume.
12. Orrit, M., and Bernard, J., this volume.
13. Timp, G., et al., *Phys. Rev. Lett.*, *68*, 580 (1992).
14. Dalibard, J., Castin, Y., and Molmer, K., *Phys. Rev. Lett.*, *68*, 580 (1992).
15. See e.g. Aspect, A., Dalibard, J., Grangier, P., and Roger, G., *Optics Commun.*, *49*, 429 (1984).
16. Mahler, G., this volume.
17. Roskos, H. G., et al., *Phys. Rev. Lett.*, *68*, 2216 (1992).
18. Datta, S., *Superlattices and Microstructures*, *6*, 83 (1989).
19. Ekert, A., *Phys. Rev. Lett.*, *67*, 661 (1991).
20. Deutsch, D., *Proc. Roy. Soc. London*, *A425*, 73 (1993).
21. Cerny, V., *Phys. Rev.*, *A48*, 116 (1993).
22. Barnett, S. M., and Phoenix, J. S. D., *Phys. Rev.*, *A48*, R5 (1993).
23. Keyes, R. W., *Rev. Mod. Phys.*, *61*, 279 (1989).
24. Ferry, D. K., and Porod, W., *Superlattices and Microstructures*, *2*, 41 (1986).
25. Brady, D., *Nature*, *344*, 486 (1990).
26. Cathey, W. T., *Optoelectronics*, *8*, 126 (1993).
27. *Nikkei Computer*, July 19, 1993.
28. Gallagher, W. J., *Solid State Technology*, Nov. 1993, p. 151.
29. Andreou, A. G., *Nature*, *354*, 501 (1991).
30. Hopfield, J. J., *Network*, *1*, 27 (1990).
31. Deutsch, D., and Josza, R., *Proc. Roy. Soc. London*, *A439*, 553 (1992).
32. Lloyd, S., *Science*, *261*, 1569 (1991).

Index